Introduction to Measurement Science and Engineering

WILEY SERIES IN MEASUREMENT SCIENCE AND TECHNOLOGY

Instruments and Experiences: Papers on Measurement and Instrument Design
R. V. Jones

Introduction to Measurement Science and Engineering
P. H. Sydenham, N. H. Hancock and R. Thorn

Introduction to Measurement Science and Engineering

P. H. Sydenham

Measurement and Instrumentation Systems Centre
South Australian Institute of Technology
Adelaide, South Australia

N. H. Hancock

School of Engineering
Darling Downs Institute of Advanced Education
Toowoomba, Queensland, Australia

and

R. Thorn

Measurement and Instrumentation Systems Centre
South Australian Institute of Technology
Adelaide, South Australia

JOHN WILEY & SONS
Chichester • New York • Brisbane • Toronto • Singapore

Wiley Editorial Offices

John Wiley & Sons Ltd, Baffins Lane, Chichester,
West Sussex PO19 1UD, England

John Wiley & Sons, Inc., 605 Third Avenue,
New York, NY 10158-0012, USA

Jacaranda Wiley Ltd, G.P.O. Box 859, Brisbane,
Queensland 4001, Australia

John Wiley & Sons (Canada) Ltd, 22 Worcester Road,
Rexdale, Ontario M9W 1L1, Canada

John Wiley & Sons (SEA) Pte Ltd, 37 Jalan Pemimpin #05-04,
Block B, Union Industrial Building, Singapore 2057

Library of Congress Cataloging-in-Publication Data

Sydenham, P. H.
Introduction to measurement science and engineering / P. H. Sydenham, R. Thorn, N. H. Hancock.
p. cm.—(Wiley series in measurement science and engineering)
Includes bibliographies and index.
ISBN 0 471 92223 4
1. Mensuration. I. Thorn, R. (Richard) II. Hancock, N. H. (Nigel H.) III. Title. IV. Series.
T50.S93 1989
681′.2—dc20

89-14776
CIP

British Library Cataloguing in Publication Data

Sydenham, P. H. (Peter H.)
Introduction to measurement science and engineering.
1. Engineering. Measurement
I. Title II. Thorn, R. III. Hancock, N. H.
620′.0044

ISBN 0 471 92223 4

Typeset by APS Ltd, Petersfield and Salisbury
Printed and bound in Great Britain by Biddles Ltd, Guildford

CONTENTS

PART B: PRACTICAL APPRECIATION

AUTHORS' PREFACE

This text has its genesis in the *Handbook of Measurement Science*, Volumes 1 and 2, which were published in 1982 and 1983 respectively. The *Handbook of Measurement Science* project collected together the basic knowledge of Measurement Science and those volumes have enjoyed wide acceptance. Volume 1 has now been reprinted and also translated into Polish; Volume 2 is now in translation into Polish; a Volume 3 has been commissioned. These, and other activities, demonstrated that this field of knowledge is now mature and that a Book Series in Measurement Science and Technology should be developed.

In particular it was perceived that a mid-level undergraduate *teaching text* was now required, one that commenced with the appreciation of instruments as information gathering and processing machines and also presented a structured approach to the general design and effective use of instrumentation systems.

The result, after some four years of development, is this volume, which the authors believe will be of lasting value to both undergraduate and postgraduate students, and instrumentation practitioners. Furthermore, application of these concepts and techniques is not restricted to the physical and engineering sciences: researchers in the biological, agricultural, medical, accounting, economics and social sciences will also be able to derive significant benefit from this book.

There is, of course, a wide range of texts, both general and specialist, published on measurement. However, these are generally concerned with reporting the *technology* involved and detailing the specific techniques commonly employed for particular measurands. They are organized in what is essentially, and perhaps unavoidably, a 'catalogue' style which makes them valuable reference manuals but, in the authors' experience, inappropriate as teaching texts. They have not been designed to be true teaching texts. Technologies come and go with increasing rapidity, making 'recipe' text solutions to measurement problems rapidly out of date.

Introduction to Measurement Science and Engineering is a first text that is unusual, and perhaps unique, in presenting a structured *top-down* approach, including examples and tutorial material. A serious attempt has been made to present a balance of both the formal and the heuristic knowledge that is typical and vital to instrument practice. The curriculum implicit in this text provides a generic methodology, applicable to any discipline needing instruction in 'how to set up and make good measurement'. Here technology is of secondary interest in examples that reinforce the main emphasis on the science of measurement and appropriate engineering design.

This book draws heavily on the *Handbook of Measurement Science*, both volumes of which should be regarded as supporting reference works. The *Handbook* will be particularly appropriate to Honours and postgraduate study.

One important area of teaching is not covered in this book, that of the necessary supporting laboratory instruction. In the authors' opinion it is preferable that teachers in this field use their own initiative to make use of locally available plant and equipment to create relevant exercises. The several publications of IMEKO on laboratory instruction are valuable sources in this regard.

Despite the current trend towards authors using wordprocessors and desk-top publishing software the publisher still has to convert an author's manuscript into a tightly edited, professionally produced and coordinated printed work. Our appreciation is extended to a patient Ellen Taylor, Editor, and to those behind the scenes who provided production, printing and promotional support at John Wiley & Sons, Ltd, Chichester.

We also wish to thank Michael Vaughan and David Harris for assisting with preparation of figures and those authors, in the *Handbook of Measurement Science*, who kindly gave their permission to make use of parts of their contributions.

PETER H SYDENHAM
RICHARD THORN
Adelaide,
December 1988

NIGEL HANCOCK
Toowoomba,
December 1988

SERIES EDITORS' PREFACE

Measurement and instrumentation technology is a key enabling tool. Its range of applications is immense. Measurement is the basic method of science, and modern natural science is wholly dependent on instrumentation. The manufacturing machines and plant of modern industry and modern transport vehicles, whether land, sea or aerospace, all require advanced automatic control, based on measuring instrumentation. Our natural environment, under great pressure, must be monitored by measurement. Medical diagnosis and care are also coming to rely to an increasing degree on sophisticated measuring equipment.

The wide application of measurement presents an extensive and diverse spectrum of measurement tasks. They involve many types of measurands embracing all physical and chemical quantities. They differ in respect of the magnitude of the measurands and their dynamic behaviour. They involve different accuracy requirements. The conditions in which the measurement instrumentation must operate range from those of the laboratory to extremely arduous environments.

To meet this great spectrum of requirements there is a large arsenal of technical means of instrumentation of high complexity and capability. There is a great range of sensors which convert external excitation into signals which can be conveniently handled. Modern information technology has developed extensive and diverse means for processing the acquired information, displaying it, or converting it into physical action. To an increasing extent modern instrumentation incorporates machine intelligence.

This great range and wide diversity of measurement and instrumentation tasks and tools means that the technology can no longer be presented in the traditional manner as a catalogue. For effective teaching and study the subject must be organised as a systematic science with a body of fundamental concepts and principles and a logical framework.

Such an organisation of measurement and instrumentation science has been developed over the last two decades. It has evolved in particular in the Proceedings of the International Measurement Confederation (IMEKO). Peter Sydenham has documented the approach excellently in the authoritative *Handbook of Measurement Science*. The systematic approach to measurement and instrumentation considers measurement as an information process and measuring instruments as information machines. In this way the subject is treated by the concepts and methods of the theories of systems, signals and information. This approach also recognises that the essence of the technology is the design of measuring systems and that a systematic methodology of design is its most essential tool.

Systematically organised measurement and instrumentation science is an excellent vehicle for the general education of engineers and scientists. It is primarily useful because of the key practical importance of the technology. However it is also significant because it combines rigour with the application of theory to practice.

The development of systematic approaches to measurement and instrumentation has not as yet resulted in the development of adequate teaching materials. In this volume the authors have taken a major step to remedy this deficiency by providing an introduction to systematic measurement and instrumentation science for the beginning student. The book lays the foundation of knowledge and understanding of the concepts and tools required to build and use measuring systems for many applications and in many implementations. Soundly based on authoritative literature the book presents the material logically, and as comprehensively as possible, but above all clearly and simply, without unnecessary mathematical complexity.

As an enthusiast for measurement and instrumentation science and technology, and for the systematic approach to its teaching, I cannot provide an unbiased judgement. I can, however, express my strong view that this is an important book which should have a major impact on how this important subject is taught and studied.

L. FINKELSTEIN
London,
July 1989

LIST OF FIGURES

PART A

PHILOSOPHICAL AND THEORETICAL APPRECIATION

In physical science a first essential step in the direction of learning any subject is to find principles of numerical reckoning and methods for practicably measuring some quality connected with it. I often say that when you can measure what you are speaking about, and express it in numbers, you know something about it; when you cannot measure it, when you cannot express it in numbers, your knowledge is of a meagre and unsatisfactory kind: it may be the beginning of knowledge, but you have scarcely, in your thoughts, advanced to the stage of science, whatever the matter may be.

Lord Kelvin (Sir William Thomson, 1824–1907), in address to Institution of Civil Engineers' meeting, England on 3 May, 1883.

The first five chapters comprising Part A develop an appreciation of the philosophical, terminological, historical and theoretical issues relevant to the professional pursuit of measurement and its instrumentation.

As the relevant and essential knowledge and the application of measurement science and engineering is multidisciplinary Part A, by necessity, must cover, to be adequately comprehensive, a considerable breadth of conceptual and mathematical material. Although much of it is already taught in other topics areas it is rarely delivered in the context of measurement systems.

It will often be possible to cover some of the material relatively rapidly, according to the other curriculum topics of the course in which this is a subject. In courses where the mathematical background developed is less penetrating it is advisable to supplement this material with additional lectures.

To be schooled only about the conceptual and theoretical issues is certainly not an adequate educational exercise for professional work in this field. Effective design and application of measurement requires a strong understanding of practical aspects, for this must be incorporated into a study. For this reason the three chapters of Part B address practical issues of the nature of the media in which instruments operate, how they are designed and how to select and commission them.

CHAPTER 1

Introduction

1.1 MEASUREMENT SCIENCE AND INSTRUMENT ENGINEERING

Simple measurement is an everyday event: the youngest of schoolchildren soon learn to measure their height, weight and age; the length of a piece of wood and the time taken in a footrace. Figure 1.1 illustrates the range of simple measurements made in keeping us healthy.

Developed societies increasingly rely on complex, multidimensional measurement using complex, often automatic equipment. To conceive, and adequately design, such equipment requires the rigorous, structured approach as has been formalized within the discipline of *measurement science* and applied in the discipline of *instrument engineering*.

The discipline of *measurement technology*, however, is generally recognized as being concerned with particular techniques and methods of implementation – it is considered further in a subsequent section.

The two disciplines of measurement science and instrument engineering share the same objective: quite simply to make good measurements of everything that needs to be measured. The measurement of length, time, for instance, appears to be straightforward. The measurement of smell, taste, quality, love, hate and intelligence, as other examples, is another matter.

Some definitions are now in order. To fix ideas in readiness to delve into the deeper aspects of measurement science and instrument engineering it is first necessary to define measurement.

Measurement *is the process of gathering information FROM the physical world.*

Measurement is achieved by means of *sensors*, also called *transducers*, and involves the gathering of *data* and its comparison with an agreed *standard*. Each of these terms will be defined in later chapters: note at this stage that *data* and *information* are not the same entity. The word *FROM* has been deliberately emphasized to stress the act of sensing and that energy (or sometimes mass) must flow between that subject and the sensor.

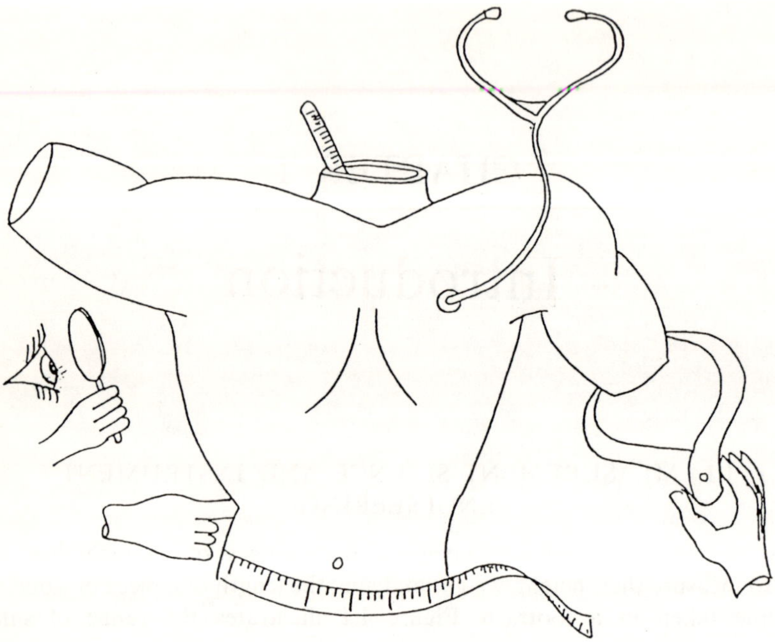

Figure 1.1 Everybody relies on measurements of some kind, some are obvious, others are not

When sensors connect the real world to computing systems (now so commonly done) they are labelled *physical world to computer interfaces* (*PWCI*) to distinguish them from visual display screens, terminals and printers which constitute *human to computer interfaces* (*HCI*). Both are *information technology machines* which will be considered further in Chapter 2.

Hence measurement science may be defined as

Measurement science *is the systematic study and organization of the methods by which information is gathered FROM the physical world.*

The process is depicted in Figure 1.2, where it can be seen that human senses are first used to extract broad information about a topic. As the study matures measurement instruments are applied to enhance man's sensing ability.

The act of measurement obviously requires more than sensors and standards. There must also exist a method of comparison with a standard (preferably automatically), a means of display and a host of other features for the convenience of the user. The result is an *instrument* or system of *instrumentation*. Only when ALL of these aspects are satisfactory is the device regarded as a good instrument. Hence instrument engineering may now be defined:

Instrument engineering *is the systematic design and manufacture of reliable, cost effective measuring instrumentation that fulfills a given measurement specification.*

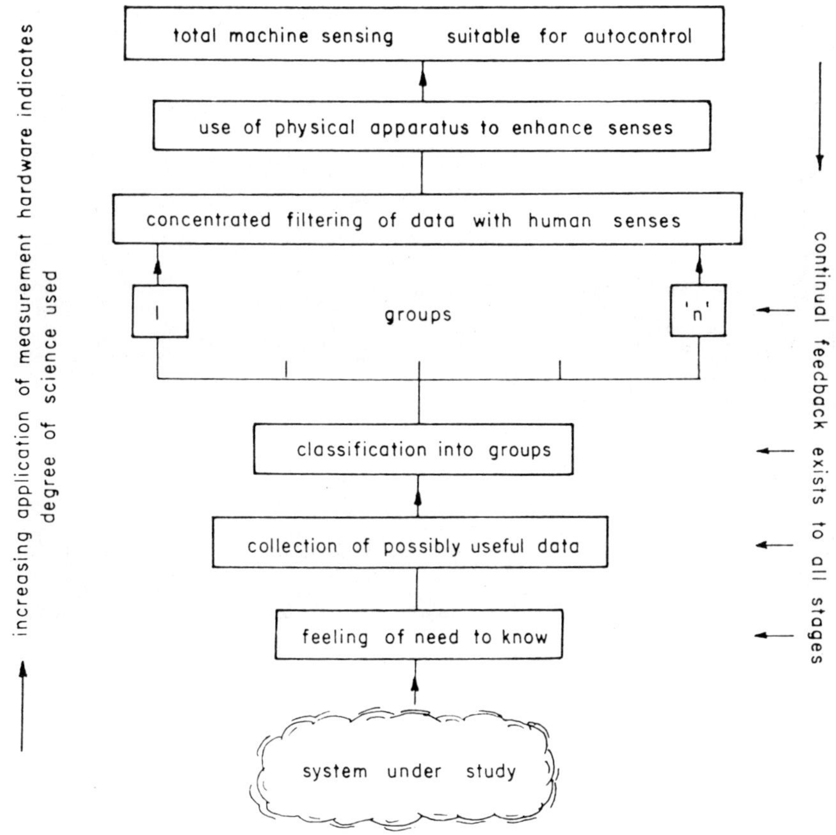

Figure 1.2 Application of measurement in study of a problem

This definition is no different from a possible definition of any other branch of engineering. However, it is important to note that *capable* instrumentation can be defined and manufactured only if the measurement specification is adequate. An example of the *engineered* hardware and its assembly, for a simple instrument, is given in Figure 1.3.

For all but the simplest of instruments this requires a proper understanding and application of measurement science. The two disciplines go together.

The reader may now wonder why the concept of *accuracy* has not been included explicitly in the definition of instrument engineering. Accuracy is just one of numerous properties of an instrument which is required, to a greater or lesser extent, depending on the application requirement. This, and other critical measurement performance parameters, are considered further in Chapter 2.

The word *metrology* is commonly used in connection with engineering measurement. This term is variously defined as both a system of, and the science of, *weights and measures*, and is identified firmly with the traditional measurement of mass, length, time, etc. as required for the everyday practice of engineering.

Figure 1.3 Inside a relative humidity, meter/sensor used in process control applications. (HBMS, figure 15.3.) (Reproduced by permission of Kent Industrial Measurements Ltd, St Neots, Cambs, UK.)

1.2 MEASUREMENT TECHNOLOGY AND MEASUREMENT SCIENCE

Measurement technology is concerned with how to make measurements with available principles, materials, manufacturing processes and plants. The techniques required and developed over the centuries span all of physics, chemistry and much of biology, and a knowledge of the appropriate techniques is regarded as essential to any worker in a particular field. The problem is deciding what techniques are appropriate. It is commonly observed that the introduction of a new measurement instrument or technique, or use of existing instruments, can herald in a new era in that science – a *breakthrough* is a common term used – see Example 1.1 and Figure 1.4.

Example 1.1 Microscopy began with a 'ballpoint pen' sized sphere of glass

Optical telescopes were in use in the early 1600s AD. They gave man the ability to see small objects in space. The technology–glass lenses, mirrors, fine adjustment devices and frameworks–existed to allow the next development of ways of seeing smaller objects than the eye could observe.

The magnifying lens certainly existed prior to 1600 AD but it was the work of van Leeuwenhoek that brought the attention of the scientific world to the benefits of the microscope–see Sydenham (1979).

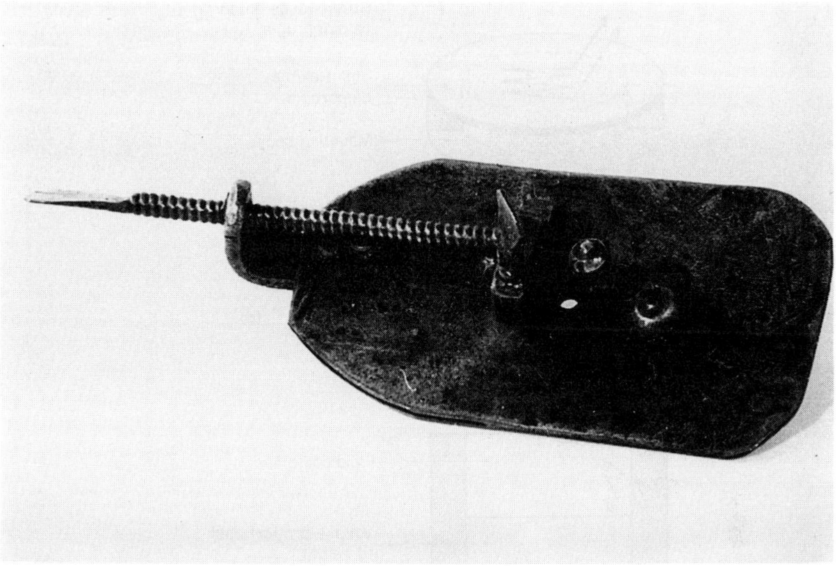

Figure 1.4 Leeuwenhoek's primitive microscope, c. 1650, is recognized as the instrument that heralded the huge field of microscopy. It made possible the observation of things that had not previously been observable. (From Sydenham, 1979.) (Reproduced by permission of Utrecht University Museum.)

His instruments were simplicity to the extreme — see Figure 1.4. The lens was a double convex system ground on each side of a glass ball about the size of the ball in the end of a ball point pen. (The small size overcame major distortion defects.) These simple systems could resolve up to 700 lines per millimetre and had a magnifying power of around 275 times, but required great skill on the part of the user. They still work as well after 240 years!

Often a technique is not new; it has just been discovered as being appropriate in the particular field. In recent years the value of this process has been recognized and the phrase *technology transfer* coined to convey official encouragement. Concepts of *lateral thinking* and *interdisciplinary approaches to problems* recognize the same idea.

There is a wealth of literature on the technology of measurement. Handbooks in any field will almost invariably contain a chapter on measurement techniques (usually the last!) and there is a range of general-purpose texts which recognize the general applicability of various measurement techniques. The material in these texts is usually organized according to the quantity to be measured, for example length, temperature, pressure, flow and so on. Some texts are listed at the end of this chapter to provide readers with places to look for examples of modern instruments.

This approach to measurement technology is essentially a simple cataloguing of experience and practice accumulated over many decades. Whilst this might

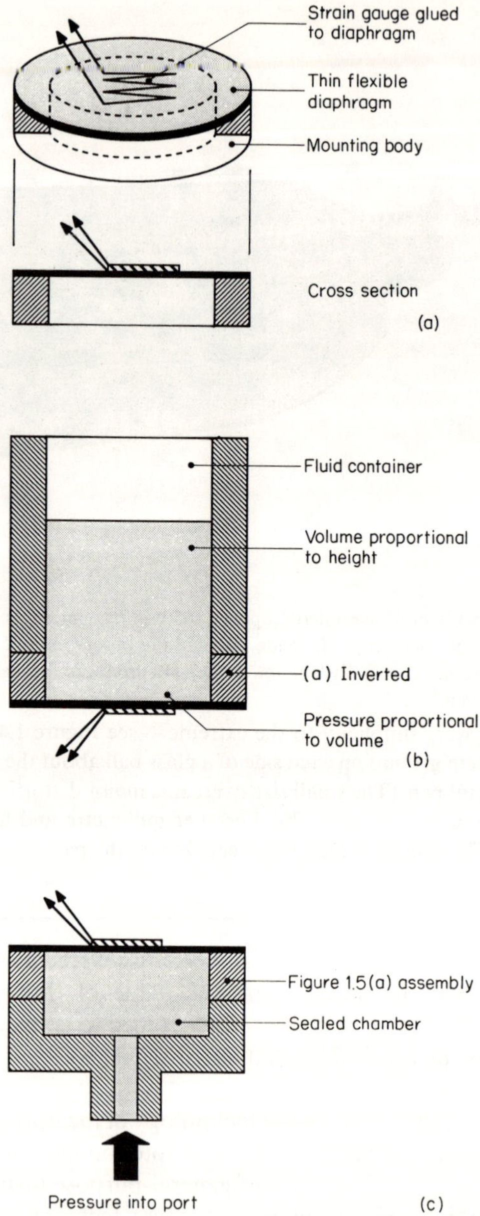

Figure 1.5 (a) Strain gauge mounted on a thin diaphragm; (b) Schematic of a volumeter; (c) Pressure gauge layout

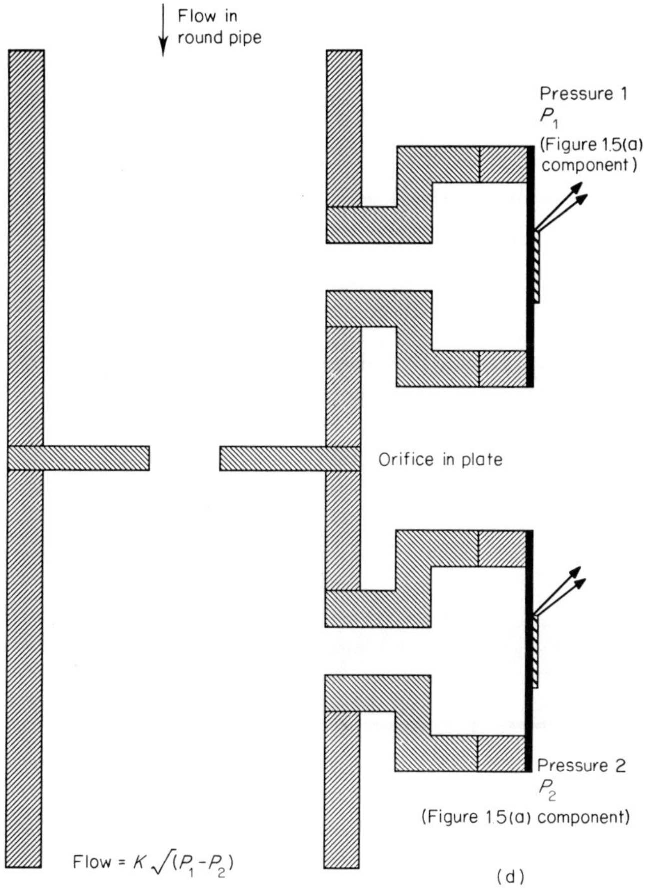

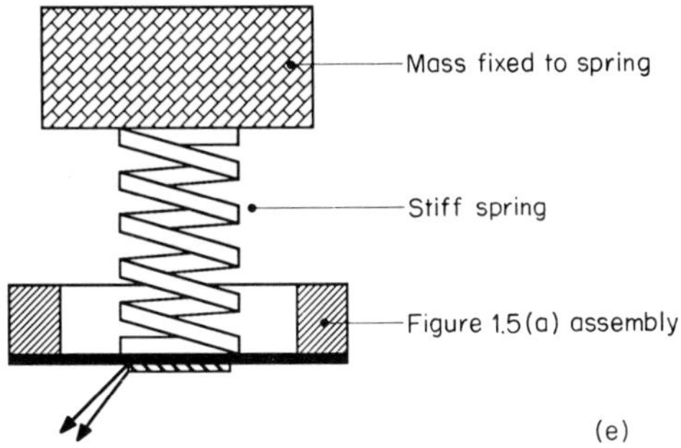

Figure 1.5 (*continued*) (d) Arrangement for flow metering with d.p. gauge; (e) Layout of accelerometer using a spring and mass

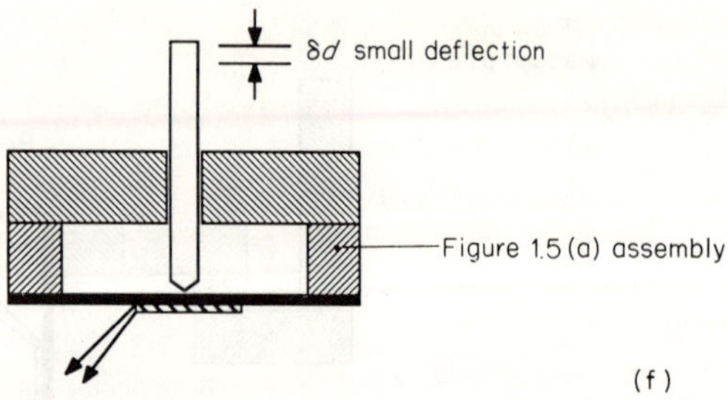

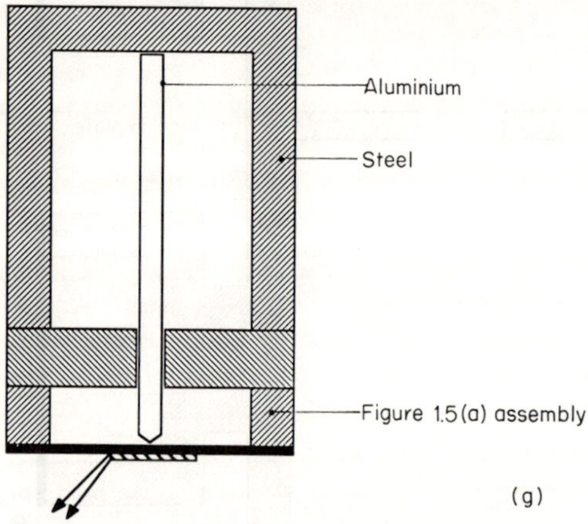

Figure 1.5 (*continued*) (f) Displacement sensor arrangement; (g) Thermometer configuration

appear a very reasonable approach, there arise two dominant problems for many designers, selectors or users of measuring instruments.

Firstly, there can be considerable overlap between the quantities that can be measured with simple modifications of a particular basic technique–see Example 1.2.

Example 1.2 How to measure almost anything with one basic concept

Consider a strain gauge, (a *sensor* that converts mechanical strain into equivalent electrical signal) mounted on a small flexible diaphragm as shown diagrammatically in Figure 1.5(a).

Mount the assembly of Figure 1.5(a) in an enclosure, Figure 1.5(b), and it will measure the change in *volume* of the enclosure; it becomes a *volumeter*.

Seal one side of the enclosure and allow gas or liquid to enter the other side and it becomes an absolute *pressure gauge* as in Figure 1.5(c). Connect both sides to different pressures and it becomes a *differential pressure* (*d.p.*) gauge.

Connect the two sides to differential pressures established by a suitable pipeline system and it becomes a *flowmeter*, Figure 1.5(d).

Add a small mass to the diaphragm, as shown in Figure 1.5(e), and it can measure acceleration, seismic activity and vibration.

Connect a rod to the centre of the flexing diaphragm and the system becomes a *displacement* sensor as shown in Figure 1.5(f).

Connect the free end of the rod, in Figure 1.5(f), and the body to a fixed datum and expansion of the rod due to temperature converts the system into a *thermometer* as in Figure 1.5(g).

These are but a few of the possibilities. The reader is invited to develop more, such as for weighing mass; measuring viscosity in bread dough; determining relative humidity; determining the proximity of an object without making contact; and measuring flow of an open stream of water.

Choice of the appropriate technique for a given need is no longer straightforward. Numerous possibilities exist and deciding which is best is a multi-dimensional problem that needs a good underpinning of knowledge and experience. Measurement science and instrument engineering provides that basis.

Secondly, the user often does not know quite what should be measured. If it is required, for example, to assess the quality of an orange by objective measurement, just how would one go about the design of an appropriate instrument? Clearly some properties must be relevant and can be referred to with ease-size, colour, firmness, smell, shape and weight - but sensing some of these is not at all easy nor is it obvious how to do some of them.

Measurement science, in part, provides a formal structure for identification of the potentiality relevant attributes, which are labelled *referents*; and for their structured decomposition into actually measurable *measurands*, linked according to *Measurement Process Algorithm* (*MPA*)-these ideas are developed in Chapter 2.

The remaining sections of this introductory chapter complete the background to the disciplines of measurement science and instrument engineering. The all-pervasive application of measurement in the daily conduct of science and engineering and the control of processes is considered further and the historical context of the discipline briefly outlined. Suggestions and recommendations are made for further reading in these areas.

1.3 ON THE UNIVERSAL IMPORTANCE OF MEASUREMENT

1.3.1 The role of measurement science

The everyday requirement to measure things has already been pointed out. A deeper consideration is now required to appreciate the responsibility carried by those who practice measurement science and instrument engineering.

Science is the procedure through which we gain knowledge about the physical world. Information is collected, organized, ordered and generalized; experiments are conceived, performed and repeated; and as confidence and understanding develops the result is expressed in the form of a *scientific law*. This process is generally known as the *scientific method*. The methodology of the scientific investigation is aptly and rigorously summarized in the 1930's statement by Bertrand Russell.

> In arriving at a scientific law there are three stages: the first consists in observing the significant facts; the second in arriving at an hypothesis, which if it is true, would account for these facts; the third deducing from this hypothesis consequences which can be tested by observation.

In recent decades the hypothesis formed has been more commonly referred to as a *model*, a conceptual framework whose validity is always open to challenge or modification as further information is gathered.

Associated with the model are several, and sometimes very many, *parameters* which have numerical values that make the general model specific to a particular application. Figure 1.6 is an example of a model involving the need for many measurements–that of a rural system converting natural energy and rain into meat. The measurements are highlighted.

Clearly, measurement is essential to both the first (observing) and third (testing and validating) stage of any scientific investigation. At the second (model-forming) stage measurements are often made to 'calibrate' the model; that is, to determine appropriate parameter values. The modelling aspect is taken further in later chapters.

The theory and philosophy of measurement and its role in scientific method is a major topic which has been explored further by several authors, including Finkelstein (1977, 1982), Jones (1988), Stein (1970) and Sydenham (1979, 1982, 1983).

1.3.2 The role of measurement in the newer sciences

Central to the practice of science is the assumption that all experimental work will be *objective*. This means that phenomena observed and results obtained will not depend on the individual making the observation. Experiments may be repeated by different observers and each will get the same result (within the limits of 'experimental error'—see Section 2.5.5).

Considerable diligence is required on the part of experimental designers and operators to ensure that objectivity is maintained: the history of science records numerous episodes where certain phenomena have been observed by certain workers as a result of poor or inadequate experimental technique – one example was the nineteeth century research into the age of the Earth.

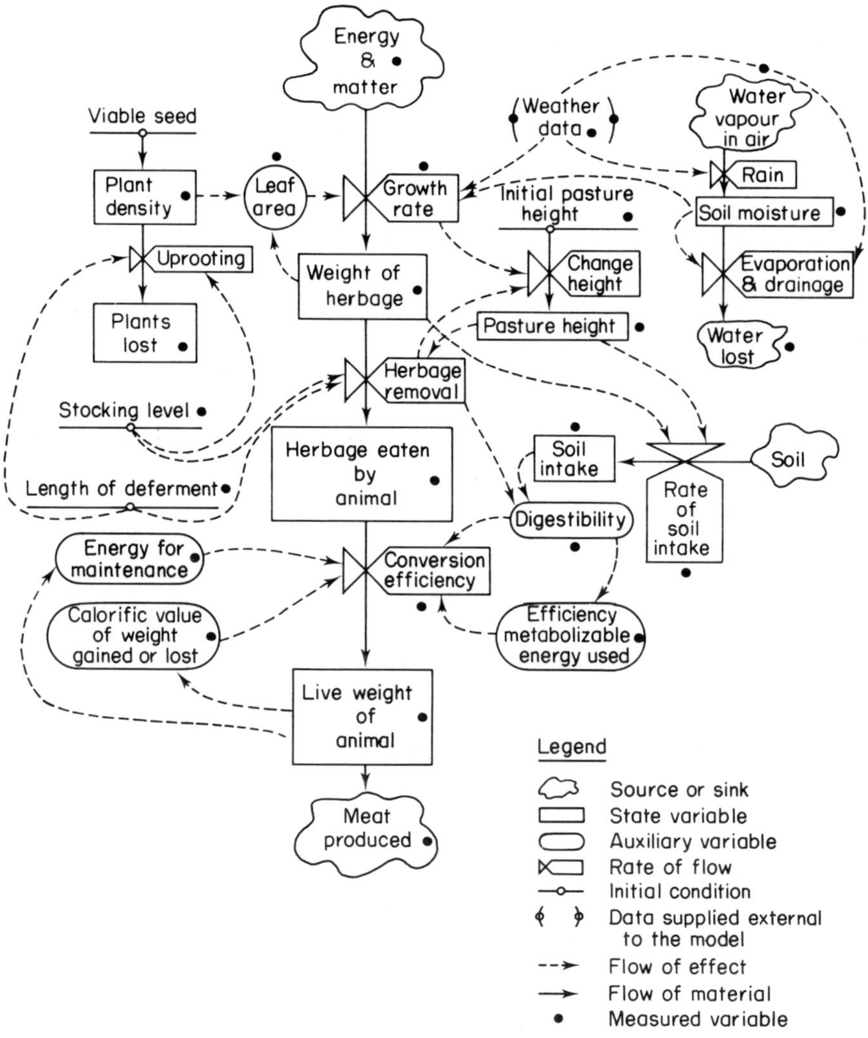

Figure 1.6 Model of system converting natural energy and rain into meat. (Reproduced by permission of the Faculty of Rural Science, University of New England, NSW, Australia.)

The use of agreed systems of measurement is one of the major ways by which objectivity is maintained in the conduct of science.

Unfortunately in some of the newer sciences (rather harshly labelled 'soft sciences') and at the frontiers of some of the physical sciences (arrogantly labelled 'hard sciences') reliable, objective measurement is very difficult or impossible at present. The reasons for this include unravelled complexity of the phenomenon and often the lack of satisfactory sensors or sensing techniques–consider for example, measurement of the taste of food or the measurement of intelligence—and see also Example 1.3.

Example 1.3 Many sensors are yet to be developed

It is not yet possible to measure the crime-rate distribution across a city with an overview measuring sensor system in the same way as we can measure crop damage in pastures, or sewage discharge in the sea, using the 'eye in the sky', remote sensing satellite scanners.

In the first case the model and its measurable parameters have not yet been realized (and thus, to some, it will be purely fanciful to even consider such an example) so apparatus cannot be devised. In the latter case the parameters for successful measurement are understood enough to enable physical apparatus to be set up that maps such variables into measured information.

To attempt to cope with these formidable measurement difficulties techniques have been developed within some of the newer sciences in which, according to some commentators, objectivity is compromised or even abandoned. This is not the place to debate the scientific validity of merit of these techniques - the reader is referred to the papers of Wild and Oliphant (1983), and Johnston *et al.* (1983) as an introduction to the debate. It is the challenge of measurement science to extend the frontiers of objective measurement, and the challenge of instrument engineering to design and construct satisfactory tools for the task. As long ago as around AD 1600 Galileo Galilei is accredited with saying.

> Count what is countable, measure what is measurable, and what is not measurable, make measurable.

1.3.3 The role of measurement in achieving control

Measurement is also concerned with providing information about the state of systems, usually man-made, that are already understood. Here the objective of the measurement is to assist exercise control over the system and the measurements are made repeatedly or continually, and often automatically. Such systems span an increasingly wide range of human endeavour from controlling processing and manufacturing plant to use within such products as motor vehicles, domestic kitchen appliances and military defence equipment.

Conception and design of such commercial measurement systems may not be as glamorous as, say, a giant telescope, but they are no less demanding in their development. Figure 1.7 shows a car factory arrangement in which numerous measuring devices and systems are used to control manufacturing processes and quality of cars.

The commercial importance of measuring systems is often high. Their design needs high quality and high reliability - catastrophic damage and loss of human life can result from failure of some of these systems. Obvious application examples are aircraft flight controls, nuclear reactor control systems and medical electronic life support systems employed in modern hospitals.

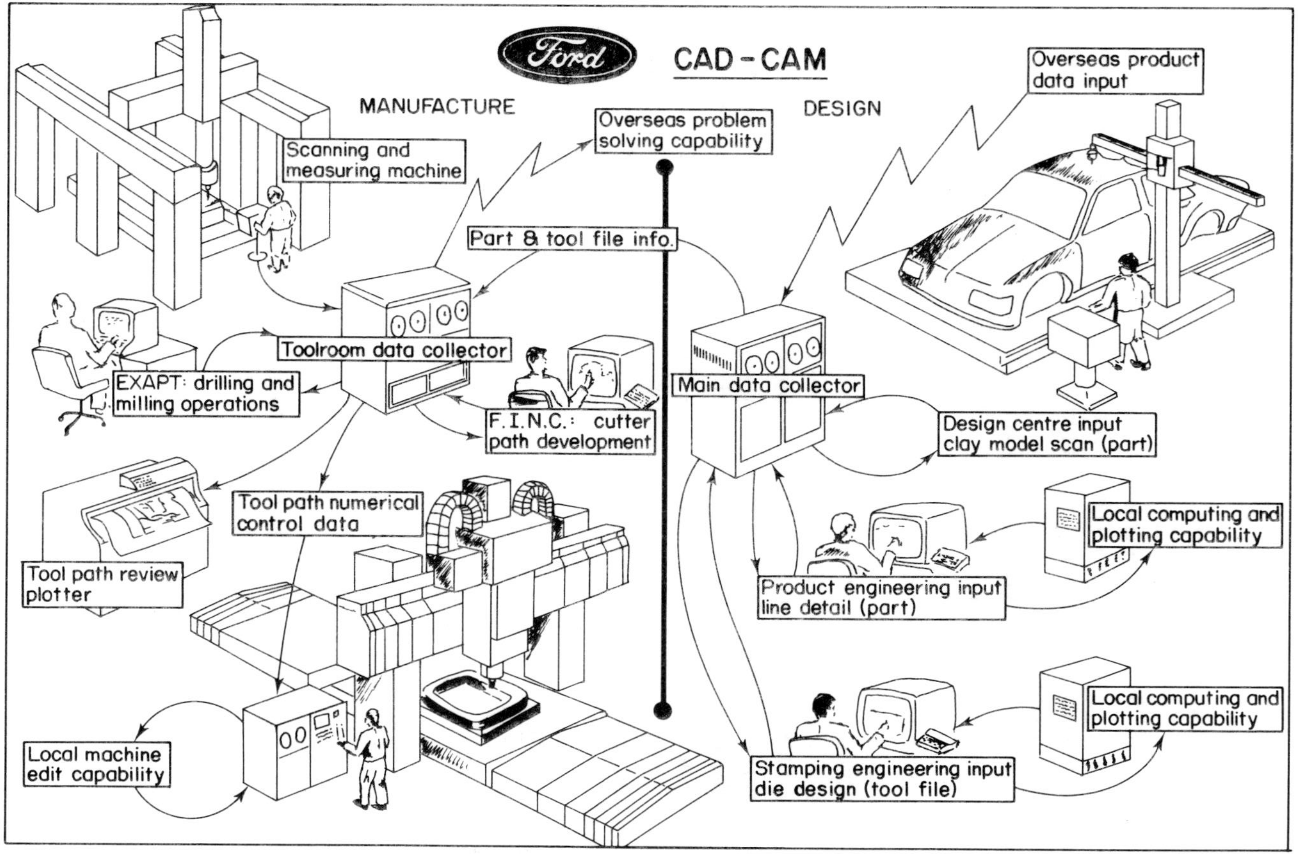

Figure 1.7 Measurements are needed to bring about control as shown here in a car manufacturing plant. (From Deverson D F, 1988, Paper presented I E Aust Bicentennial Electrotechnology Conference, Melbourne, May 1988.)

It is important to remember that the quality of control achieved can only be as good as the information derived from the measurement systems employed.

1.3.4 The need for a National Measurement System

Several countries have recognized that intimately interwoven into their national activity is the ever-increasing presence of the need for, and application of, measurements. Some governments have made specific studies of their *National Measurement System* (*NMS*). These studies have shown how the various national endeavours each need measurements and how they impact on the various sectors of a nation.

Study of the National Measurement System of the USA commenced around 1965, reaching major reporting stages in the mid 1970s. Key reports such as Huntoon (1967), which explains the concept, and NBSIR (1975), which overviews the findings, demonstrated that the economic value of the NMS of a nation lies in the range 0.1–5% of the nation's Gross National Product (GNP) and depends on the state of its economic development. In each it is a very large amount of resource! In the USA, for example, the Sandia Corporation alone spent, in 1975, around US$10 million to maintain its ten calibration laboratories which were estimated to impact directly on investment worth over US$1 billion.

Finkelstein (1976) has shown that around 5% of the capital cost of the UK process industry is in its instrumentation. The defence industry is likewise heavily dependent on instrumentation systems. Modern front-line equipment in particular employs a very wide range of instrumentation (too vast to list here) and in the case of a military submarine, or submarine search aircraft, a considerable part of the gross cost is directly attributable to sensing and instrumentation systems installed.

1.3.5 On the need to improve the quality of measurement

It is an obvious platitude to suggest that the quality of measurement needs to be continuously improved. However, quality is not simply the accuracy of measurement. Quality of instrumentation also includes:

- its appropriateness to the particular measurement task;
- knowledge of the uncertainties associated with the instrumentation;
- reliability of the instrumentation.

It is important, therefore, to understand the *capabilities* and *limitations* of instruments. Only then can the user adequately make selection from *off-the-shelf* equipment, or adequately specify the required performance to the instrument designer.

The penalties for poor quality instrumentation range from financial loss to major catastrophe. Consider, for example, the financial loss to the generating authorities of, say a 1% lower measurement metering of electricity or gas flowing into consumers' homes. Consider the loss of production when components are manufactured out of tolerance such that they have to be rejected. Consider the risk to human life due to the poor measurement of anaesthetic and drug dosage.

In the practice of science researchers have long realized they need measurements of the highest quality. This is not only to be able to observe new phenomena but also to test agreement with, or departures from, their theoretical predictions. It is the small difference between observation and theory that often constitutes a breakthrough in fundamental knowledge – see Example 1.4.

Example 1.4 How small differences in measurements can alter the path of progress

In the seventeenth century the astronomer Joannes Kepler had a thorough understanding of the capabilities of the (crude) astronomical instrumentation in use at the time. As a result he was given confidence to seek a better explanation of planetary motion than the current model of circular orbits.

He inherited a large set of astronomical observations made by the meticulous observer, Tycho Brahe. Consistent departures from values expected from the circular orbits model, of a mere one minute of arc magnitude, existed and had been put down to being within instrument uncertainty. Kepler was able to show that whilst the instruments of the day were simple (sighting rods with angular scales) they did have an uncertainty level less than these discrepancies. He sought, and found, a more complex motional model to which the data had a better fit: this was of course, the model of the planets moving in elliptical paths.

Herschel, in the eighteenth century, wrote about the importance of seeing small differences between theory and observation. He coined the term *residual phenomena* (see Jones, 1968).

The application of measuring instrumentation in the practice of science has risen dramatically over the past century and today scientific discovery can seldom be made without using the most sensitive measurement systems available. Hence the size of the residuals awaiting verifications are continually decreasing so that it is essential to be constantly improving measurement capability.

1.4 A BRIEF HISTORY OF MEASUREMENT SCIENCE AND INSTRUMENT ENGINEERING

1.4.1 A sense of history is needed

The current sophistication of instrumentation can create the impression that development has followed an ordered, scholarly development path. This has not been the case. A comprehensive history of measurement has been published (Sydenham, 1979) from which the following brief account has been extracted. It is, of course, inextricably bound up with the history of science itself.

1.4.2 First instruments–ancient times

As the earliest people moved to live in groups they saw the need for ordered trade. To sell and buy commodities and levy taxes they required some simple measurements. Weighing balances and weights, measuring containers for volume, rods representing lengths and simple timekeeping devices have been described in the classical literature. Some artifacts still exist today, the oldest example being from around 6000 BC. Such trading needs led to *legal metrology* at least as early as 3000 BC. Standards of units were drawn from daily experience–the carat, a unit of weight, was a seed of the date palm. The inch, foot, cubit and yard were from the size of parts of the adult human body (a thumb bone, foot, forearm and stride length, respectively.)

Early instruments were dominantly mechanical in form. Figure 1.8 shows some instruments representing those of the ancient world.

Simple mechanical computing devices, such as the astrolabe, were made to predict planetary movements; astronomy was the earliest science to be practised.

Some measurement was undertaken for scientific investigation but this was rare. A few, surprisingly complex, instruments are dated to this period but they are untypical and embodied principles associated with much later times. The Antikythera mechanism, for example, employed planetary gearing and has been dated to 80 BC.

Philosophical considerations about measurements have come down to us from scholars of Ancient Greece, but these records provided little material on better ways to measure.

Around AD 100 recorded knowledge passed to the custodianship of the Islamic nations. As surprisingly little activity was recorded in the Western world from that time to around AD 1100, this period has become known as the *Dark Ages*. However, Islamic contributions to clocks and beam balance design are recorded. Clock, magnetic compass and direction-finding instrument developments also existed in Ancient China at this time.

Overall the science of measurement developed little during this 7000-year period – a slow beginning indeed.

1.4.3 Experimental science begins – Middle Ages to AD 1800

The reawakening of interest in secular intellectual thought, from around AD 1000 in Europe, led to an increase in the pace at which sensing techniques were developed. This pace of development has continued to accelerate up to the present day.

The 700-year period to roughly AD 1800 saw development from primitive measurement through to the first electrical transducers. It was during this period that man began to use sensors to enhance his natural sensing ability and to practise observational science. The settling out of the modern scientific method, most noticeably from about AD 1600 onward, provided systematic research

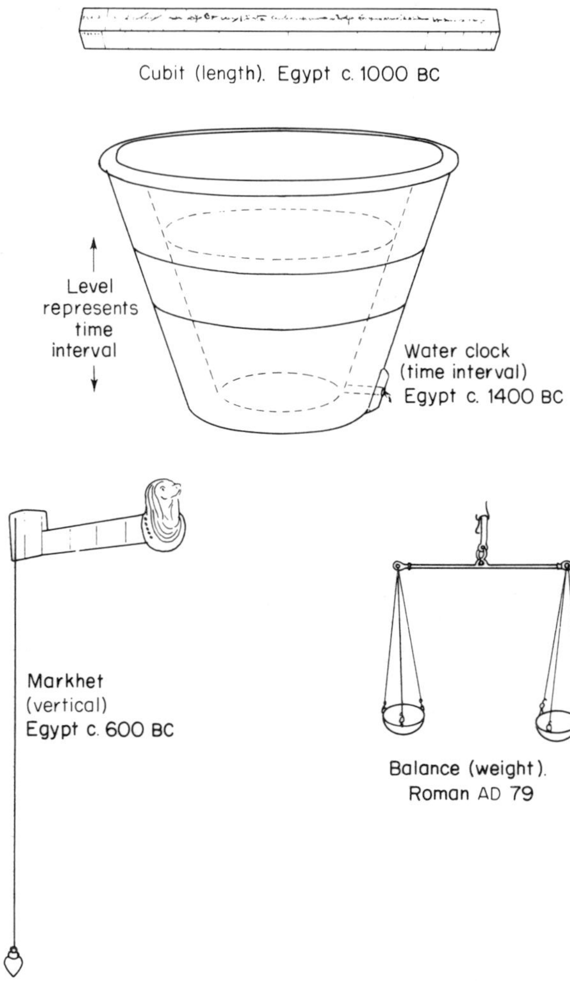

Figure 1.8 Typical instruments of the ancient world. (Drawings of objects in the Science Museum, London.)

methods. These brought with them the need for increasingly more sensitive instruments.

Optical science moved from unaided sensing to devices that provided ability to see things far away and close up – the telescope and the microscope. The nature of light and its composition gradually were realized through experimentation and theorizing. At the pragmatic level spectacles were invented to give people improved sight. Astrolabes, quadrants and sextants gradually improved as sea navigation increased.

The mechanical aspect of instruments continued to improve in parallel with the advancement of the mechanical sciences. The clockwork mechanism providing a constant-velocity drive for recorders was perhaps the most obvious manifestation. Mechanically controlled time pieces were made using gearing and a range of innovative mechanisms to give constant rotational speed.

From their birth in cathedral clocks, which needed to be of quite large proportions, mechanisms were later miniaturized to give pocket watches, and increasing quality of design and manufacture led to the magnificent chronometers that could keep time to within a few seconds over many months – the accuracy needed to navigate from Britain to the West Indies. Principles of compensation for the thermal expansion of components were well understood and incorporated in many clocks.

Tensile-testing machines developed to support a growing understanding of the importance of the strength of materials in mathematical approaches to design. As all instruments of the day needed mechanical technology advances were many and varied.

Magnetic effects were observed by oldest man, but it was Gilbert, in AD 1600, who has become known as the father of this field of science. Naturally occurring magnetic minerals, the Earth's field and electrical storm activity invited experimentation. Magnetic geophysical instruments developed including compasses to help miners dig better-directed tunnels and navigators to travel about more confidently.

The generation of magnetic effects by electricity soon became of interest and electromagnetism became a recognized field. The torsion suspension used in magnetometers led to instruments for measuring static electric charges, the only kind of electricity then known. Figure 1.9 shows Coulomb's apparatus for investigating magnetic effects – clearly a forerunner of several electrical and magnetic indicating instruments.

Static electricity was of great interest for it was spectacular in its effects. Until AD 1800 electricity was only known in its electrostatic form. Charge generators, from small to huge, were constructed, gradually pushing up their voltage level and ability to sustain a steady stream of charge. These machines were often the pride of the wealthy who used them as party games in their drawing rooms and, since they were used as status symbols, the finest craftsmen were commissioned to build them.

The electrostatic era began with pith balls and rubbed cats fur and ended with knowledge of insulators, capacitors, polarity, objective charge measurement, pyroelectricity and how to apply this unlikely energy source as an electrical telegraph system. However, only a few simple scientific laws were deduced from this mass of experimental results. Electrical circuit laws, for example, did not exist and until quite late in the period it was still uncertain whether it was the colour, size, type of material, or other properties that kept electric charge in a cable.

The fields of *thermometry*, the measurement of temperature, and the establishment of temperature scales; *hygrometry*, the measurement of humidity; *psychrometry*, the measurement of atmospheric humidity using two thermometers; and *anemometry*, the measurement of fluid velocity, particularly wind speed, have their foundations in this period.

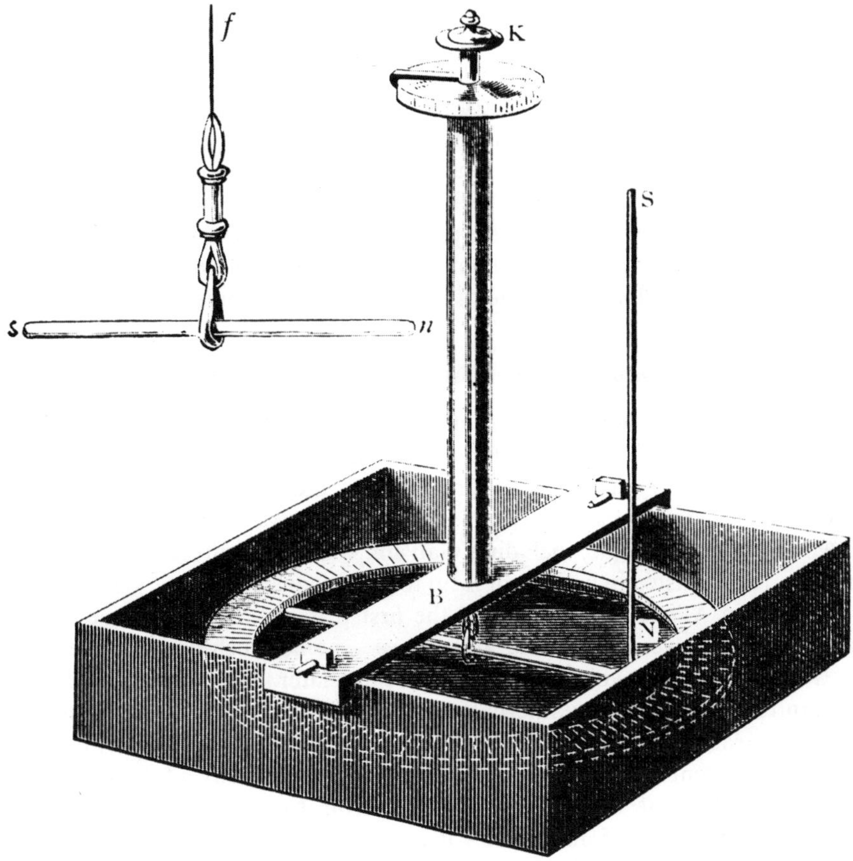

Figure 1.9 Coulomb's torsion balance apparatus. (From Mullineux-Walmsley R, 1910, *Electricity in the Service of Man*, Cassell, London.)

During these 700 years weaving machines were invented and toward the end of this period they were developed with a high degree of automation. Many forms of mechanical 'automata' were constructed, and associated with this endeavour was the advent of modern computing. Pascal, Moreland and Leibniz built mechanical calculating machines. However, such acts were seen as carrying out a 'God-like' function and so their construction, in common with pursuit of much of the seventeenth-century scientific research, incurred the wrath of the established Church. Opposition was extreme and often put the lives of their inventors and researchers in serious jeopardy.

By the end of the period numerous scientific laws had been discovered with reports being placed regularly in a growing number of scientific journals.

Scientific research was, however, largely undertaken by those with independent financial means – the wealthy, the clergy, the tax collectors and those patronized by the rulers and the wealthy. Tycho Brahe, the sixteenth-century astronomer mentioned earlier, was supported this way. Harrison, a carpenter, took many years to make his famous chronometers and worked in the hope of winning a great monetary prize – which he eventually did.

Accumulation of wealth outside of the Church toward the end of this period allowed science to advance more freely. An instrument manufacturing industry, of private firms, grew up in which classical craftsmanship for its own sake and embellishment with ornate and supernatural imagery gradually gave way to the need for performance at a reasonable price. However, each instrument was made individually and its manufacture could take several years. Mass production of cheaper, more affordable instruments, however, was to come much later.

1.4.4 Foundations of electrical method – nineteenth-century AD

In AD 1800 a simple device was invented that greatly speeded up the progress of science and of instrumentation – the electrochemical Voltaic cell. This was the forerunner of the dry cell, today's torch battery and calculator battery. Figure 1.10 shows the original apparatus.

Although many people, including Galvani, were gradually becoming aware that electricity existed in a different form to that from electrostatic generators, it was Volta, in AD 1800, who formed the first primary cell from brine-soaked paper placed between silver and zinc plates – names such as electrodes, ion, electron and polarity, voltage and current came later. The device provided (in modern terminology) a low internal resistance voltage source that was easy to construct, reasonably cheap, used commonly available materials and produced an electric current for a reasonable time duration of typically a few minutes. Its importance was that, being simple to make from easily available materials it provided for widespread and reliable use.

The immediate result of this discovery was that the gentlemen of the time could now experiment more easily with batteries, wires, coils of wire and with simple systems built from these components. When asked by a visitor to show him his laboratory Wollaston, the optical experimenter, had his butler carry in a tray of components – that was all he needed. Until this time in history it had been easy to conduct optical and mechanical experiments – now electricity could also be conveniently researched, whether for curiosity or simply for entertainment.

Commercial gain also began to enhance the pace of development. By 1830 the need for better communications led rapidly to the invention of workable electric telegraphs in which information was conveyed over considerable distances by the on–off switching of electric current in a complete circuit loop formed from a pair of wires. This was one of the earliest examples of a digital process using binary (on–off signals) information and necessitated the definition of symbolic codes, the most famous being that invented by Samuel Morse and demonstrated in 1844. Invention such as this used the developing laws of electromagnetism to build relays, switches, sounders and, about fifty years later, telephones which could convey speech signals directly in analogue signal form.

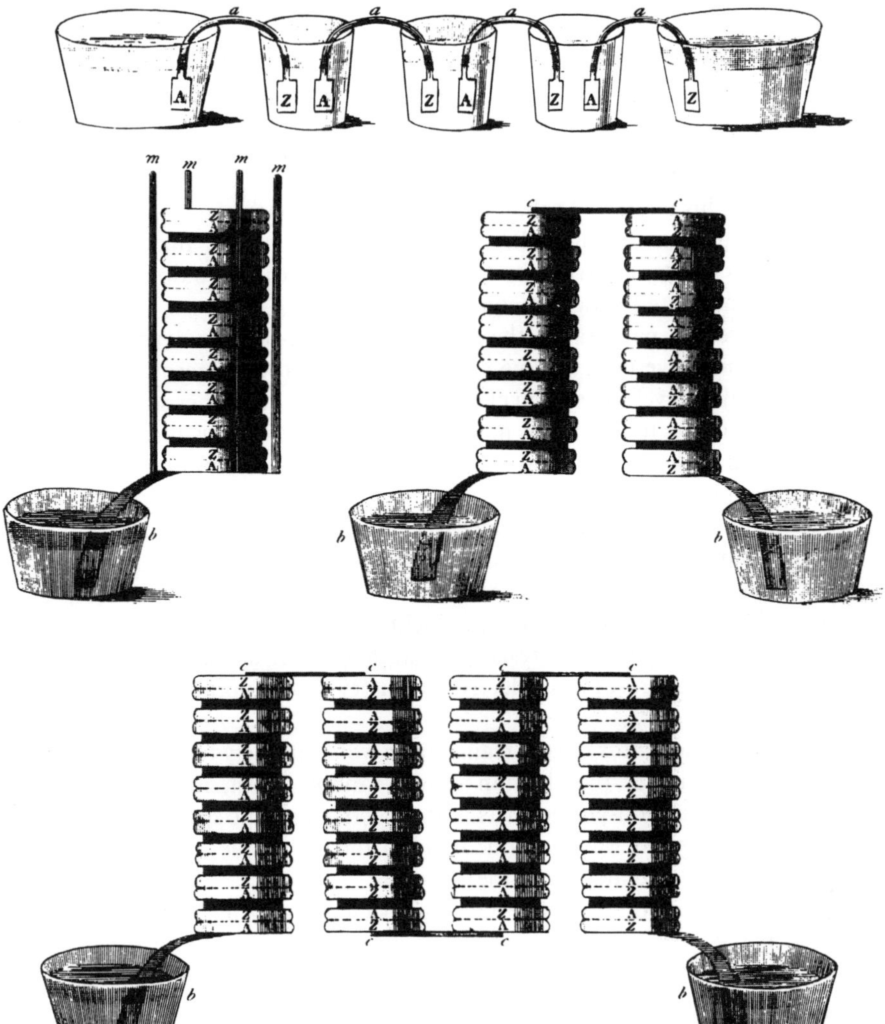

Figure 1.10 Volta's original illustrations of his voltaic cell. (From James-King W, 1962. The development of electrical technology in the 19th century, *US National Museum Bulletin 228*, Smithsonian Institution, Washington, DC.)

In parallel with these developments electrical measuring instruments progressed from gold-leaf electroscopes to sophisticated galvanometers, electrostatic voltmeters, wattmeters, current balances and the forerunners of modern moving-coil indicating meters.

The nature of the electron and other charged particles was gradually unravelled and a host of other phenomena observed which led directly to the establishment of so-called 'modern physics' in the early twentieth century. The basis of the cathode ray oscilloscope also developed in cathode discharge research. Radio was invented using simple, brute force, electrical methods. X-rays were discovered by chance during study of radioactive materials. The circumstances for electronic devices to appear were in place.

The design of calculating-engine machines advanced to incorporate stored-program concepts based on Babbage's work. His mechanical computational machines were to be driven by a small steam engine but did not become operational in his lifetime. At that same time philosophers, such as Boole, were devising the algebra of logic (Boolean algebra), formalism that later would permit logic circuits, electrical and later electronic, to be combined to make calculators and stored-program computers.

In the preceding paragraphs the major nineteenth-century developments in physical science have been outlined. In parallel, and too numerous to describe here, were major developments in chemistry, materials and other sciences; in the engineering of structures and of machines to produce power, mechanical and electrical; and in the ideas of feedback control. Limited automation of processes became possible.

In addition, and of great significance to sensing and instrumentation, a host of other miscellaneous physical phenomena were observed and studied during this period. These phenomena form the basis for many present-day sensors and actuators. Well-used transduction laws include the thermoelectric effect (hence, thermocouples); the electrical resistance/mechanical strain interaction (hence, strain gauges); electro-optics; electro-chemistry laws; opto-magnetism; and piezo-electrical effects.

By the end of the century many transducers were routinely being used to sense and measure variables that human senses cannot detect, most significantly perhaps the various non-visible parts of the electromagnetic spectrum. Infra-red radiation, for example, was quantified and at the end of the century Callendar measured the heat radiated from the Moon.

Significant technology transfer had also occurred and instrumentation had begun to make extensive use of existing technology developed for other fields. Devices used in everyday telegraphy, and later telephony, were converted into stream-flow gauges, metal detectors and crack detectors.

Toward the end of the nineteenth century the practice of science and engineering had become increasingly internationalized. The need arose for international standards of physical measurands; this led to world-wide agreements and the emergence into practice of the metric system, in many variants. National Measurement Laboratories (Section 1.3.4 above) were created to support this standardization.

In summary, by the end of the nineteenth century there was widespread, if not universal, use of objective measurement techniques and quite sophisticated instrumentation in the daily practice of science and engineering. It is perhaps surprising, therefore, that the philosophy of measurement attracted only scant attention. A few individuals, including Lord Kelvin and Helmholtz, sought to formalize and structure some aspects of the process of instrument design. However, in general interest was sparse and did not gain significant recognition as a useful academic or commercial activity. Instrumentation continued to be developed by referring to the experience of others and by adding a little amount of science and a lot of intuition and innovation.

1.4.5 Electronics and information technology – the twentieth century

The pace of instrumentation development in this century has continued to accelerate to the present day and shows no signs of slowing. Today's instrumentation products and techniques are often overtaken by dramatically updated versions in a mere two-to-five years.

It is not possible in the space available here to do justice to the wide range of major developments that have already taken place in this century. However, there have been two dominant influences in the form and direction of instrumentation development. These are firstly the emergence of *electronic engineering* at the turn of the century; and secondly, later, the explosive growth in the area of *information technology* (*IT*).

Electronic engineering has its origins in marine radio communication systems. Transmission from ship to ship and ship to shore required better detectors, amplifiers and generators of radio-frequency signals and led to the invention of the thermionic diode (1904) and then the amplifying triode (declared in a 1907 patent for detection). Developments in radio communication were rapid and spectacular and have tended to cloud the fact that instrumentation was also starting out in the form we see today.

Early uses of these vacuum tube devices (called valves) for instrumentation included telemetry between freight trains and signal poles, radio picture transmission (fax in today's terminology), film sound track (see Figure 1.11) chart recorders with kilohertz response, industrial controllers of furnaces and the like, aircraft and ship instrumentation, medical and industrial X-ray, electrocardiographs, direction finders and gun-position locators.

The convenience (and later the low cost) of electronic systems rapidly displaced mechanical instrumentation systems. Complex arrangements of springs, levers, gears with their problems of stiction, friction and maintenance were replaced with electronic equivalents wherever they could be devised. The first working stored-program computers used electromechanical parts – these were displaced by valve-based computers in the 1940s. In the 1950s the trend was so strong that interest in fine machines declined to the all-time low of today.

The invention of the solid-state amplifying transistor in 1947 opened up possibilities to process information that were hard for many to comprehend. One famous physicist Sir George Thomson, in 1955, made a forecast that the transistor might find some limited use in 'walkie talkies' – two-way radios – but that they were unlikely to find many other applications!

Military conflict has always provided incentive and effort to exploit new ideas and new technologies. The First World War (1914–1918) hastened the development of radio communications and the Second World War (1939–1945) saw the development of radar systems, gun aimers, homing bombs, flight controllers and

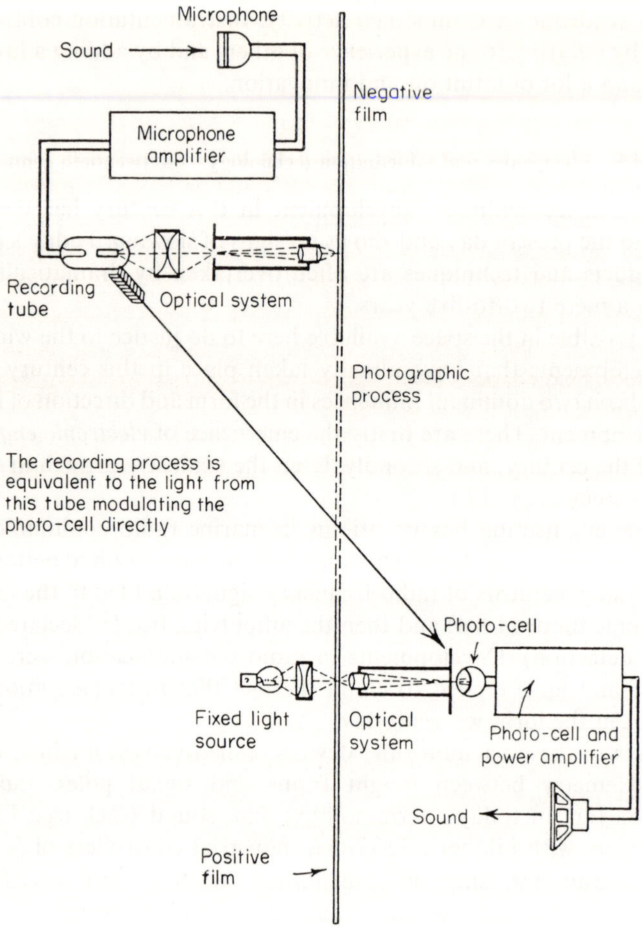

Figure 1.11 Talking picture film sound-track by optical means, c. 1930. (From Walker R C and Lance T M C, 1933, *Photoelectric Cell Applications*, Pitman, London. Reproduced by permission.)

other devices in which instrumentation was a critical part. The Second World War is credited by many as being the event that took warfare into an era where it became a battle of technologies - see, for example, Jones (1978). These inventions demonstrated that electronic devices could be reliable and cheap enough to be of economic value to industry and medicine as well as to the military.

The miscellany of physical 'effects' and interactions discovered in the nineteenth century found use as sensors when electronic amplification became available from around 1920. The resistance strain gauge (suggested by Faraday) was first used in engineering measurement in the 1930s; the light sensitivity of selenium was used in picture transmission and in light intensity control by that time. Thermocouple and platinum resistance thermometry became routine in process control from around 1900.

As electronic engineering progressed – and is still progressing – the possibilities for enhanced sensing and hence better measurement science continue to broaden. Ever smaller sensing signals can be detected and amplified. Sophisticated techniques such as phase-sensitive detection and lock-in amplifiers permit signal measurement even in the presence of high levels of extraneous noise. As a result further potential sensor techniques have appeared.

Semiconductor effects are now employed in a range of applications, for example, as strain gauges, thermometers and radiation detectors. In the last decade superconductivity effects have been employed to measure extremely small magnetic field changes and offer the possibilities of earthquake prediction and enchanced medical diagnosis.

The development of the integrated-circuit manufacturing technique has led to a range of silicon sensors manufactured by chemical 'micromachining'. Laser and optical fibre technology developed for communication purposes is employed to make gyroscopes which are entirely without moving parts and to form a range of sensors capable of operating in hazardous places such as where high radiation, high voltage and flammability exist.

Clearly one of the most spectacular applications of electronics has been in the construction of calculating engines (or as we now call them, computers). Compared with prior mechanical arrangements or vacuum tubes, the dramatically lower power requirements, higher reliability and smaller size of electronic components required for each logic function permitted machines of increasing power and complexity to be built.

However, it was not widely appreciated until the 1970s that computers handled, stored and processed *information* rather than just numbers; and that computers, their interfaces and intercommunications were all aspects of information technology. Humans, of course, interfaced with computers via punched cards and paper tape initially, and more recently via keyboards and video display units (VDUs). Sensors could also be interfaced to computers, usually via an analogue-to-digital converter (ADC) – the era of modern instrumentation was born.

The use of digital electronics in measuring instruments was at first restricted to inherently digital tasks such as counting circuits for frequency and nuclear particle measurement. In the 1960s the appearance of the light-emitting diode (LED) and later the liquid crystal display (LCD) led to cheap digital voltmeters (DVMs) and these were rapidly employed as digital panel meter (DPM) modules to replace the traditional moving coil and pointer movements. The internal design of DVMs and DPMs permitted remote digital output of the displayed voltage, usually in a binary-coded-decimal (BCD) format, at virtually no extra cost. Hence the direct digital interfacing between test equipment and computer systems became commonplace.

A further major change occurred in the 1970s when microprocessor integrated circuits became available. Microprocessors and complete microcomputer boards rapidly became incorporated within instruments and were permanently programmed to provide sophisticated facilities for the user such as auto-ranging, auto-calibration, averaging and peak detection.

In the late 1970s the semiconductor memory essential for microprocessor systems tumbled in price, and large-capacity, microprocessor-controlled data loggers became available at relatively low cost. The user of instrumentation was thus encouraged to sense and log more and more variables and, as the information was stored in digital form, employ a desk-top computer to perform signal processing and data reduction.

By the 1980s microcomputers (or 'personal computers') had increased markedly in speed and power such that much signal analysis could be undertaken 'on-line'. Keystroke or program control could, for example, initiate software to compute a frequency spectrum from a time sequence of measurements made with a data logger; perform correlations between different signal sequences or make decisions dependent on the information acquired (i.e. a sensor signal) or on results calculated. Instruments with these types of facility built-in were rapidly labelled 'intelligent' or 'smart'.

During these recent decades it was hoped that the microelectronic and microprocessor 'revolutions' might be matched by a similar 'sensor revolution' in which cheap solid-state, general purpose, sensors would be available. This is still to occur, although there have been significant advances, notably in silicon, opto-electronic and chemical sensors. Such a change of availability is now also being driven by the needs of the various industries, notably in automobile manufacture.

Hence there has been a tendency in recent years to emphasize software and signal-processing aspects of instrumentation. This is perhaps unfortunate, particularly for the non-specialist user, for two reasons.

Firstly, the user is increasingly 'separated' from the information input (from the sensor or sensors) by the signal processing being performed between the input and the final display presented to the user. The user has little choice but to trust the instrument.

Secondly, there is a tendency to believe – often erroneously – that the signal processing can make up for inadequacies in the siting of the sensor, or in the performance of the sensor itself. Clearly the result of any measurement, just like the result of any calculation, can only be as good as the initial information sensed or the data supplied. It is hoped that the increased appreciation of the fundamentals of measurement sciences will help to counteract the tendency.

In parallel with the dramatic technological developments in measurement capability and technique so far this century there have been more significant developments in the philosophical aspects of measurement science. Attempts to measure, for example, the effectiveness of education (around 1900) opened up interest by the emerging 'social' sciences. Texts on measurement theory have been published.

To date over 700 books have been published on topics related to measurement – mostly of the catalogue style. Professional societies have emerged for instrumentation matters. Consultancies, laboratory accreditation schemes, journals, conferences and many more infrastructure mechanisms now exist.

However, despite there being increased activity in applications of measurement, only a handful of workers have made lasting, foundational contributions. In the spectrum of measurement activity it is the academic area that has been the last to develop. It is hard to explain why this field has attracted so little theoretical interest compared with control engineering, its partner its industrial progress.

Today there is a growing awareness that measurement science and instrument engineering can provide scholarship and intellectual challenges equal to any other discipline. Furthermore, and perhaps more importantly, there is also recognition that a more rigorous. formalized approach to instrumentation design and use can avoid perpetual rediscovery of pitfalls and 'reinvention of the wheel', can yield better instruments designed more quickly and can lead to better measurements made with instruments to hand.

It must be recognized, however, that the discipline, although well established, still has far to go.

TUTORIAL EXAMPLES

1.1 Keep a diary for a day recording the measurands that are involved in your activity. List the measurands, the instruments used and the time taken to make the measurements. For each measurand develop the 'cost of measurement' by estimating the time involved in making the measurement and amortizing the capital cost of the instruments at 10% of their purchase price per annum.

1.2 Two metal plates placed close together, but not touching, form an electrical capacitor for which the capacitance is given by $C = \varepsilon_r \varepsilon_0 \, A/d$ where

- ε_r is relative permittivity of the gap volume;
- ε_0 is permittivity of free space;
- A is area of overlapping plates;
- d is distance the plates are held apart.

Draw the physical arrangement, for each case, by which this simple transducer could be used to sense displacement, area, permittivity of liquid, temperature, flow, acceleration, and acidity.

1.3 Locate and list the various measurement and standards agencies in your country. Study their activities and draw a systems diagram showing these and their relationships.

For your country, where would you take a surveyor's tape (that used to set up field tapes) for calibration?

Who is responsible for checking that petrol pumps deliver the correct amount?

1.4 Compile a short account of the development of one of the following instruments:
(a) chart recorder; (b) signal amplifier; (c) micrometer; (d) timer.
(Sydenham, 1979 is a source for all cases).

REFERENCES

Finkelstein L, 1976, Preliminary study of the economic significance of industrial measurement and instrumentation technology for the United Kingdom. Internal Report DSS/LF/110, The City University, London.

Finkelstein L, 1977, Instrument Science - introductory article. *J. Phys. E: Sci. Instrum.*, **10**, 566-72.

Finkelstein L, 1982, Theory and philosophy of measurement. In: Sydenham P (Ed), 1982, *Handbook of Measurement Science*, Vol 1, Wiley, Chichester.

Johnston R, Encel S and Hooker C, 1983, The soft sciences. In: Williams R (Ed), 1983, *The Best of the Science Show*, Nelson, Melbourne.

Jones R V, 1968, More and more, about less and less. *Proc. R. Inst.* **43**, 202, 323-45.

Jones R V, 1978, *The Most Secret War*, Hamish Hamilton, London.

Jones R V, 1988, *Instruments and Experiences*, Wiley, Chichester.

Huntoon R D, 1967, Concept of a national measurement system. *Science*, **158**, 67-71.

NBSIR, 1975, Final summary report study of the National Measurement System 1972-1975. NBSIR report 75-925, National Bureau of Standards, Washington, DC.

Stein P, 1970, Sensors as information systems. *Research/Development*, June, 33-40.

Sydenham P H, 1979, *Measuring Instruments; Tools of Knowledge and Control*, Peter Peregrinus, Stevenage.

Sydenham P H (Ed), 1982, *Handbook of Measurement Science*, Vol 1, *Theoretical Fundamentals*, Wiley, Chichester.

Sydenham P H, (Ed), 1983, *Handbook of Measurement Science*, Vol 2, *Fundamentals of Practice*, Wiley, Chichester.

Wild P and Oliphant M, 1983, The hard versus the soft sciences. In: Williams R (Ed), 1983, *The Best of the Science Show*, Nelson, Melbourne.

BIBLIOGRAPHY

In addition to those texts listed amongst the references above, the following are recommended general works. However, there exist many other worthy and valuable publications whose omission from this brief bibliography should not be taken as a criticism.

Bentley J P, 1983, *Principles of Measurement Systems*, Construction Press (Longmans), London.

Doebelin E O, 1975, *Measurement Systems Application and Design*, McGraw-Hill, New York.

Jones B E, 1987, *Current Advances in Sensors*, Adam Hilger, Bristol.

Mansfield P H, 1973, *Electrical Transducers for Industrial Measurements*, Butterworths, Sevenoaks, UK.

Noltingk B E, 1987, *Instrument Technology*, Vols 1-4, Newnes Butterworths, Sevenoaks. Also as *Instrumentation Reference Book*, 1988.

Norton H N, 1969, *Handbook of Transducers for Electronic Measuring Systems*, Prentice-Hall, Englewood, Cliffs, NJ.

Sydenham P H, 1984, *Transducers in Measurement and Control* (3rd edn), Adam Hilger, Bristol.

CHAPTER 2

Fundamental concepts

2.1 INTRODUCTORY REMARKS

Many of the concepts and ideas relevant to measurement science and instrument engineering are commonplace – so much so that they are seldom rigorously stated or discussed. Indeed many of these ideas were simply assumed in the discussions of the previous chapter. Other concepts, however, especially those concerned with the nature of the measurement process, have been adequately articulated only in recent years. Both categories are considered in this chapter: their adequate appreciation is essential for sound instrument design.

2.2 DEFINITIONS

Definitions are stated and discussed for five terms fundamental to the measurement process, namely – signal, data, calibration, information and knowledge. Four of these words are part of common English usage; however, standard dictionary definitions are less rigorous than those necessary for measurement science.

2.2.1 Signal

A *signal* is that which flows through the conduit formed, by the measuring system, between the system being observed and the eventual output of the measuring system.

These days a signal within an instrument, or as part of a measurement system, is often envisaged as an electrical phenomenon. However, any disturbance such as a mechanical movement or vibration, also constitutes a signal. Likewise an emission of, or change in, the transfer of light, sound or other energy can also act as a signal. In each case the energy flow is labelled the *carrier* and the disturbance, that is, the change from its normal energy state, is labelled the *modulation*. Transportation of mass can also be used to convey signals but that is less commonly used than the energy carrier.

The conduit from the system being observed and extending through the measurement to its output may be wires, cables, levers, gear trains, pipes (for mass flow), solid, liquid or gaseous media (which convey vibrations, e.g. acoustic or ultrasonic). The conduit can also be empty space (a vacuum) involving no physical contact between the system being observed and the measuring system. Radio waves, visible light, ultra-violet, infra-red and other electromagnetic radiations freely traverse empty space.

The energy or mass flow that constitutes the signal may change form as it passes through the measurement system, sometimes several times. In each case the transformation is performed by a *transducer*.

Finally we must question what the signal means or what it can tell about the system under observation. Hence it becomes apparent that a signal is also a *symbolic representation* of some attribute (or combination of attributes) of the system under observation. Ideally the symbolism should be unique and unambiguous but this is not always possible. Chapter 3 explores this concept further and sets out a theoretical basis for satisfactory symbolic representation.

2.2.2 Data

Data are the numbers associated with a signal, usually in some recorded form.

These numbers are often envisaged as a tabulated computer printout and if the signal is changing with time (or distance) the numbers, being snapshot values of a continuum, will constitute *sampled data* unless the signal has been processed in some way, perhaps integrated or totalized. The term *raw data* is often used to indicate that no form of processing has been undertaken.

In recent years the term *continuous data* has been coined to indicate that sampling has not taken place and in this case the record of the signal will be in the form of an ink line on a paper chart or perhaps a recording on magnetic tape. Strictly these are not numerical forms of the signal but simply further transformations of the signal into position (on the chart) or amplitude (of magnetization) respectively.

2.2.3 Calibration

Calibration is the act of turning data into information. This is achieved by comparison with some agreed measurement standard and the result is to give *meaning* to the symbolism inherent in the signal. Figure 2.1 shows a calibration of a flow-measuring instrument in progress: a known quantity of water is passed though the flow meter in a defined time. This definition immediately demonstrates the need for agreed measurement standards, the ramifications of which are explored further in Section 2.2.6. It also illustrates the clear distinction between data and information.

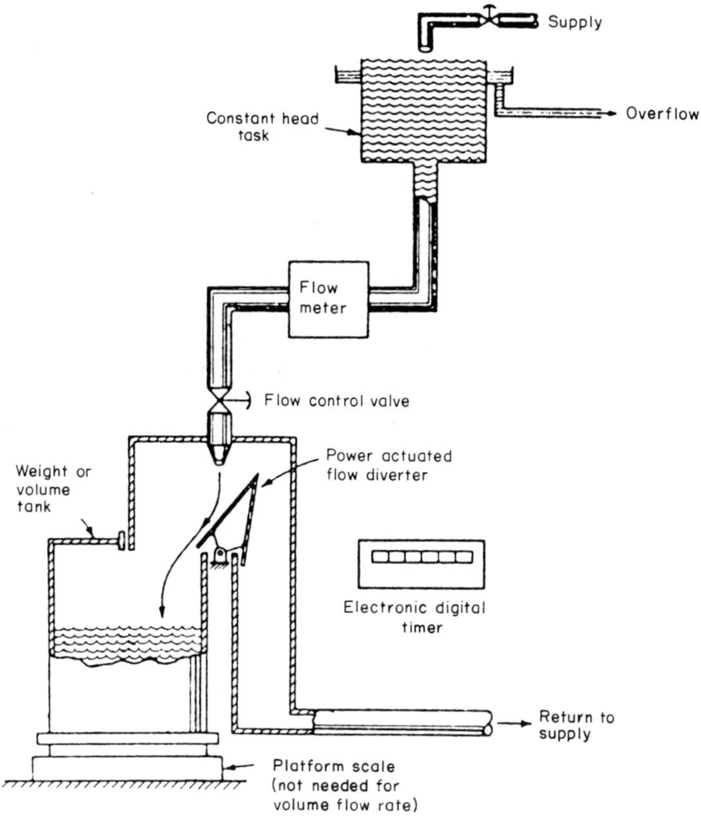

Figure 2.1 Calibration of a flow-measuring instrument. (From Sydenham P H, 1984, *Transducers in Measurement and Control* (3rd edn), Adam Hilger, Bristol. Reproduced by permission of IOP Publishing Ltd.)

2.2.4 Information

Information, as acquired using a measurement system, is the representation of attributes of the event or object being measured by a known and agreed symbolism. This follows directly from the previous definitions of signal and calibration and shows that information is the result of a complete measurement process. Furthermore, this does accord with the general understanding of the word information as a set of meaningful statements (for example, an encyclopaedia is 'full of information').

At this point it is also useful to mention the discipline of *information theory* pursued by communications engineers. This is the study of the quantity conveyed in a message transmitted between two points without concern for its actual meaning: it is thus a very different use of the word 'information'. Clearly, however, information theory is of direct relevance to the design of instruments and is considered further in Chapter 3.

2.2.5 Knowledge

Knowledge is generally recognized as an organized body of information. This simple definition is adequate for our present purpose, that is the clear understanding of the process of measurement, although a consultation of any dictionary will reveal that the word has much wider philosophical implications. The process whereby knowledge is extracted from an observed system by measurement is illustrated in Figure 2.2. Discussion of these topics is not appropriate in this book and the reader is referred to Stein (1970).

It is perhaps sufficient to appreciate that fully calibrated measurement systems (or, more simply, 'instruments') gather information and not knowledge. Knowledge is a matter of *interpretation* of the information.

A measurement or instrumentation problem is often first posed as a requirement for knowledge about some particular event or object. In such a case the first step is to determine the specific information required, as far as possible, and this may be a very difficult task. Once the information requirement has been specified, or guessed, measurement science can assist in the conception and design of appropriate instrumentation. A rigorous procedure for doing this is discussed in Section 2.3.

2.2.6 An everyday example

During the foregoing definitions and discussion the reader will probably have conceived measurement systems as complex, multidimensional and electronic. Many systems are indeed like this requiring meticulous design and manufacture (and very many dollars!) if they are to give reliable meaningful information. However, simple instruments and everyday measurement activities embody all of these ideas – see Example 2.1.

Example 2.1 Measuring the size of a table

The length of a dining table must be measured. Unfortunately there is no ruler available at the time so that the length of the table cannot be measured directly. Rather than waste time getting a ruler an instrumentation system is to be devised with available materials.

A piece of string, held taut across the table's length, is marked off at two places corresponding with the two ends of the table. It is folded up and taken to the ruler where it is held taut against the rule and the required measurement carried out.

Here the *signal* is one of mass transfer (of the string). The string plus the human transport constitute the *carrier* and the two marks are the *modulation* of this carrier. The *conduit* through which the signal flows is formed by the human transport. The *symbolic representation* is the distance between the two marks along the string, in this case at a scale of one to one. *Data* relating to the length of the table may be said to exist in the form of the marked piece of string (although it is not in numerical form it is certainly recorded). The ruler is obviously the *calibration standard* and the data

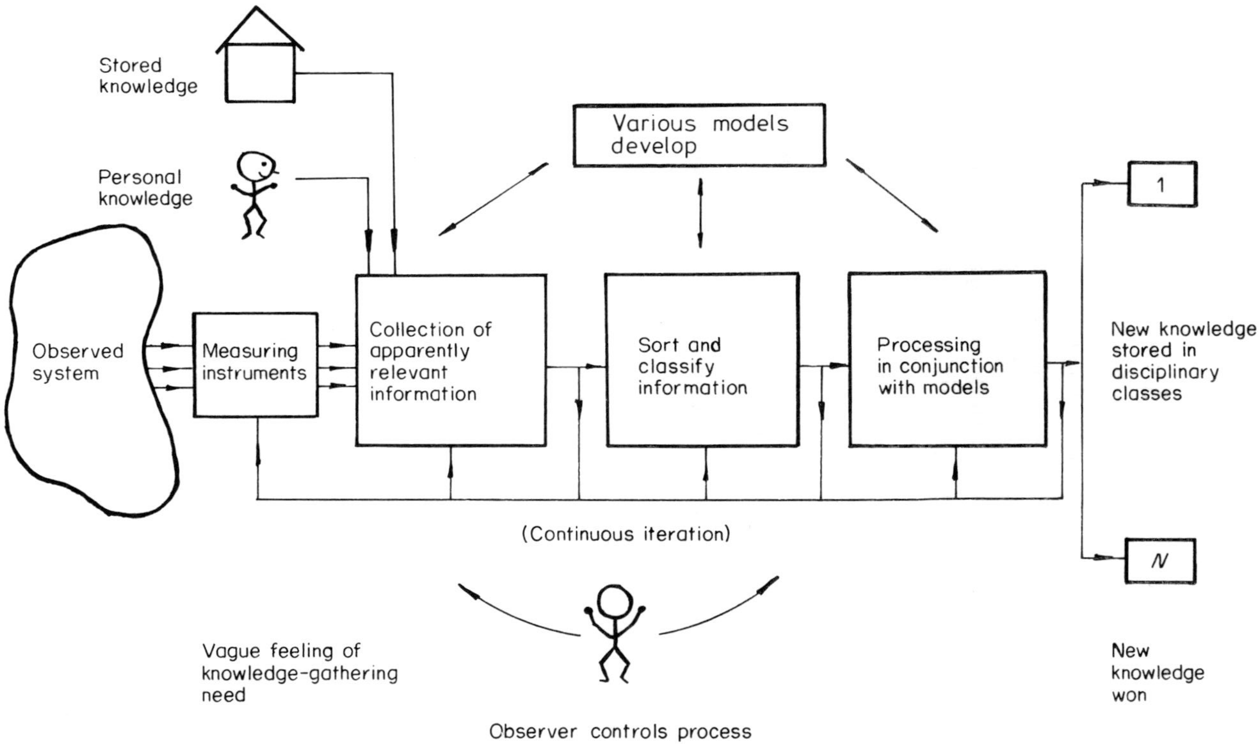

Figure 2.2 How measurements about an observed system result in new knowledge. (From Sydenham, 1985. Reproduced by permission of The Institute of Measurement and Control.)

becomes *information* when we read off the length in millimetres. Whether anyone then has significant *knowledge* about the table is uncertain but it is clear that a piece of information has been obtained.

2.3 REFERENTS, MEASURANDS AND THE MEASUREMENT PROCESS ALGORITHM

We shall now consider the *general* process of designing a measurement system – note, *any* measurement system. There are a number of stages, as set out in the following subsections, which together have been termed a measurement process algorithm (MPA). This algorithm is a conceptual design procedure applicable to any measurement task from the (apparently) straightforward through to the seemingly impossible. To fix ideas it is helpful to bear in mind Examples 2.2–2.4.

Example 2.2 A builder measuring the height of a brick wall

Measuring the height of a wall on a building site is a straightforward measurement situation because the concept of 'height' is both well understood and simple. Several alternative measurement techniques would immediately spring to mind – the piece of string transferred to the tape measure; a tape measure held to the wall; running a calibrated wheel over the distance counting the revolutions; and more.

Now consider a less obvious measurement need.

Example 2.3 A mechanical engineer measuring the efficiency of a Diesel engine

Here the concept of 'efficiency' is well understood and may be rigorously defined as regards a Diesel engine. However, it is not such a simple task to derive the information required – simultaneous measurements must be made of fuel input, heat output and mechanical power output; and the results combined as required by the definition of efficiency.

The resulting assembly of sensors and signal processing could then be regarded as a single instrument, perhaps labelled a Diesel-engine efficiency meter. How does one determine the appropriate sensors to use? And what is the appropriate way to combine the information from each sensor? How does one ensure that the choices are indeed appropriate and the signal processing correct so that the derived result, the efficiency measurement, is not erroneous? This measurement need normally requires special expertise to set up such a system, special plant to test the instrument and agreed definitions of some complexity.

Example 2.4 A fruit grower measuring the quality of oranges

This is clearly a significantly more difficult measurement problem than that given in Example 2.3. Indeed many would question whether such a measurement can be made at all. The concept of quality is, however, well understood and there would be a large measure of agreement (at least among orange growers) about the quality of any particular orange. The measurement requirement cannot be simply dismissed as a matter of opinion unless there is a 'standard human observer' who is prepared to be the yardstick for this measurement. This is not practical and experience has shown such human standards change unaccountably and unexpectedly. Hardware sensing is needed.

But how do we then determine what constitutes a measurement of quality? What aspects of the orange must be sensed and how must the information derived be combined to provide a measurement of quality?

2.3.1 The system to be measured and its model

Consideration of the above three examples suggests that two basic requirements must be established before commencing the design of a measurement system. Firstly, there must be an adequate definition of the quantity to be measured (in the examples height, efficiency, quality); and secondly an adequate understanding must be established of the system to be measured (in the examples wall, Diesel engine, orange).

Our understanding of the system to be measured must be structured in some way. The structure is normally referred to as a *model*, just as in any other branch of science (Section 1.3.1). Models can be of many kinds. Some major types are:

- *linguistic*, in which the relationships are expressed in words and sentences;
- *manufactured scale model*, in which a physical replica (for example a 'scale model') of the system is constructed;
- *iconic*, also called *graphical*, in which the relationships are expressed in terms of graphs and charts, etc.;
- *mathematical*, in which relationships between aspects of the system to be measured are expressed in terms of equations

Each represents only *aspects* of the observed system and their information content varies widely. They each have a role to play in representing the real system. Mathematical models generally contain the most information about performance but are also the most difficult to construct, particularly for imperfectly understood systems. However, the adequate construction of a model, or models, is essential if the resulting instrument design is to be successful. Hence various aspects of model construction and usage ('modelling') are considered, in progressively more detail, throughout this book (particularly so in Chapter 5).

The development of knowledge passes from use of relatively subjective, initial linguistic and physical models, through an *icon* stage to become objective mathematical representations. (Chapter 1 of Sydenham (1979) and Chapter 2 of Sydenham (1982) contain a wider discussion of these general modelling considerations.)

The process of model construction and recording reveals to the observer that there exists a conceptual 'boundary' of system extent beyond which events and occurrences do not influence the observed system behaviour.

Within this boundary the model must be able to account for the effect of *all* events and occurrences. Ambient air temperature, for example, will have only a very minor effect on our measurement of the height of a brick wall (although there will be thermal expansion of both the wall and our measurement scale which may be significant if high precision is required).

Conversely ambient air temperature may have a major effect on the efficiency of a Diesel engine. In that case (and perhaps also for the assessment of the quality of an orange) temperature will be a significant *influence variable* and our model must incorporate a description of this effect. The resulting instrument will then probably require an air-temperature sensor, and signal processing to incorporate the temperature information into the final output (that is, in our examples: height, efficiency, quality.)

2.3.2 Referents

To commence the process of measurement system design, particularly for difficult tasks such as that of the orange, it is first necessary to identify those attributes (of the orange, or Diesel engine) which appear relevant. These are known as *referents.*

There are, of course, a vast number of attributes that may be listed for even a simple physical object. (Try listing *all* the attributes of a house brick.) Hence it is a valuable first step to distinguish those which appear, at least at first sight, to be relevant to the measurement requirement. The attribute of *colour*, for example, is likely to be relevant to the quality of an orange but it is hard to connect it to the efficiency of a Diesel engine.

The result of this process, diligently undertaken, will be at least a clear indication of the applicable *boundary* of the system to be measured. The concept of quality, for example, will have been researched in relation to the orange, and some possible methods of measurement may be already apparent in the reported literature.

2.3.3 Measurands

Some of the listed referents will be obviously measurable with physical apparatus (such as temperature, mass). Such measurable referents are labelled *measurands.*

Others (for example, efficiency, colour, taste, love) are more complex and as yet there may not be any obvious hardware sensing system in existence for the particular measurement task. Further consideration is then necessary before each can be broken down into one or more measurands.

A **measurand** *is a* **referent** *that is physically measurable by a known equipment.*

The objective at this stage in a measurement system design is to render all referents into measurands: once this is achieved an instrument design is possible. Until that step has been established a satisfactory measurement is *not* possible unless a human observer is applied to carry out the sensing and processing.

2.3.4 The measurement process algorithm

Once a complete set of measurands has been assembled and an adequate model expressing their inter-relationships formulated it is possible, in principle at least, to write a formal specification for each sensor that will be required and the signal processing necessary so that a complete instrument design may be produced. This is, of course, highly idealistic. It is much more likely that serious difficulties and inadequacies will remain – some essential referents will not convert easily into measurands and the relationships (and hence the signal processing necessary) between some measurands will remain unclear. Further research and perhaps direct experimentation may be necessary.

Hence the process of new instrument design is likely to be an iterative process involving the revision of the system model or models, the system boundaries, the referents and measurands. The overall process is, therefore, best presented in block-diagram form (a model of the process!) as shown in Figure 2.3. This procedural model has been designated the Measurement Process Algorithm (MPA) by Sydenham (1985). The MPA was constructed to capture that general process by which science is advanced, commonly referred to as *scientific method.*

2.3.5 An MPA application example

To bring the above concepts and procedures into perspective the steps of a typical measurement situation are taken in an example. The task chosen, Example 2.5, is reasonably straightforward to conserve space, but the methodology applies equally well to all measurement including the very complex cases. Clearly complex tasks will require correspondingly more time to develop and document.

Example 2.5 Establishing how to measure the quality of a loaf of bread

Bread is being baked in a factory bakery. There exist legal and consumer reasons to measure each loaf as it is made to guarantee the quality of each loaf that is passed into sale.

As shown in Figure 2.3 the first need (box 1) is to establish what has to be known and make a record of this requirement. The attribute needed here is *quality*, the item is *bread*, for which the *particular* loaf is that arriving at the inspection point at a defined time. No attempt should be made at this stage to immediately define what to actually measure with apparatus.

The next step (box 2) is to learn about such situations in order to be able to list the referents that appear to give a reasonable definition of quality. This is done by

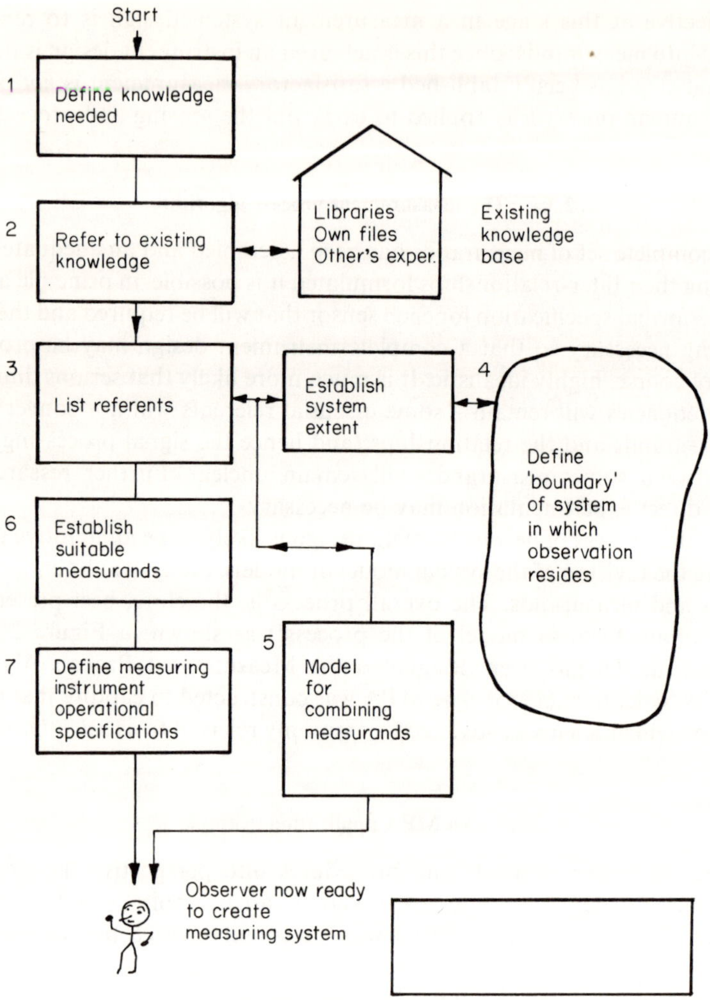

Figure 2.3 Block diagram of the measurement process algorithm (to stage of observing instrument being specified). (From Sydenham, 1985. Reproduced by permission of The Institute of Measurement and Control.)

studying a typical loaf, the place where it is to be measured, library articles and industry reports on similar tasks of quality assessment.

Feeling confident that enough has been learned about the problem a set of referents might then be established (box 3). In this case these would at the least be *degree of cooking*, *size*, and *weight*.

This is now the appropriate time to establish the *boundary* of the system for design purposes. This, here, is in an automatic baking plant where each loaf is inspected a short time after being taken from the oven. The boundary would be defined by the place in which the loaf is to be measured, that is, an inside place with hot, humid, and dusty conditions. There would not be any influence on the measurements by far distant events although a non-climate controlled bakery would have its environment influenced by the heat of the day.

This is also the time (box 5) to begin to decide how the referent-based measurements be combined – *all in a given range* would be reasonable.

Each referent has then to be converted into a suitable measurand (box 6). *Size* and *weight* obviously can be measured with three displacement calipers and a weighing balance. As size is three dimensional, three length measurements are needed. *Degree of cooking* is another matter; this referent needs to be broken down into other referents because it cannot be measured directly as such. A reasonable basis for assessing it is the compressibility of the loaf, that is, its *compliance* at a certain *position* on its surface. These two parameters are clearly measurands. All referents chosen can be measured so a system is practical.

Actual measurement requires specific *metrological* performance from each measurand instrument; the next task (box 7) is to decide performance specification for each measurand. These might be (only some are listed here) that the three size lengths be nominally 140 mm, 140 mm and 350 mm and they must lie within the range ± 5 mm. The weight must be at least 0.680 kg measured to within ± 0.08 kg. The degree of cooked condition is defined as compliance of 1 mm when a 0.1 kg weight is applied at the centre of a side – and so on.

After finalizing the combinational model (box 5) the set of measurands and their parameters are compiled as a list along with a statement of the task (box 1 definition), the combinational model (box 5 definition) and the system extent (box 4 statement).

The statement, for the above example, is now presented.

System for Measuring Quality of Bread

Task

The instrument system has to measure the quality of each loaf produced in the Top Bakery's automated oven.

Measurands Tree

The following sensing system is to be used.

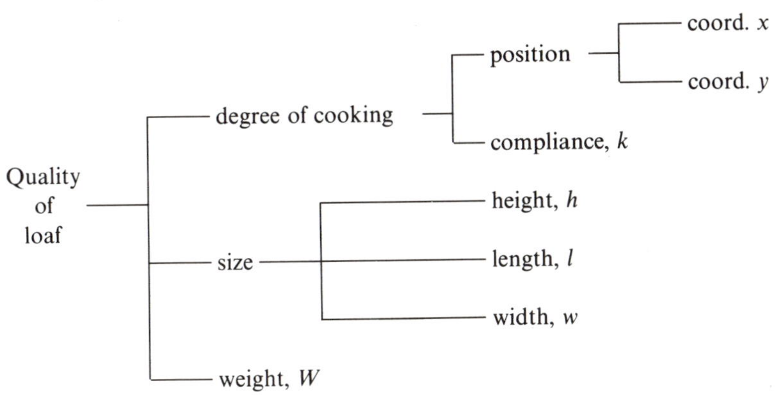

The various sensor outputs are to be combined to map the many measurands into a 'quality' measurement according to this connectivity.

Operating environment for the sensing system

The sensing system is to operate next to the automatic baking plant, which presents only close-by environmental influences as follows:

Temperature: hot
Pressure: ambient
Humidity: high
Vibration: blenders, assembly plant and intermittent motors are in close proximity.

It is permissible to control the environment for the loaf inspection to avoid the need to compensate for influence parameters.

Sensor measurement performance list

MEASURAND	x	y	k	h	l	w	W
Nom. value	175 mm	170 mm	1 mm	140 mm	350 mm	140 mm	0.68 kg
Min. value	170	165	0.5	135	345	135	0.60
Max. value	180	175	1.5	145	355	145	0.76
Discrim.	1	1	0.1	1	1	1	0.01
Repeatbty	1	1	0.1	1	1	1	0.01
Error	0.5	0.5	0.5	0.5	0.5	0.5	0.005

All sensors to respond to 90% of any new value within 1 second.

Combination of sensor information

The mapping process will be according to the above tree diagram which translates into the following logical combination of signals.
IF x lies within 170 mm and 180 mm
AND IF y lies within 165 mm and 175 mm
AND IF k (at position defined by x and y, for applied load of 0.1 kg) lies within 0.5 mm and 1.5 mm
AND IF l lies within 345 mm and 355 mm
AND IF w lies within 135 mm and 145 mm
AND IF h lies within 135 mm and 145 mm
AND IF w lies within 0.60 kg and 0.76 kg
THEN the quality of the loaf is satisfactory
[End of report]

In this way a sound strategy has been created that will facilitate loaf measurement. This operational specification statement produced can then be used either to purchase a system or guide a designer to build one – in practice a hybrid system will usually be constructed consisting of both purchased standard components and a few non-standard components which have to be specially manufactured.

2.4 MEASUREMENT SYSTEMS AS INFORMATION MACHINES

A machine is generally understood in common English usage to be a device that performs some useful function. In science and engineering a machine is defined as a device which can do work. Implicit in this definition is the idea that a machine handles energy - see Example 2.6.

> ***Example 2.6 Energy conversion in machines***
>
> (a) An internal combination engine transforms chemical energy (of the fuel) into mechanical energy.
> (b) An amplifier increases the strength (energy level) of a signal by converting the energy of the power supply.
>
> (In such 'power' situations high efficiency of the conversion is vital as losses are expensive and give rise to engineering problems of removing waste heat.)

Similarly an information machine is defined as one which handles information. A computer is the obvious example, and operates by transforming symbolic input information (much of which is unhelpfully labelled 'data') into symbolic output information.

To achieve these transformations energy or mass movement is used as the vehicle for symbolic representation inside the machine - see Example 2.7.

> ***Example 2.7 Symbolic representation in information machines***
>
> (a) A mechanical calculator uses the angular position of a gear wheel to represent a number.
> (b) An electronic calculator (e.g. a computer) uses the voltage at a point in a circuit to represent a (binary) number.
> (c) A mercury-in-glass thermometer transforms 'hotness' into a length which is rendered symbolic by comparison with a scale etched to to the glass stem.

Example 2.7(c) above is, of course, an instrument. Hence it is clear that:

An **instrument** *is an information machine which transforms physical world information into symbolic output information.*

In each case an energy supply is required to make the machine operate. However, it is the symbol conveyed and *not* the energy (or mass) transferred that is important in the measuring machine; this often leads to very different designs for seemingly similar devices - see Example 2.8.

> ***Example 2.8. The gearbox – a work or information machine?***
>
> A motor-vehicle gearbox transforms the rotational energy at its input shaft into a different form of rotational energy (angular velocity, torque) at its output shaft. A key design requirement is that as much as possible (high conversion efficiency) of the input energy is transferred to the output shaft and this is achieved by appropriate gear wheel and bearing design.

In contrast, a servomechanism (closed-loop controller) gearbox transforms the angular position information at its input shaft into a different angular position at its output shaft. A key design requirement here is accurate shaft positioning for which quite different gearwheel and bearing design is required, for here *fidelity* of movement conversion is far more important than conversion energy efficiency.

Information-gathering machines can be automatic in the same way as work-performing machines, and may also be to some extent *intelligent* in that they might be designed to respond differently according to the information gathered.

Information machines, in themselves, are not knowledge-gathering machines but tools that assist knowledge procurement. The process of knowledge creation requires human participation, as discussed in Section 2.2.5 above.

2.5 BASIC MEASUREMENT TERMINOLOGY

Measurement is so common in everyday life that people inherently assume they understand all there is to be known about it. In reality common knowledge is only superficial and limited. A particular area of constant misuse is the terminology associated with measurements. As with any mature discipline, measurement science also has its own special words to give precise meaning to its hardware, processes and results.

It is not appropriate in this introductory text to present a full discussion of the many terms that are used across the breadth of measurement practice. These have evolved throughout the history of science and technology and many have only narrow, specialist application. However, there are some basic terms of universal application that must be used rigorously when specifying instrumentation or describing instrument performance.

The definitions given here are taken from authoritative and international documents; the reader is referred to Section 3.2 of the *Handbook of Measurement Science* (Sydenham, 1982). (Small differences exist between documents and these may be important in exacting applications.) The value of these definitions and their use in rigorous, formal description is illustrated by Example 2.9.

2.5.1 Discrimination

Discrimination is the quality which characterizes the ability of the measuring instrument to react to small changes of the quantity being measured – the measurand. It is used to state this ability in quantitative terms such as 'The thermometer has a discrimination of 0.05°C'.

The word *resolution* is still commonly used to describe this feature but is now no longer the preferred term and should not be used in this context. (This word is now reserved for a specific phenomenon in the field of optics.) Likewise the word *sensitivity* conveys the same general concept but should not be used in formal documentation as it is used in connection with transduction performance.

2.5.2 Repeatability and reproducibility

Repeated measurements of the same value of a measurand will not, if the discrimination is small enough, be exactly the same. The closeness of sets of measurements made in the *short term* is termed the *repeatability* of the instrument.

When sets of measurements are repeated, in similar circumstances, over *long periods* of time the closeness of the sets of measurements is termed the *reproducibility* of the instrument.

2.5.3 Precision

The above three performance parameters – discrimination, repeatability and reproducibility – are often generally referred to as the *precision* of the instrument. However, the term precision is used only in the general sense and cannot be quoted as a numerical characteristic of an instrument.

2.5.4 Accuracy

The closeness of the measurement value provided by an instrument to the *defined true value* is termed the *accuracy* of the measurement made with the instrument.

Accuracy, as will be explained below, must usually be defined into a measurement system for it is not inherent in most instruments. The ability of an instrument to measure accurately must be decided through comparison – the act of *calibration* – with a declared reference value for the measured unit concerned. (A very few routine instruments do not need this comparison because their internal operation can be shown to give accurate measurements to declared levels. One example is a particular type of commercial laser interferometer used for length measurement.)

Example 2.9 Formal description of measurement of a darts player's performance

Consider the performance of a darts player attempting to place all of his set of darts into the central bullseye.

Discrimination

The discrimination of position for lodgement of a dart in the whole dart board area is decided by the diameter of the point of the dart. Narrowing the shaft would improve the discrimination but there would be no real value in doing this beyond a certain thinness.

(In the design and construction of a measuring instrument the discrimination is usually set to just allow the repeatability to be witnessed as slightly varying values for each measurement event.)

Precision factors

The darts player now throws all of his darts to see if he can place them into a single central location. They 'group' as a scattered set of different positions. The *smaller the group diameter the better the precision* of the throwing; a quantitative statement of this grouping would talk of the *repeatabillity*.

(As values commonly vary in a random manner, precision parameters are usually quoted in terms of a statistical manipulation of the set of values. The *standard deviation* parameter is often used to enable one quantity to be quoted, thereby avoiding the need to provide a set of values.)

Accuracy

If the *centroid* of the set of dart positions lies in the bullseye the player has thrown an accurate set, for accuracy is judged by how close the throws are placed to the *defined desired place*. The degree of accuracy is decided by how close to the centroid of the set comes to the bullseye, the declared place to hit.

Relationships between terms

The following cases illustrate unexpected relationships.

- A highly precise player need not be an accurate player: the centroid of a small grouping (highly precise) could lie well away from the bullseye (low accuracy).
- It is possible to be highly accurate (that is, place the centroid of the throws in the centre of the board) with a very imprecise and scattered set of throws.
- The smallness of the discrimination is not related to the precision or the accuracy of the throws provided it is small enough to allow each placement to be separated.

From Example 2.9 it should be clear that the concepts of discrimination, accuracy and precision are very different. It is vital that they are not confused when referring to measurement or instrumentation.

2.5.5 Error and uncertainty

It follows from the above that an ideal perfect instrument would have discrimination of zero, perfect repeatability and reproducibility without any spread of values and be exactly coincident with the values defined as the standard. This is neither achievable nor economically necessary. The cost-effective instrument design is one which is just adequate to meet the required specification.

Any measurement produced by any instrument will always, to some extent, be both imprecise and inaccurate. Departure from perfection is referred to as the *error* of the measurement. It is quoted in either fractional units of the quantity (for example, ± 0.2 kg), as a percentage of the measurement itself (for example, $\pm 0.1\,\%$) or as a relative measure (for example, 1 part in 10 000)

When referring to a particular measurement value the term *uncertainty* is preferred to use of *error* as there is no implication of a 'mistaken' or 'wrong' value associated with the word 'error' in general English usage. Furthermore, the word 'uncertainty' does suggest a band or margin surrounding the measurement in which the true value of measurement can be reasonably assumed to lie. The actual error of a particular measurement from its true value is virtually impossible to ascertain for many reasons, not the least of which is the impossibility of finding the true value, for that requires existence of the mythical perfect instrument!

In any measurement procedure, however simple, there exist many sources of error. Reconsider, for example, the piece of string used to measure the length of the table in Example 2.1 and try listing *all* of the potential sources of error in this measurement procedure.

The design of measurement systems is primarily concerned with keeping errors under control

The design objective is not to minimize every source of error but to keep them in control to an adequate level required by the specification for the instrument.

The error aspect in instruments is developed further in Section 4.6 and throughout many later sections and examples – such is its importance.

2.6 STANDARDS, UNITS AND THEIR MAINTENANCE

As has been explained, for measurements to be ascribed accuracy levels there must exist a method of defining each of the measurement units used in widespread practice and there must also exist a means to compare the instrument used in everyday practice with this standard of the unit.

At the global level this is done using an international network of national physical standards laboratories. Well-known examples are the National Bureau of Standards (NBS), now within the National Institute of Science and Technology (NIST), of the USA, the National Physical Laboratory (NPL) of the United Kingdom, and the National Measurement Laboratory (NML) of Australia. A treaty between most of the world's national laboratories regulates the international activity and coordinates inter-laboratory standards development and acceptances and their intercomparisons.

Development and maintenance of these international and national level standards is a specialist branch of measurement science, one that is characterized by the need to reach the lowest possible uncertainties for the particular units covered. The equipment required is very complex and extremely expensive. When a defined and agreed apparatus is adequately proven it acts as the *primary standard* for that physical quantity. An example is shown in Figure 2.4.

Due to the great cost of maintaining this level of unit not all are covered in each country. The most important standards at this level are those of the *base units of the SI metric metrology system* – metre, kilogram, second, ampere, kelvin, candela – and the hierarchy of major derived units as set out in Figure 2.5. Many

Figure 2.4 Apparatus used to maintain the primary frequency standard of the second. (HBMS, figure 3.1.)

other locally important units are also to be found maintained in standards laboratories, an example being that of *density*, which is important in the *custody transfer* of natural oil products.

Below this primary level governments and industry maintain a variety of *secondary standards*, each of lower accuracy and much lower cost. These are those used for the day-to-day calibration of the nation's instruments.

A widespread practice is to establish a national accrediting agency whose task it is to bring about assessment of measurement capability for measurand units variously maintained in laboratories across the country. These laboratories can then be used to calibrate workshop and field measuring systems used in everyday practice. Examples are the National Association of Testing Authorities (NATA) of Australia and the British Calibration Service (BCS) for the United Kingdom.

It is essential to ensure that measurements made with the everyday laboratory and factory instruments each give results that are consistent across the nation; that is for the same value of measurand they each give the same value within their uncertainty range. That value must be the nationally declared value obtained from the national or international standard.

If this were not done then similar parts or components made in different plants may not fit together, and scientific results from different laboratories could not be used in combination with any real confidence. This process is called *traceability*, and is achieved by ensuring that each level of calibration can be

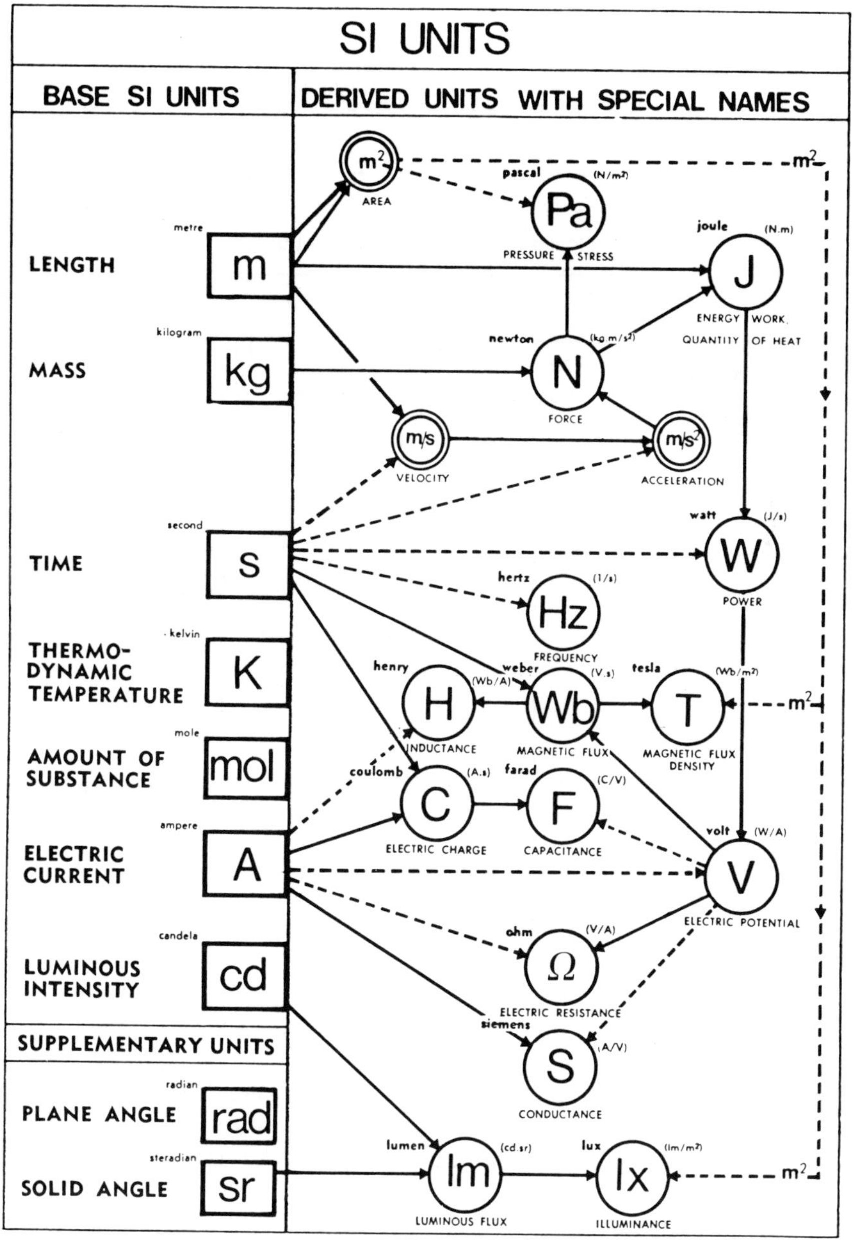

Figure 2.5 Chart showing relationship of SI measurement units. Dashed lines indicate division; full lines indicate multiplication. (HBMS, figure 3.2.) (Reproduced from MHI-1972 Metric Conversion in Building and Construction. © Standards Australia.)

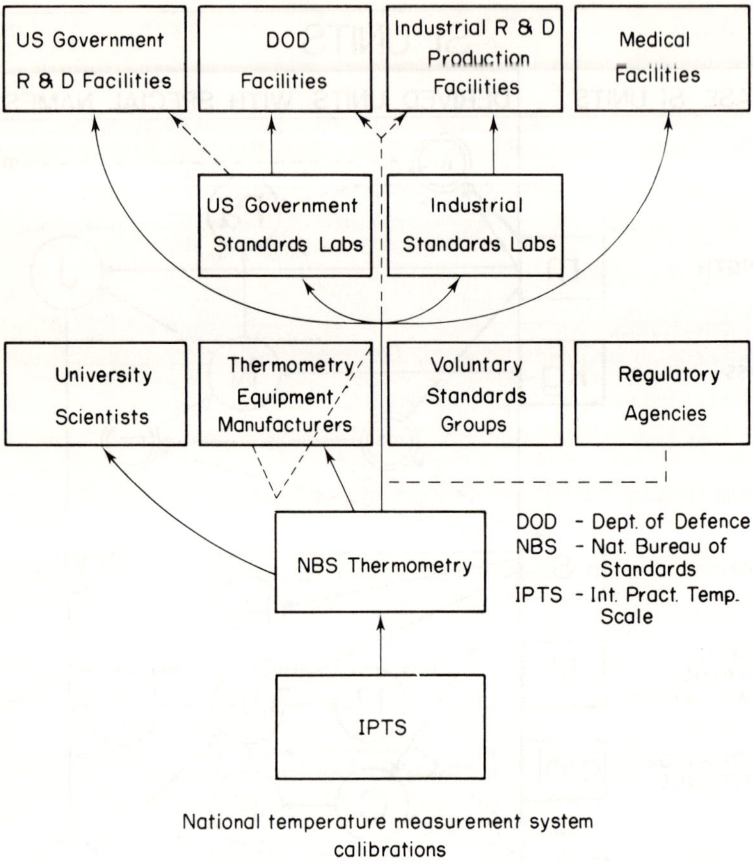

Figure 2.6 Concept of traceability – USA example. (From 1975 *NCSL Newsletter*, **15**, 31. © 1975 NCSL. Reproduced by permission.)

legally traced to the next through a system of *certification*. Figure 2.6 illustrate the concept of traceability of a measurement calibration for a nation's instruments.

As each level cannot be used to calibrate the next lowest without increasing the uncertainty it will be clear that the prime standards must be much better than the laboratory and industrial field calibrations. A rule of thumb is that a factor of uncertainty increase of around 3–10 times can arise at each step. As an example the base unit standard for length has uncertainty 1 part in 10^{11} whereas toolroom laboratory slip-gauge standards can only be used to an uncertainty of around 1 part in 10^6.

This is, of necessity, only a very brief introduction to this vital area of measurement science. A fuller account of standards, units and traceability may be found in Section 3.4 of the *Handbook of Measurement Science* (Sydenham, 1982).

TUTORIAL EXAMPLES

2.1 Describe the role of measurements in a process or equipment that is commonly available to you. Comment on why they are vital to its operation and how they are used in it.

2.2 Prepare a commentary on the problems of measurement in one of the following fields: (a) economics of a country; (b) learning performance; (c) the suitability of an item of clothing; (d) the medical investigation of living people.

2.3 Using the measurement process algorithm approach establish the measurement system specification for a knowledge-seeking task of your choice. Examples might be:
(a) Are trucks passing on a fast highway within legal road limits?
(b) When is a hamburger steak cooked?
(c) When does a machine-tool cutting bit need changing in an automatic lathe?
(d) Control of the quality of water in a fish-farm aquaculture pond.
(e) When to cut a cricket pitch wicket.

2.4 Tests are made of lengths of seat belt as they are cut from a roll by a production machine. The values are automatically measured and recorded by a data logger that presents a chart of the results as shown in Figure 2.7.
From this chart assess:
(a) the discrimination of the measurement instrument being used;
(b) the repeatability of this sequence expressed as the standard deviation - assume Gaussian distribution;
(c) the accuracy of the measurements;
(d) the length of the batch expressed as a mean length and uncertainty.
(Rigorous answering of this question assumes basic statistical knowledge - which is given in Section 3.3.1.)

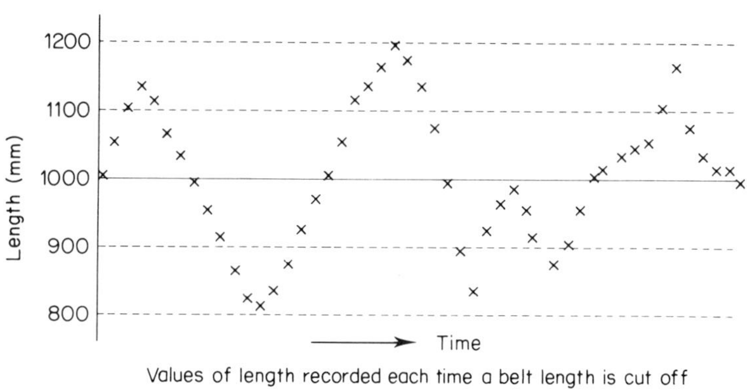

Figure 2.7 Test results of lengths of seat belt cut by an automatic cut-off machine - see tutorial example 2.4

2.5 Explain why the national standards for measurement must be considerably better than those used in everyday practice. Record the improvement in uncertainty as the standards, for a measurand of your choice is traced from the field instrument up to the international definition or definitions needed.

2.6 Establish the calibration procedure and how it would be carried out, in your city or country, for one of the following instruments:
(a) mercury-in-glass thermometer;
(b) beam balance set of weights;
(c) sand depot weigh bridge;
(d) petrol pump in service station;
(e) laboratory voltmeter.

REFERENCES

Stein P K, 1970, Sensors as information processors. *Research/Development*, June, 33–40.

Sydenham P H, 1979, *Measuring Instruments: Tools of Knowledge and Control*, Peter Peregrinus, Stevenage.

Sydenham P H, 1982, *Handbook of Measurement Science*, Vol 1, Wiley, Chichester.

Sydenham P H, 1985, Structured understanding of the measurement process. *Measurement*, **3**, 115–20, 161–8.

CHAPTER 3

Signals and information

3.1 INTRODUCTION

Development of theoretical appreciation of how signals can be modelled within measurement systems is essential for their effective engineering – the added rigour helps design and appraisal to be undertaken more efficiently. Sole use of the intuitive approach to estimating and analysing instrument behaviour cannot adequately predict what will happen to signals as they are modified by *signal-processing* stages and corrupted by unwanted *noise disturbances.* A few hours spent rigorously modelling a proposed system can lead to performance optimization that would take days, and sometimes months, of work to establish by experimental means.

Development of the ability to describe the behaviour of signals in a rigorous manner using mathematics has been a most important part in the development of electronic engineering. The mathematical description was initially confined to analogue signals but has now progressed to include digital forms, as commonly employed by other branches of information technology.

Development work began in the early 1900s, driven by the success, and often lack of success, of early experiments in telegraphy, telephony and radio communication. Although the mathematics has mostly been expressed in terms of telecommunications requirements it also applies to measurement systems because they, too, are communication links that interface an observed object to an actuator. However, measurement systems also have to convey meaning, and to ensure that extraneous noise does not alter the message fidelity. This chapter provides an overview of relevant signal theory from the measurement systems, rather than telecommunications perspective.

The following material is presented as a condensed but more fully illustrated version of Chapter 4, by Professor E. G. Woschni, from the *Handbook of Measurement Science* (Sydenham, 1982). It is linked into that chapter in terms of order of material, symbology and illustrations so as to allow more detail to be readily obtained where the level of mathematical experience can make full use of its content.

The material presented in this chapter may well appear to be difficult and demanding upon first reading. Many mathematical ideas are covered without full justification of development due to space limitations. This has been done because the relevant foundational mathematics have usually been dealt with in engineering and physical science degree courses by the end of the second year, the earliest year to which this book is directed. This chapter puts that theory into a measurement-science and instrument-engineering perspective. Readers are encouraged to make use of the examples presented, consulting Woschni (1982) and other texts as referenced when and where they need to explore the theoretical bases underlying these principles.

3.2 CLASSIFICATION OF SIGNALS

3.2.1 Types of signal

Within signal-bearing systems there will be both the wanted signals and extraneous noise signals. As both are carriers of information both can be treated by the same method. To convey information the signal-bearing channels must be capable of having some property changed in accordance with a measurement-mapping process.

Two dominant types of signal type exist – the first is the continuously varying, non-quantized, *analogue* form. Measurement information is conveyed as a change of the *amplitude* or *phase* of the signal bearer.

The second form, once less common but now in use in more applications than the analogue form, is the quantized *discrete* or *digital* form. Information is carried by a host of methods which vary the relative *timings* in digital signals. By far the most common form of the latter is the two-state, *binary* signal.

Figure 3.1 provides a summary of the various methods by which information (parameter I in each diagram) is conveyed. Each have their particular virtues; each can be analysed using well established theory.

Signals that have all properties known, including those about their future, are said to be *deterministic* (test signals are examples); the information content of these does not change with time as they do not acquire or lose information.

A signal to be measured is, by definition, one still in a largely unknown state, and is called a *non-deterministic* signal. When a signal is known to vary in a probabilistic manner it is called a *stochastic* signal; its future state can only be determined in terms of statistical measures about a group of possibilities.

Signals can be combined, with appropriate apparatus, according to mathematical combinatorial rules. Simple examples are addition and subtraction; more complex combinations are multiplication, division and convolution.

Within a signal-bearing system the signal flow can be portrayed using *signal-flow graphs* and signals can be represented in either the *time* or *frequency domain.* Many of these concepts are brought together in Example 3.1.

Characteristic			Without quantization of time (continuous)	With quantization of time (discontinuous)
Without quantization of the information-parameter	analog		Continuous analog signal 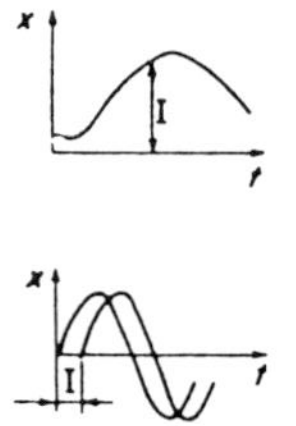 information parameter *I*: signal amplitude information parameter *I*: phase shift or frequency	Discontinuous analog signal 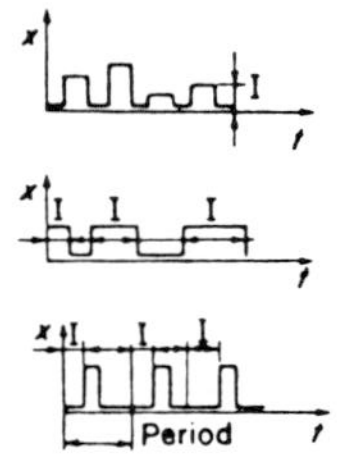 information parameter *I*: amplitude of the pulse sequence information parameter *I*: width of the pulse sequence information parameter *I*: phase shift of the pulse sequence
With quantization of the information parameters	discrete	multipoint-signal	Continuous multipoint signal 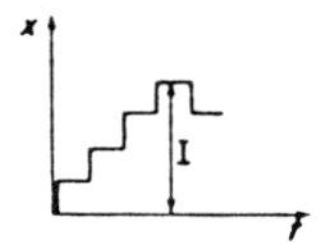 information parameter *I* signal amplitude can have *n* discrete quantities	Discontinuous multipoint signal 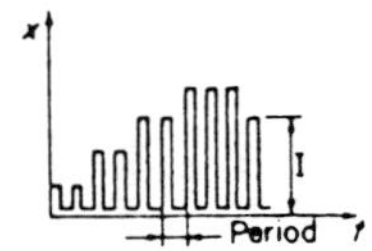 information parameter *I* signal amplitude can have *n* discrete quantities
		binary	Continuous binary signal 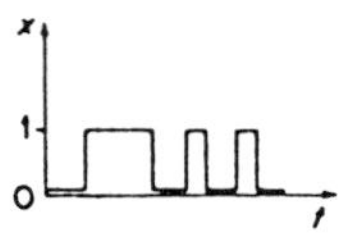 information parameter *I* can have only quantities 0 and 1	Discontinuous binary signal 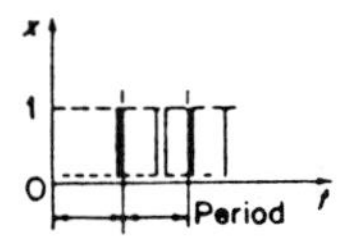 information parameter *I* is given as the code word of the 0, 1 signals of one period

Figure 3.1 Types of signal used to convey information. (HBMS, table 4.1.)

Example 3.1 Measurement of radioactivity in granite rock

Geophysicists are interested in measuring the level of weak radioactivity in granite bedrock; this assists them in understanding geological processes. A typical measuring system is shown in Figure 3.2.

From previous scientific work it is known that the radioactivity signal to be measured is stochastic. The properties of radioactivity are also well known so that a suitable detector can be made that converts the rapidly occurring radioactive pulses of energy converted into electrical energy pulses. These are fed to a high-performance *pulse counter* with the special facilities needed to capture pulses that can arise very close together.

Within the counter the randomly arising pulses are converted into a digital equivalent form of signal that usually has applied to it some degree of *averaging* of the stochastic signal.

Whilst digital signals are convenient when counting events they are not suited to the human observer; the counter input signal is then converted, with a digital-to-analogue converter (DAC), into an analogue signal that drives a scale-and-pointer form of display. Whilst the output is a continuous analogue signal it is also a

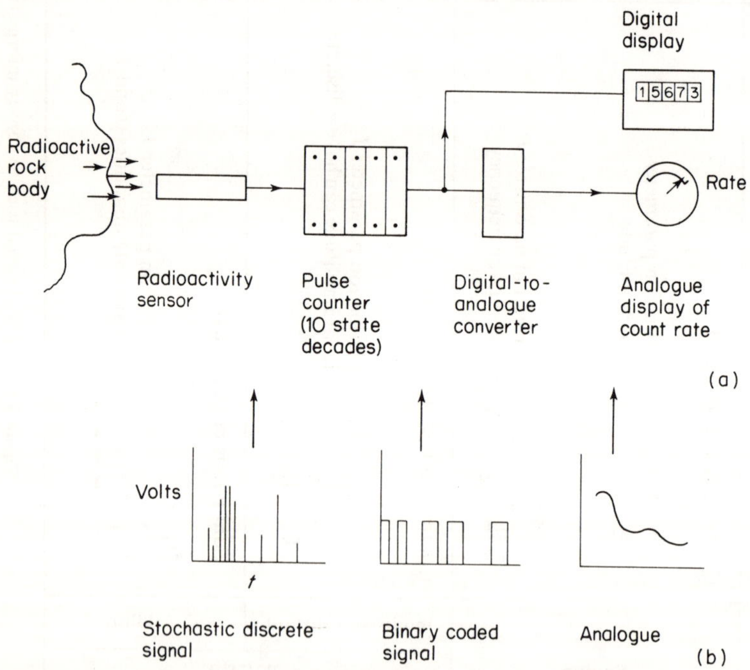

Figure 3.2 Radioactivity-measuring instrument system: (a) signal flow through organic blocks; (b) signal appearance with time, as seen on oscilloscope screen, for various stages

stochastic signal, for the next signal-level increment cannot be predicted other than on a probability basis. The original, digital, signals must be used to drive a digital panel meter (DPM).

Regardless of the form of the signal present at each stage it should still bear the information about radioactivity levels that is needed by the observer. It may, however, have been purposely *filtered* to remove unwanted aspects of the original physical process – a user-directed process – or it may have picked up unwanted signals arising from inadequate instrument design – designer- and user-controlled if the system is adequately understood.

When first devised and constructed, and periodically thereafter, the system needs to be *verified* to ensure it is mapping radioactivity levels into expected voltage representations. Verification of the counter is achieved by feeding a *deterministic signal* into the counter from a generator of random pulses whose properties are known. (Today such random pulse sequences are usually computer generated.) However, direct verification of the radioactivity detector is not possible because sources of radioactivity are essentially non-deterministic – but we do know they have been performing according to known models since they were identified in the 1890s, and this physical theory predicts the detector's performance.

3.2.2 Signals in the frequency domain

Most people find it easy to visualize and comprehend a signal that varies over time. To be told that signals can also be portrayed in terms of a continuum of varying sets of sinusoidal signals having different amplitude and phase is usually bewildering. Mathematically, however, this latter *frequency-domain* form is the better version for description.

The most important determined fundamental signal form is the harmonic oscillation, or *sine wave*. Figure 3.3 represents this generic signal, which is formalized as:

$$x = \hat{X} \sin(\omega t + \phi) \tag{3.1}$$

where

$\hat{X}$ = peak amplitude
ω = angular frequency (equal to $2\pi f$)
f = cyclic frequency
t = time
ϕ = phase angle

The importance of the sine wave is that addition of a set of different sine waves, each having their individual amplitudes and phases relative to a common datum, allows the synthesis or analysis of *any complex* waveform. Example 3.2 shows how a square wave can be generated in this way.

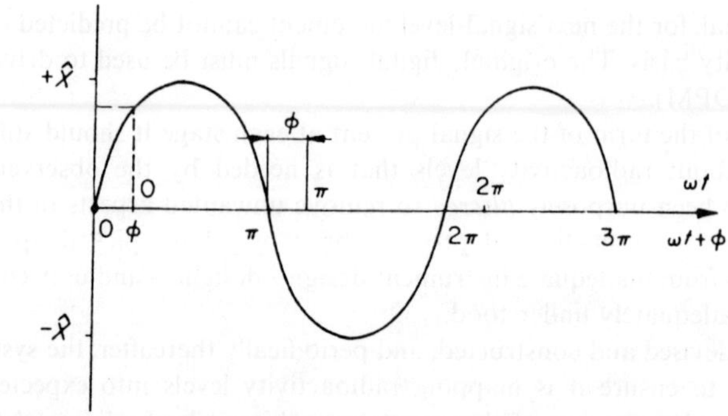

Figure 3.3 Representation of the generalized sine-wave signal in its time-dependent form. (HBMS, figure 4.4(b).)

Example 3.2 Generation of a square wave by addition of a set of sine waves

Signal synthesizers are used to generate complex wave shapes for testing such equipments as signal processors, heartbeat monitoring instruments, communication links and radio systems. They are also used to predict parameters of such harmonic events as the height of sea tides.

Within these equipments the appropriate set of sine and cosine waves are generated (by analogue or more commonly today by computer-controlled digital methods) each with the required amplitude and phase. These are then added together to form the output signal.

The process of generating a continously recurrent square wave shape by graphical means is shown in Figure 3.4. For reasons of clarity only two sine waves of the set are shown; some twenty such components are needed to form a reasonably square

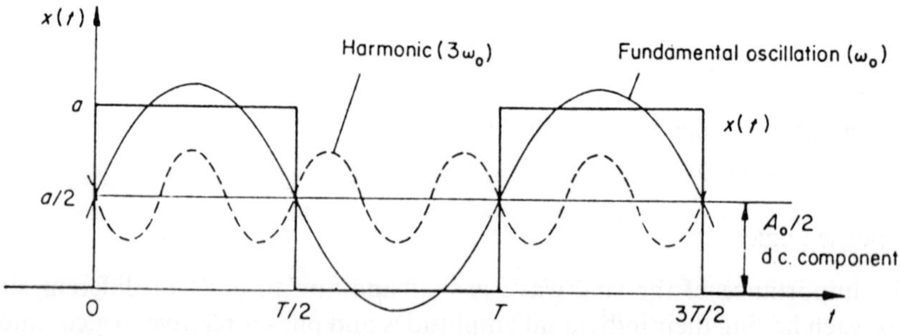

Figure 3.4 Generation of a square-wave, time-varying, signal wave shape by vertical addition of a set of sine-waves – see text. (HBMS, figure 4.5.)

shape. Musical instrument *synthesizers* and *electronic organs* also use this technique to produce their wide range of sounds, as do *speech synthesizers*.

Whereas it is possible to investigate wave shapes by laborious graphical means, such as shown in Figure 3.4, it is far more expedient to use the mathematical methods devised by Fourier. His, and other work, was carried out in the early nineteenth century to explain harmonic motions such as water waves; signal theory and electronic applications came much later.

In general a complex wave shape $x(t)$ can be represented by the expression:

$$x(t) = \tfrac{1}{2}A_0 + \sum_{n=1}^{\infty} [A_n \cos(n\omega_0 t) + B_n \sin(n\omega_0 t)] \tag{3.2}$$

where

A_0 = amplitude of the zero frequency term (d.c. component)
A_n and B_n are *Fourier coefficients*, see below.
ω_0 = angular frequency of the *fundamental* sine wave or cosine wave
n = integer of recurrence of wave
t = time.

The two Fourier coefficients are given by

$$A_n = \frac{2}{T}\int_{-T/2}^{+T/2} x(t)\cos(n\omega_0 t)\,dt \tag{3.3}$$

$$B_n = \frac{2}{T}\int_{-T/2}^{+T/2} x(t)\sin(n\omega_0 t)\,dt \tag{3.4}$$

where

T = period of signal.

The amplitude of the combined sines and cosines is given by

$$C_n = (A_n^2 + B_n^2)^{1/2} \tag{3.5}$$

It is also possible to express the Fourier series in terms of complex notation; this form is often more appropriate to use. (A relatively recent development has been the *Walsh functions* equivalent of the Fourier method which has certain practical advantages – see Woschni (1982) for an introduction.)

Depending upon the symmetry of a given time sequence about the zero time point there will arise different forms of expression for the same wave shape.

Because a complex signal can be formed from a set of frequencies it follows that the instantaneous form of a complex signal can be portrayed as a *frequency spectrum* graph – see Figure 3.5 for that of a square wave. If the wave form is continuous, such as an ongoing speech pattern or square-wave clock signal, the spectrum will comprise discrete frequency lines. Alternatively, if the signal is

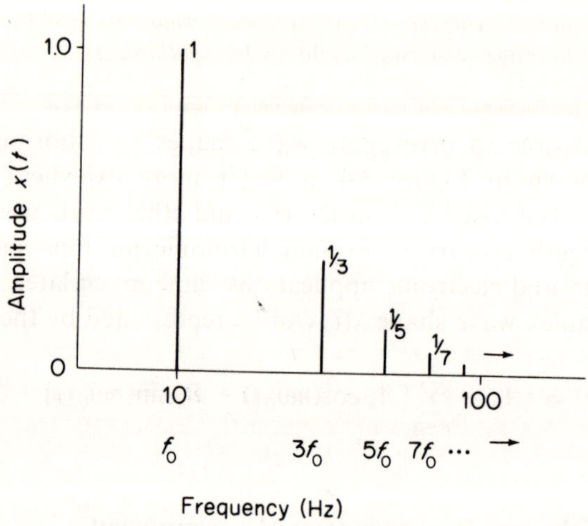

Figure 3.5 Frequency spectrum representation of a square wave

discontinuous, such as a short burst or a single pulse, *then the spectrum will be a continuous curve.*

It is necessary to develop a 'feel' for the frequency spectra of the commonly met signals and for how they vary as the period and wave shape vary. It is helpful for the reader, at this stage, to compare this form with that of the time-domain representation, shown in Figure 3.4.

As information is conveyed as amplitudes and phases of the various component frequencies present and represented by their spectral presence, it can be seen that processing of a signal to alter its frequency spectrum essentially alters the information content of a signal.

Example 3.3 Where does information reside in a radio transmitter output?

A common form of radio transmission system is *amplitude modulation* (*AM*). In this system the essential message information, that of the audio signal (frequencies in the approximate range 100 Hz to 10 kHz) is fed into the microphone by the pressure waves from the mouth of the announcer.

It is not practical to transmit these very low audio frequencies as radio waves directly but it is possible if they are converted to a much higher frequency in the electromagnetic spectrum. The amplitude modulation procedure (multiplication of two signals in fact) performs this conversion. It can be shown, using the mathematics of modulation theory (Section 3.5.4 of this chapter), that after multiplication of the audio signal (frequency f_a) with a high-frequency (HF) *carrier* wave (frequency f_c) there then exists a set of high-frequency terms in which the carrier sits between two *sidebands* of frequency $f_c + f_a$ and $f_c - f_a$. This is depicted in Figure 3.6.

The carrier, therefore, still exists after modulation has taken place but now the audio information is available in both of the sidebands. At the broadcast signal

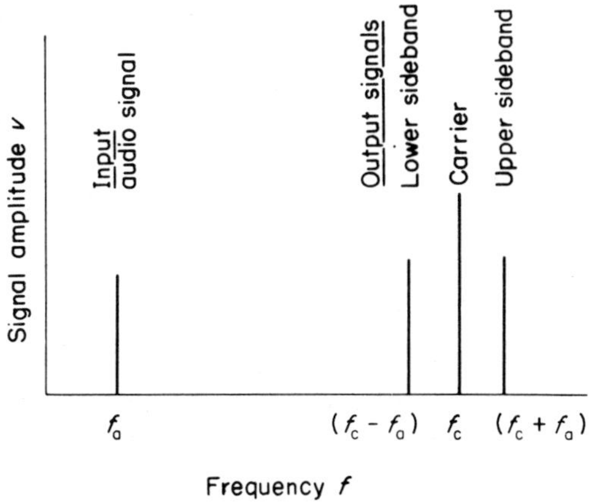

Figure 3.6 Frequency spectrum for input and output of an AM radio system

receiver the incoming HF signal is *demodulated* to extract the audio signal out of the HF sideband signal by bringing it back down to the audio region.

The audio signal is thus represented in two sidebands but has no presence in the carrier. It is usual to filter out (that is, not transmit) the carrier and one of the sidebands as this will conserve use of the broadcast spectrum.

This example is typical of many signal-processing arrangements in that *multiplication* of a reference signal takes place with the desired signal – as will be seen later in this chapter. In such cases it is important to fully appreciate where the original message information resides. Only then can we avoid inadvertently distorting or, at the worst, completely losing, the information.

So far this discussion has dealt with the frequency spectra of signals that are continuous over time. The much used digital signal, however, has discontinuities in its time flow, as also do short-lived signals, such as single pulses for testing equipment. These *non-periodic* signals can also be analysed using Fourier methods in which the mathematics is extended – but the theory is necessarily more complex and is usually covered near the end of engineering and science courses. For this material the reader is referred to Chapter 4 in Sydenham (1982), or to general signal processing texts, such as Oppenheim *et al.* (1983).

White noise is an important instrumentation signal (also called *Johnson noise*) for this is generated by resistance components of all circuitry devices and sensors. Its energy occurs uniformly at all frequences from zero to infinity; the spectrum is therefore a uniform amplitude at all frequencies. This means that intended use of a wide-band filter to remove the noise from a signal present with the noise would also take out wanted signal. Special methods can, however, be employed to improve this kind of signal-recovery situation – see later, Example 3.4.

Fourier methods have also found extensive use in *optical image processing*. In this application the frequency signal is formed in space rather than in time (the 'time domain') for electrical signals. This appears as the periodicity of lines that collectively form an image when it is scanned in one dimension; or as the separation (in two dimensions) of a matrix of spots on a photographic plate when the image is captured on a surface. As physical space has three dimensions Fourier analysis of optical systems has to be extended to cover three dimensions rather than a single dimension which is adequate for all time-domain signals.

3.2.3 Signals in the time domain

Signals can be studied in the time domain by making use of a sequence to establish the periodicity. Various features can be of interest – the period, average value, root mean square (r.m.s.) value, peak value, mark–space ratio, d.c. bias level and others.

In instrumentation the process of *averaging* leads to a number of powerful and valuable signal-enhancement techniques. In general the signal $x(t)$ may be raised to the power n before taking the average. Hence for *periodic* signals the nth-order average is defined as

$$\overline{x^n} = \frac{1}{T}\int_{-T/2}^{+T/2} \overline{x^n(t)}\,\mathrm{d}t \tag{3.6}$$

where

$\overline{x^n(t)}$ = time average of nth order at time t.
T = time period of the signal.

For *non-periodic* signals no simple time period T exists (by definition) over which averaging will be satisfactory. Hence the nth-order average must be defined as

$$\overline{x^n(t)} = \lim_{T\to\infty} \frac{1}{T}\int_{-T/2}^{+T/2} x^n(t)\,\mathrm{d}t \tag{3.7}$$

Simple averaging ($n = 1$)

For $n = 1$ the resultant output of the averager is the *arithmetic mean* or the zero-frequency component (if averaged for an infinite time period). It is to be noted that a truly random signal (such as white noise) will average to zero whereas a repeated identical signal will not diminish and vanish – this is a useful basis for separating wanted signals where they are accompanied by random noise.

Simple averaging, in many forms – see Example 3.4 for one – provides the basis for many signal-recovery methods. For these to work satisfactorily it is essential that the signal and its noise can be repeated with the noise sequence not being related to the signal in any systematic way.

Example 3.4 Principle of the box-car integrator

Consider recovering the wave shape of a microsecond duration radiation pulse (see Figure 3.7(a)) emitted from a material made to fluoresce when pulsed by a powerful laser source. The sought-after emitted radiation wave form is present in random noise that is thousands of times larger in magnitude than the wanted signal. The time-domain record of the unprocessed signal will appear as a burst of noise, as shown diagrammatically in Figure 3.7(b)).

In this type of observation it is possible to repeat the laser excitation event at regularly synchronized times, say at 100 times a second, Figure 3.7(c). Each of the signals resulting from these pulses will have the same wanted fluorescence signal wave shape, but as they are individual, unrelated events the noise in each will be a different white noise sequence adding randomly varying levels of noise to the same level of signal of each repeated excitation as shown in Figure 3.7(d).

To extract the signal a processing system is needed that will average the energy of a small time period (such as one hundredth of the full wave period) of the signal over, say, 100 output occurrences. This will average out the noise component of that small increment of the wave shape whilst enhancing its signal amplitude.

Once a system has been devised to recover the value of a particular point of the hidden wave form, the next step is to add a procedure to move the sampled point on to the next incremented position as shown in Figure 3.7(c). To ensure the same spot is sampled each time a synchronizing pulse must be generated by the laser excitation system.

Analogue box-car systems are not moved stepwise but are scanned very slowly through the whole wave-form period.

As the wave shape is recovered quite slowly due to the need to repeat many excitations there is also a need for some method that will store the recovered shape. Analogue systems make use of *storage oscilloscopes*; the digital alternatives use *memory*.

Why are these equipments known as *box-car* integrators? The oscilloscope display of the repeated sampling pulse, moving slowly through the wave-shape period, is like a freight train 'box car'; the system is averaging the energy of the samples, and this, of course, is integration.

How well do they work? Gaussian random noise (Section 3.3.1) averages toward zero as the square root of the number of times a signal is averaged. Thus, for example, for 10 000 repetitions of the signal at a point (at 200 Hz this would take 50 s) the noise is reduced to one hundredth of its original value. The signal, however, has now grown in magnitude to 10 000 times its original amplitude. Thus the *signal-to-noise ratio* (*SNR or S/N*) for that sample has improved by 1 000 000 times. Well-designed systems can extract wanted picovolt level signals in the presence of volt-level noise.

The price paid to extract such small signals is that it takes a considerable time to obtain the whole wave shape. If each sample in this example requires 50 s to recover, then it follows that to recover 200 of these samples (in order to obtain reasonable definition) will need nearly 1.5 h to recover. Note that increasing the number of repetitions at each point provides diminishing benefits (square root function), so the

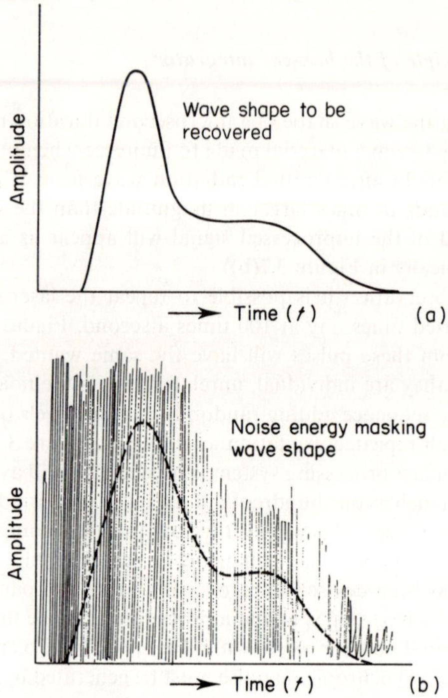

Figure 3.7 Box-car integrator time-domain signals: (a) wave shape to be recovered; (b) wave shape buried in white noise; (c) sampling pulse passing through wave shape period; (d) a few of the samples taken at a point

method cannot provide perfect recovery and there is little point in averaging too many samples.

Box-car integrators are used to investigate many types of optical effects, to observe neurological signals and to recover any signals that can be repeated in a synchronized manner for long periods. To be effective the wave shape to be recovered must be stable over the observing period.

Mean square averaging ($n = 2$)

Using $n = 2$ in Equation (3.7) yields the resultant signal $\overline{x^2(t)}$, which represents the *mean-square average* of the input signal. This finds application when assessing power levels of radiation, and other electromagnetic signals.

One of the most common uses of the mean-square average is the form of the *root mean square* (r.m.s.) value, algebraically $\sqrt{\overline{x^2(t)}}$. It is particularly useful for a description of 'symmetrical' signals (such as the sine wave) whose arithmetic mean is zero. For example, the voltage measured across a heating element connected to the a.c. mains will average to zero, thus telling nothing about the heat produced. Use of the r.m.s. value gives the heating level (when multiplied by the resistance constant).

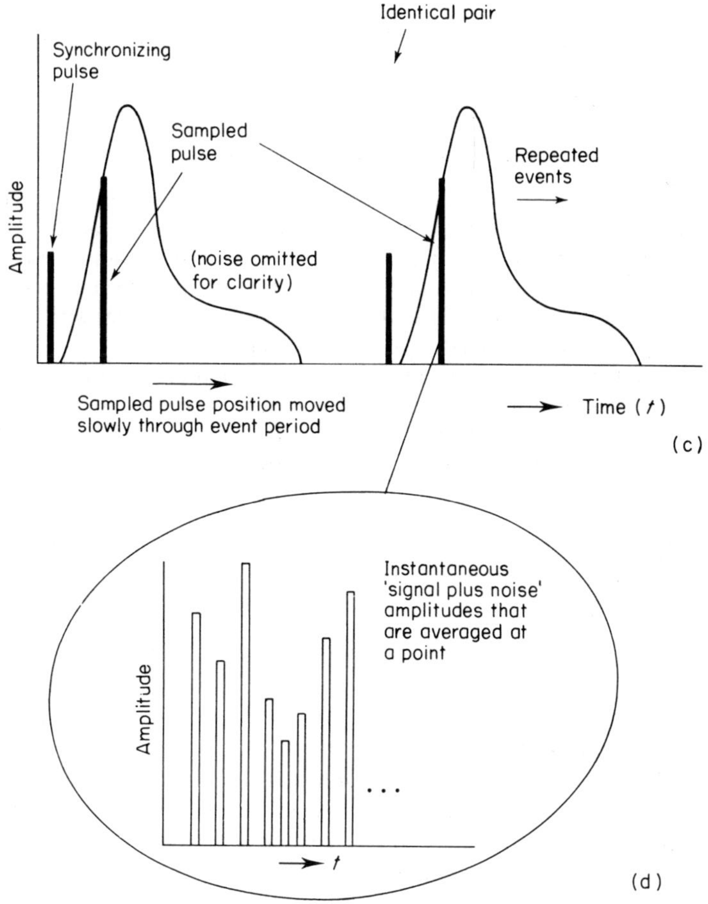

Figure 3.7 *(continued)*

Time-shifted averaging – correlation

The averaging formulation of Equations (3.6) and 3.7) are subclasses of the more general *correlation* process. If the signal is compared with *itself*, but with each comparison carried out with a different time shift τ, between them, it is called *autocorrelation*. The result is the *autocorrelation function* written as $\psi_{xx}(\tau)$ given by

$$\psi_{xx}(\tau) = \lim_{T \to \infty} \frac{1}{T} \int_{-T/2}^{+T/2} x(t)x(t+\tau)\,\mathrm{d}t = \overline{x(t)x(t+\tau)} \tag{3.8}$$

Compare this equation with Equation (3.7) and note that autocorrelation is simply a specialized kind of averaging process. An easy way to visualize this process is to consider two identical transparent films, each having close writing upon it. When the two are exactly coincident and overlapping, looking through them shows that correlation is at a maximum. As the films are moved relative to each other the words become indistinct. Patterns buried in noise can be retrieved in this manner, as explained in Example 3.5.

Example 3.5 Discovering new astronomical radio sources in space

Autocorrelation is used extensively to process radio signals received from outer space. This is because it can detect periodicity (which can be generated by rotating galactic radio sources) in noisy signals such as are received by radio telescopes.

A string of incoming signals is first recorded. This record is then autocorrelated (i.e. correlated with itself) but with each processing run taking place with the record displaced by a known time relative to the preceding set.

If a deterministic signal is present the correlation process output (called the *correlation coefficient* will rise to a maximum value according to the agreement between the two signals. The correlation coefficient function that results can give information about the spectral nature of such a signal.

In some applications two independent signals , $x(t)$ and $y(t)$, are compared and the *cross-correlation* function is then given by

$$\psi_{xy}(\tau) = \lim_{T \to \infty} \frac{1}{2T} \int_{-T}^{+T} x(t)y(t+\tau)\,\mathrm{d}t = \overline{x(t)y(t+\tau)} \tag{3.9}$$

Here the two time-domain signals are $x(t)$ and $y(t)$ and the result called a *cross-correlation function* $\psi_{xy}(\tau)$.

This has found use as the basis of such measuring instruments as flowmeters, ground-speed meters for aircraft and to measure the velocity of hot moving steel in steel mills. It is also used to assist in identifying the control characteristics of process plant. Example 3.6 explains how the *correlation flow meter* operates.

Example 3.6 Measuring flow of liquids in a closed pipe using correlation

Consider the need to measure the flow rate of waste water in a closed pipe. A source of heat is applied as shown in Figure 3.8. This heats the passing fluid with 'thermal noise' variations; these heating variations are purposely made to be close to random as this assists the correlation detection process. The resulting temperature changes are quite small and do not affect the fluid.

The stream now has imposed on its already existing temperature an additional time sequence of random temperature which flows along with the fluid. Whilst this is randomly generated the pattern stays much the same over a small interval of travel and it is, therefore, a marker pattern that is *time invariant*.

As the stream passes each sensor it generates noise-like signals. These are then cross-correlated (after some amplification) in the processor, as shown, to find when the two patterns are best matched. The delay of the correlator, which varies the alignment of the two patterns, is adjusted until the computed cross-correlation function is at a maximum. When this occurs it can be reasoned that the 'marked'

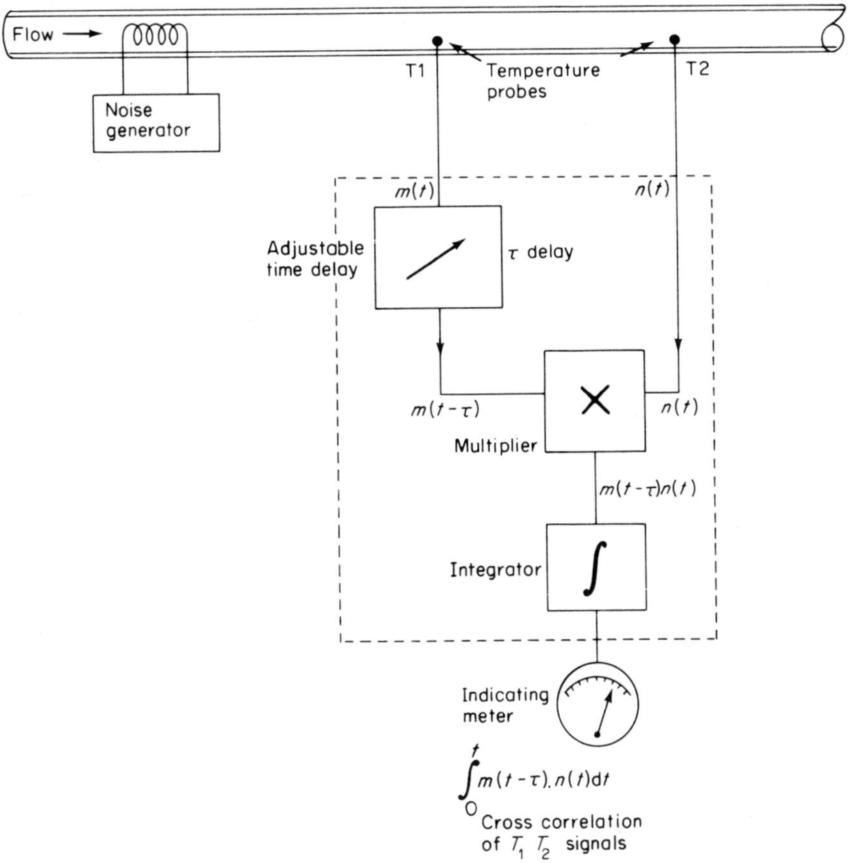

Figure 3.8 Block diagram of signal generation and processing components of a correlation flow meter. (From Sydenham P H, 1984, *Transducers in Measurement and Control* (3rd edn), Adam Hilger, Bristol. Reproduced by permission of IOP Publishing Ltd.)

sequence must have taken that time delay to pass from the first temperature detector to the other; flow rate is thus measured. Random noise is used to 'mark' the stream as this gives the sharpest correlator output.

Summary of averaging

It will be seen that the averager, box-car integrator, auto- and cross-correlators (and phase-sensitive detector discussed in Section 3.5.4 below) can each be analysed using variations of the general correlation expression, Equation (3.8).

In practice it will also be seen that they each make use of the same basic building blocks – multiplier, averager and delay elements. Processing equipment can be constructed in using either digital or analogue circuitry.

3.2.4 Transforming signal information between frequency and time domains

It can now be seen that the features of a signal can be expressed either in the time or the frequency domain; if one form is available the other can usually – there do exist constraints – be derived by use of the *Fourier transform* (*FT*). The task of conversion requires considerable computer power, but that need has been eased by use of the *fast Fourier transform* (*FFT*) computer algorithm now commonly available as commercial software.

Measurements will make use of one form or the other according to which is more practical. Mechanical system vibration measurement, for example, usually makes use of continuously scanned cyclic excitation (frequency-domain analysis); whilst the transfer function of a sensor is often best realized by exciting it with a single short-duration pulse or step function to obtain its dynamic response (time-domain analysis) – as explained in Example 3.7.

Example 3.7 Obtaining the frequency response of a sensor by application of a step change in the time domain

The frequency response is needed for a new model of rapid-response, electrical-output thermometer. Attempts to measure this by subjecting it to a source of fixed amplitude, variable frequency (0–200 Hz) temperature cycles is not practical because such a *temperature signal generator* would be expensive to construct, unlikely to be sufficiently accurate and certainly hard to calibrate.

On the other hand it is relatively easy to subject the sensor to a suitable amplitude step change in temperature by simply plunging it into a fluid of a different temperature to its ambient.

The resulting step-response output is recorded, digitized, and using, for example, a personal computer with FFT software, transformed into the required frequency response form.

3.3 CHARACTERISTICS OF SIGNALS

3.3.1 Deterministic and randomly varying signals

Deterministic signals

Where the signal level can be predicted with practically absolute certainty it is called *deterministic*. For example, given knowledge of the period, amplitude and a reference phase point for a sine-wave signal generator, it is possible to predict the value at any time in the future.

This has important relevance to the *calibration* of instruments. It allows an inaccurate instrument that has adequate *reproducibility* (see Section 2.5.2) to be calibrated by comparison with a standard. Once the errors, at suitable values across its working range, have been so established, they can be added to or subtracted from future measurement readings to obtain the accurate value.

Stochastic signals

Where the signal is *stochastic* in nature it will only be possible to predict future values on the basis of knowledge of the statistical properties of previous values. In general the greater the number of previous values the less the uncertainty there will be about the statistical properties of the signal and the better the predictions that may be made of future values. Note, however, that no matter how perfectly the statistical properties may be known the future values can only be predicted in terms of the *probability* of each having a certain value.

Furthermore, it has been assumed that the statistical properties of the signal are *stationary*, that is constant, so that they do not change with time.

The statistical properties of the signal are described in terms of a *probability density function* $w(x)$. This function expresses the probability (a value between zero and one) of x taking a value between x and $x + \Delta x$ (where Δx is a vanishingly small increment). Alternatively if x is restricted to fixed values x_i, as is the case for digital signals, the probability function is $w(x_i)$. In each case the probabilities must sum to unity over all possible x or x_i values; that is, for analogue signals:

$$\int_{-\infty}^{+\infty} w(x_i)\,\mathrm{d}x = 1 \tag{3.10}$$

and for discrete, digital signals:

$$\sum_{t=0}^{\infty} w(x_t) = 1 \tag{3.11}$$

The probability density function mostly assumed to be applicable is the *Gaussian distribution* given by

$$w(x) = \frac{1}{\sqrt{(2\pi)}\sigma}\exp\left(\frac{-(x-\bar{x})^2}{2\sigma^2}\right) \tag{3.12}$$

where

$a = \bar{x}$ is the *arithmetic mean value* (indicates centre of the distribution) and
σ = *standard deviation* (*s.d.*) (indicates spread of distribution) given by

$$\sigma = \sqrt{[\overline{x^2} - x^2]} \tag{3.13}$$

Figure 3.9 shows, with parameters normalized, how this function varies for different standard deviations (s.d). It will be observed that standard deviation is a measure of the preciseness of the data set; the smaller the s.d. the less the spread of values. Note that the area under each curve is unity, as required by Equation (3.10).

Whilst the Gaussian distribution is commonly encountered other distributions exist and it is wise to test data (using, for example, the chi-squared statistical test) to see if that distribution is appropriate to use. Another distribution, for example that found in measurement involving radioactivity, is the *Poisson* distribution; the shape of this function is not symmetrical about the

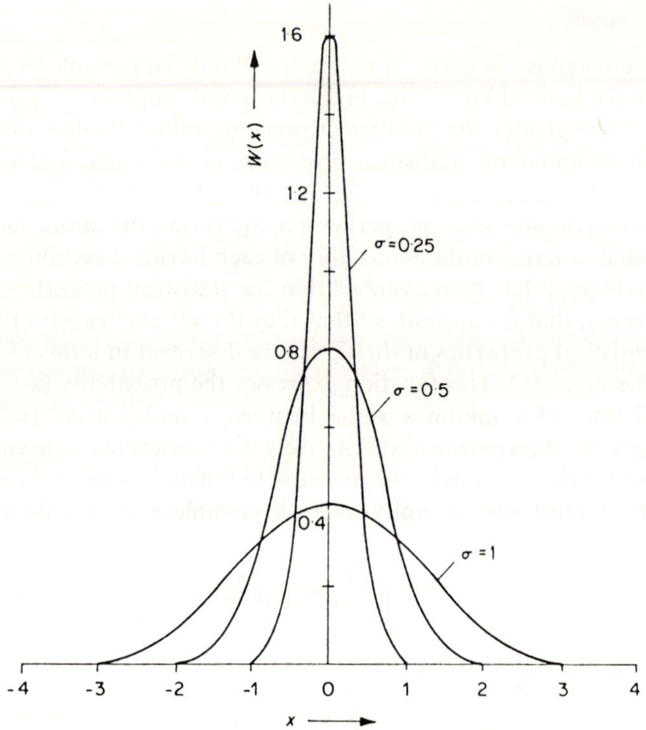

Figure 3.9 Characteristics of the Gaussian probability function. (HBMS, figure 4.16.)

mean and can be said to be *skewed.* It arises when both probability of occurrence and non-occurrence cannot both be defined. (Others are the *exponential* and *Weibull* distributions, mainly used in a failure-rate analysis.)

So far all mention of random signals that arise as noise energy has been as though it is always an unwanted signal that degrades systems performance. There exist many applications where its characteristics and known stochastic properties can actually be put to good use, Example 3.8 being one application area.

Example 3.8 Measuring temperature using noise generated in resistance

Thermodynamic effects within any resistive component of an electrical system generate stochastic noise energy which has the mean-square voltage $\overline{v^2}$ appearing across the resistor terminals given by:

$$\overline{v^2} = 4kTRB \tag{3.14}$$

where

k = Boltzmann's constant;
T = absolute temperature of resistor;
R = resistance of electrical component; and
B = bandwidth of observer.

(This phenomenon is considered further in Section 4.6.2.)

If the bandwidth and resistance value are held constant by using appropriate circuitry, it is possible to calculate the temperature of the resistor and, hence, that of its surroundings: the absolute temperature will be directly proportional to the mean-square noise voltage observed.

This concept has been used to determine temperature ranging from the cryogenic range of liquefied gases through to those of burning gas. Relevant experimental data, such as that published by Horiuchi (1977), indicates that output sensor signal levels are low and reasonably linear.

Other applications of noise theory reported by Horiuchi (1977) include measurement of the wear of trolley wire used in electric trains; of plating process quality; and of flow of solid particles in pipes. Noise is also a good test signal to use in studies seeking to identify the parameters of systems. It is used in Example 3.6 to 'mark' the flow stream in order to allow effective correlation to take place. Theoretical considerations show that the most effective code marker in that case must be that of a random noise sequence.

Other examples use noise to determine 'tin' roof mechanical properties from the impact sounds made by falling rain and to detect the onset of unwanted and potentially dangerous boiling of the sodium used in nuclear reactors. (Further consideration of the use of noise analysis is, unfortunately, beyond the scope of this text – see for example, Harris and Ledwidge (1974) for a fuller treatment.)

3.3.2 The description of multidimensional signals

The most elementary form of analogue signal is one having only one component, that of its magnitude. Examples are *scalar* quantities such as temperature and pressure or partial information about a *vector* quantity such as the magnitude of a force.

Multidimensional signals arise from the need to convey the full information about a vector quantity, for example the three components of force expressed as the x, y, and z *Cartesian* components. These signals clearly cannot be simply conveyed by a single voltage on a wire or as the pressure in a hydraulic line. Either three separate channels must be used or the three components *multiplexed* onto a single channel. The concept of multiplexing signals is covered in Section 3.5.1.

Multidimensional signals can, of course, exceed three components and the sequence can be extended as the order, n, rises to give the formulation for signals with n components x_1, x_2, ... x_n represented by a signal vector $\mathbf{x}$ in the n-dimensional space as

$$\mathbf{x} = (x_1, x_2, x_3, \ldots, x_n) \tag{3.15}$$

The end point of the vector determines the corresponding signal.

Euclidean signal space

A commonly used way to represent analogue signals is in the *Euclidean signal space* where the Pythagorean 'square root of the sum of squares' theorem applies. The magnitude $|x|$ of an n-dimensional vector $\mathbf{x}$ is given by

$$|\mathbf{x}| = \left(\sum_{r=1}^{n} x_r^2 \right)^{1/2} \tag{3.16}$$

and the distance $d(\mathbf{x}, \mathbf{y})$ between two vectors $\mathbf{x}$ and $\mathbf{y}$ is given by

$$d(\mathbf{x}, \mathbf{y}) = |\mathbf{x} - \mathbf{y}| = \left[\sum_{r=1}^{n} (x_r - y_r)^2 \right]^{1/2} \tag{3.17}$$

Non-Euclidean signal space

Where discrete signals are involved, as in the now extensively used digital systems approach to signal engineering, a *non-Euclidean* signal space is adopted. This is so-called because the magnitude $|\mathbf{x}|$ of the vector $\mathbf{x}$ is given by

$$|\mathbf{x}| = \sum_{v=1}^{n} x_v \tag{3.18}$$

and hence the distance between two such vectors is given by

$$d(\mathbf{x}, \mathbf{y}) = |\mathbf{x} - \mathbf{y}| = \sum_{v=1}^{n} |x_v - y_v| \tag{3.19}$$

In both Equations (3.18) and (3.19) note the absence of the 'square root of the sum of the squares' which is essential for the description of physical (three-dimensional) space.

Non-Euclidean space cannot be used to represent the three components of a force but it is ideal for representing the state of digital signals as will be seen below.

Binary signals

The most important discrete signals are *binary* signals as these are used as the basis of digital computing and other processing systems. Binary signals can assume only two values – 0 and 1. For these the signal space constitutes an n-dimensional hypercube having edge length l, and where the various edges are occupied by possible signal states. Figure 3.10 illustrates the first three levels which can be visualized.

The discrete signal can also be represented pictorially by the *codegraph*. Figure 3.11 shows two cases. Figure 3.11(a) is for the single-stage decimal signal with its ten alternative states. Figure 3.11(b) is for the 3-bit binary signal where several (for $n = 1, 2, 3$) stages of the codegraph *tree* exist simultaneously. Following down a path of the tree reveals the state of each bit in the binary word. For example, the lowest path passes from the left-hand originating point in the sequence 1 (blacked in node), 1, and 1, to give the binary word 111.

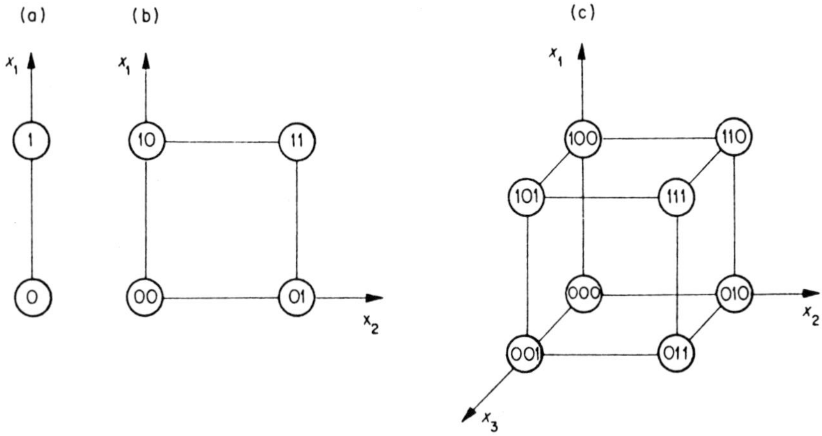

Figure 3.10 Representation of binary signals in the non-Euclidean signal space: (a) one bit; (b) two bits; (c) three bits. (HBMS, figure 4.18.)

3.3.3 Some important signals

Several signals are commonly met in instrument systems. Their properties are summarized in Figure 3.12; mathematical formulations are given in Table 3.3.

Harmonic oscillations (sine waves) are used to continuously excite a system, examples being to provide a carrier wave in a communication system (Example 3.3) or to excite a phase-sensitive detector set-up (Example 3.15).

The *unit step* is used where time-domain testing is preferred as explained in Section 3.2.4.

In some applications it is not practical to provide a step change; a *unit pulse* (also called a *delta* or *Dirac* function) can be used instead. An example is the use of a short explosion to excite seismic waves in geophysical prospecting.

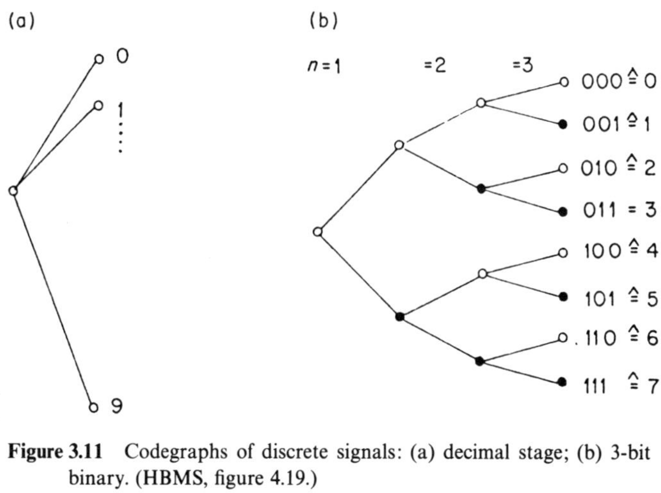

Figure 3.11 Codegraphs of discrete signals: (a) decimal stage; (b) 3-bit binary. (HBMS, figure 4.19.)

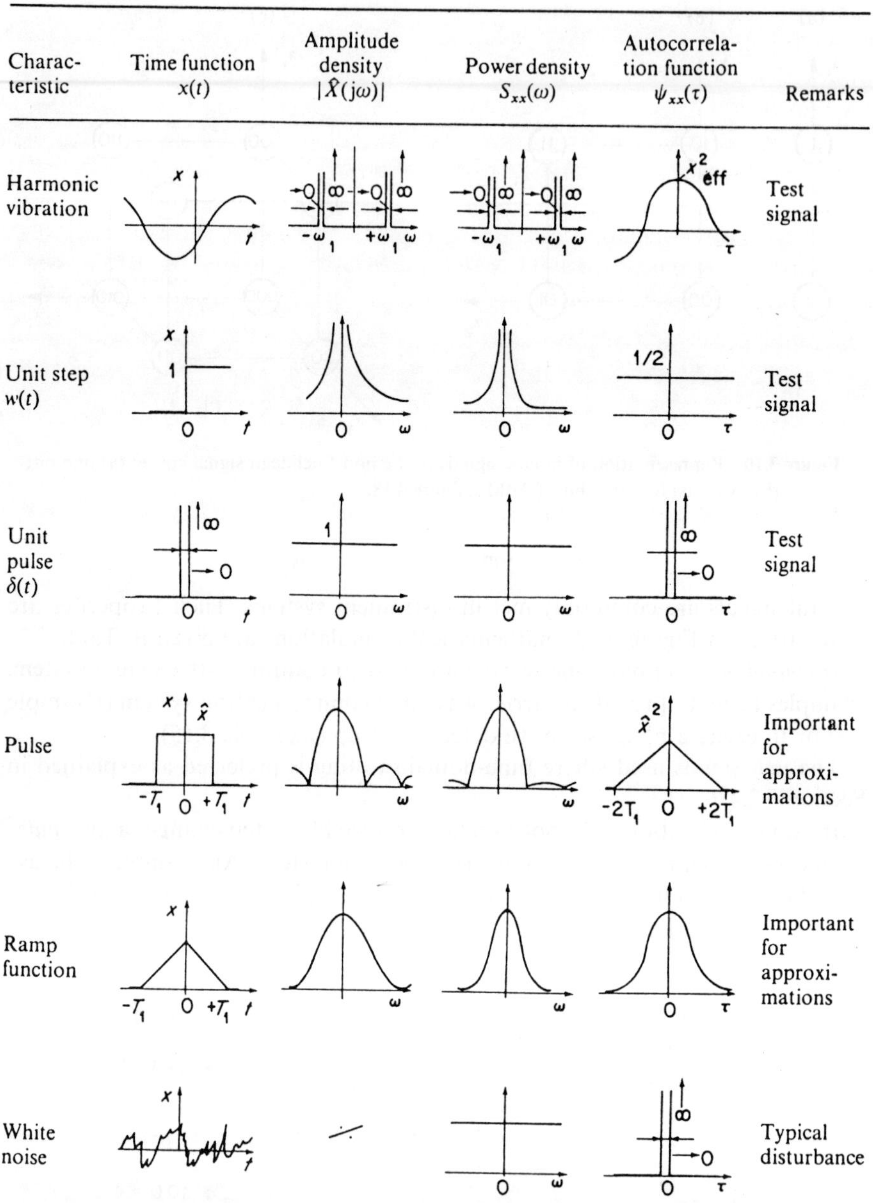

Figure 3.12 Properties of signals commonly met in instrument systems. (HBMS, table 4.4.)

The *pulse* and *ramp* are used where a closer approximation to a less than the ideal test situation applies. An example is the use of a ramp force applied to test a steel structure – it would not be practical to test it with step forces for the accelerations involved would destroy the structure and it would be very costly (if even possible) to provide large step-force levels.

An important signal, not shown in Figure 3.12, is the *pseudo-random binary sequence* (*PRBS*) test, or excitation, signal composed of a *sequence* of binary bits whose values are generated to form a randomly varying binary code. It is the simplest digital equivalent of the analogue noise signal (introduced in Section 3.2.3); conversion to analogue form (with a digital-to-analogue converter) would create an analogue signal having random-noise-like properties.

Whilst the sequence may appear to be truly random it is not so, for it is generated by a systematic algorithm – hence the reason for the term 'pseudo' being applied. Formulation of this signal is complicated, its theoretical consideration requiring mathematics well beyond the level of this text – see for example, Harris and Ledwidge (1974).

3.4 SIGNAL BEHAVIOUR WITHIN SYSTEMS

3.4.1 On system linearity

The nature of the signals within instruments and their internal processing blocks can be described by *mathematical models*, a topic of such importance to instrument engineering that the whole of Chapter 5 is devoted to it. This section presents the foundational mathematical description of systems viewed in the *time domain* and in the *frequency domain*. The necessary mathematical formulations of the *Laplace transform* and of the *Fourier transform* are also introduced; and by means of examples the concept of system output *response* to an input *stimulus* is illustrated. (Commonly met system responses are considered in more detail and formally classified in Chapter 4.)

Table 3.1 Comparison of linear and non-linear instrument systems. (HBMS, table 4.5.)

System	Linear System	Non-Linear System
Mark	Parameters are constant, independent of amplitude. Superposition law is valid	Parameters are a function of amplitude. Superposition law is not valid
Parameters are time invariant	Examples: most of measurement systems with small input levels: amplifiers; filters; transducers	Examples: systems with large input levels: output amplifiers; driver stages. Often linearization is possible
Parameters are functions of time	Examples: controlled amplifiers with multiplicative properties and small input levels: modulators; frequency multipliers; parametric amplifiers	Examples: controlled amplifiers with multiplicative properties and large input levels: modulators, frequency modulators; frequency multipliers

For the present discussion a system can be interpreted as a 'black box' having input and output signals that are related internally by some strategy to carry out the required information-processing task.

Before proceeding to describe signal behaviour it is first necessary to consider system *linearity*.

Where the mathematical models are *linear* the *law of superposition* applies and a considerable body of mathematical knowledge can be applied to great effect. If it is *non-linear* in operation then the task is not as straightforward; special techniques must be used on a largely case-by-case basis. Table 3.1 summarizes the essential; classes of system and provides examples.

3.4.2 Time-domain response

To develop appreciation of how a linear system is analysed a commonly met system is analysed, step by step, in Example 3.9. Section 4.4 delves more deeply into the various classes of response that arise.

Example 3.9 Analysis of the spring–mass–damper linear system, such as used in accelerometers, moving-coil meters and Bourdon tube gauges

The combination of a *spring* element with a *mass*, the movement of which is *damped*, is commonly met. A car suspension, a moving-coil meter movement and an accelerometer are examples. Analysis of the dynamic behaviour is important so that the required values of spring compliance, mass and damping action can be set to obtain the response required when the system is excited.

This type of system is described by the second-order differential equation for which the displacement y of the mass over time, after it has been displaced from equilibrium, by force x, is

$$m\ddot{y} + k\dot{y} + cy = x \tag{3.20}$$

where

m = mass
k = damping coefficient
c = spring compliance

(The derivation of this equation may be found in any elementary textbook on mechanics.)

Whilst this example is dealing with a spring–mass–damper system numerous other systems also are described by the same equation – the inductance–capacitance–resistance electrical circuit and many acoustic/hydraulic systems are examples. Mixed energy-domain transducers can also be similarly analysed by transforming all of the various domain elements into a common form, usually the electrical one. (Figure 4.3 is an example where the mechanical and electrical elements of the speed-measuring tacho-generator are formed into their electrical equivalent circuit for purposes of analysis.)

Once the differential equation has been formed, and characterized by giving dimensions to the coefficients, it can be 'run' in a computer to yield the dynamic behaviour. Of course a straightforward differential equation such as Equation (3.20) above can be solved directly to yield a result for y in terms of x. However, many common dynamic systems similarly expressed as differential equations cannot be easily solved mathematically and must be *simulated* or *modelled* to display their behaviour.

The original method for running a model was to use the *analogue computer* or *simulator* in which the mathematical functions were realized with *operational amplifier* and other analogue circuitry. The spring–mass–damper system is shown up in this form in Figure 3.13.

The classical differential equation of Example 3.9 can also be expressed in the *state-space* vector equation form

$$\begin{bmatrix} \dot{y}_1 \\ \dot{y}_2 \end{bmatrix} = \begin{bmatrix} 0 & 1 \\ -\frac{c}{m} & -\frac{k}{m} \end{bmatrix} \cdot \begin{bmatrix} y_1 \\ y_2 \end{bmatrix} + \begin{bmatrix} 0 \\ \frac{1}{m} \end{bmatrix} x \qquad (3.21)$$

in which the suffices 1 and 2 denote two successive states of the system.

This form is highly amenable to the programming structure and needs of the digital-computer alternative for running the model. Figure 3.14 shows the flow chart used to do this.

In the time domain two particular input signals are valuable in displaying the performance of a system. These are the *unit step*, which reveals the *transient response* $h(t)$ of the system, and the *unit energy pulse*, or Dirac *delta* function input, which reveals the *weighting response* $g(t)$ of the system.

The transient response function, $h(t)$, of the system, arising when it is subjected to a unit step displacement of the mass, is as shown in Figure 3.15. That shown is just one of a continuum of variations depending on different system coefficients m, k and c. (Alternative forms of this response are considered in Section 4.4.6.)

In the general response case (that lying somewhere between the extremes of the step or the pulse excitation boundaries), and as superposition applies, it is possible to model any excitation function in terms of a combined set of delta functions having the appropriate amplitudes and time delays.

3.4.3 Frequency-domain response

The response of a system at a particular frequency can be determined by applying a sinusoid to the input and measuring the output amplitude and phase, which for a linear system, will be at the same frequency and *synchronous* with the input. To obtain the *frequency response* $G(j\omega)$ the input frequency is scanned or 'swept' through a range which encompasses the bandwidth of the system and the result will be the *amplitude response* $|G(j\omega)|$ and the *phase response* ϕ. When plotted as graphs both will have frequency as the horizontal axis and they are often referred to as 'spectra'.

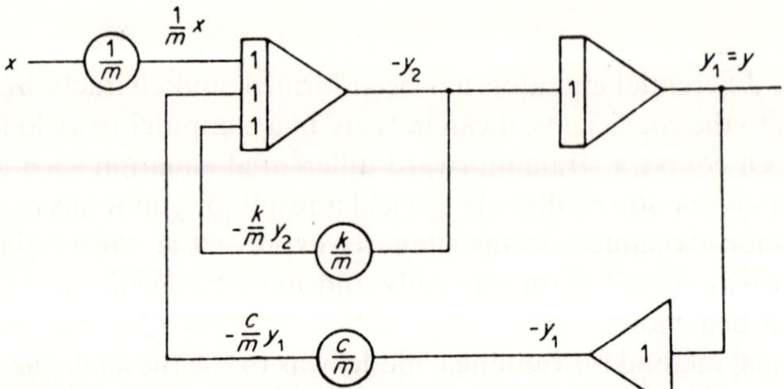

Figure 3.13 Analogue simulation of spring-mass-damper system. (HBMS, figure 4.26.)

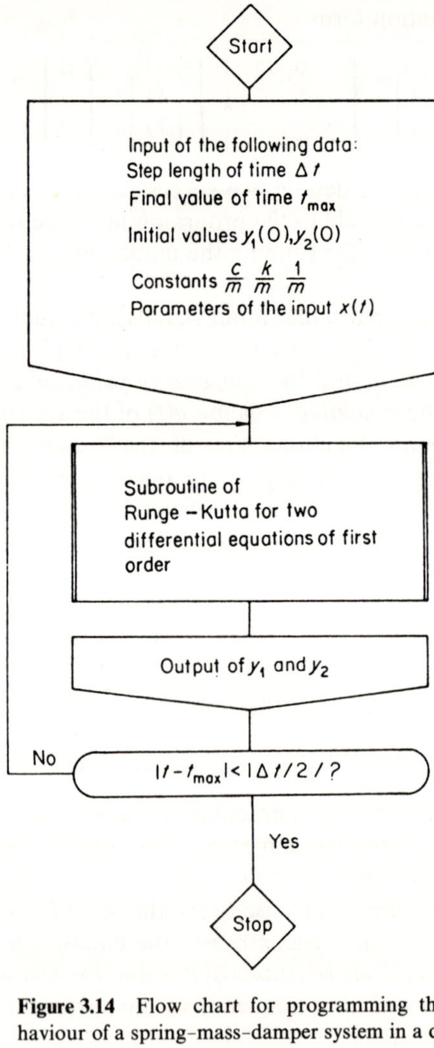

Figure 3.14 Flow chart for programming the behaviour of a spring-mass-damper system in a digital computer. (HBMS, figure 4.27.)

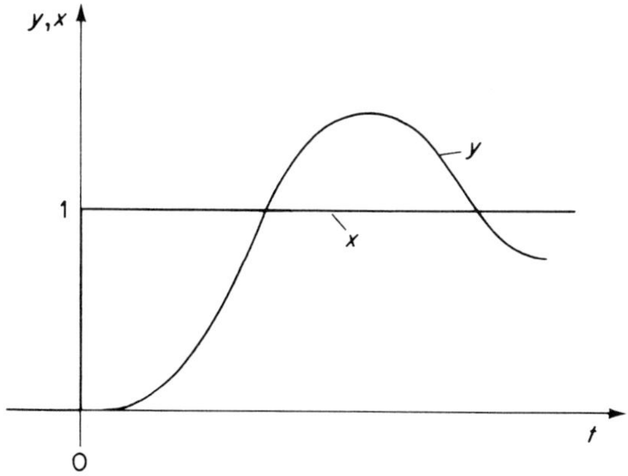

Figure 3.15 Step response of a spring-mass-damper system. (HBMS, figure 4.28.)

The frequency scan may be performed experimentally if the system already exists or it may be derived by the use of its mathematical model – as will be done here.

If the output $\hat{Y}$ is normalized to the input $\hat{X}$ and phase ϕ, then the

$$\text{amplitude response is } |G(j\omega)| = \hat{Y}/\hat{X} \tag{3.22}$$

$$\text{phase response is } \phi(\omega) \text{ formed by angle } y, x \tag{3.23}$$

After expression in its complex plane form, as

$$G(j\omega) = P(\omega) + jQ(\omega) \tag{3.24}$$

the magnitude and phase of the input–output relationship, $G(j\omega)$, at a stated frequency are given by

$$\text{magnitude } |G(j\omega)| = \sqrt{[P^2(\omega) + Q^2(\omega)]} \tag{3.25}$$

$$\text{phase } \phi(\omega) = \arctan\left(\frac{Q(\omega)}{P(\omega)}\right) \tag{3.26}$$

Graphical representations of these parameters find extensive use. Figure 3.16 shows commonly met representations; the example is of a response typical of an extended-range audio amplifier.

Another well-used graphical depiction is the *Bode plot*. This is derived from the frequency response by plotting the amplitude function (that represented in Figure 3.16(a)) with vertical and horizontal *logarithmic scales* and the phase (that of Figure 3.16(b)) with phase as a vertical linear scale and horizontal logarithmic frequency scale. These are known as the *logarithmic frequency characteristics.*

The advantages of the use of Bode plots are that the overall amplitude response of a system of stages can be readily formed by simple addition of the responses of the several stages in cascade. The amplitude Bode plot for three cascaded example stages, $|G(j\omega_1|$, $|G(j\omega_2)|$ and $|G(j\omega_3)|$, are shown in this form

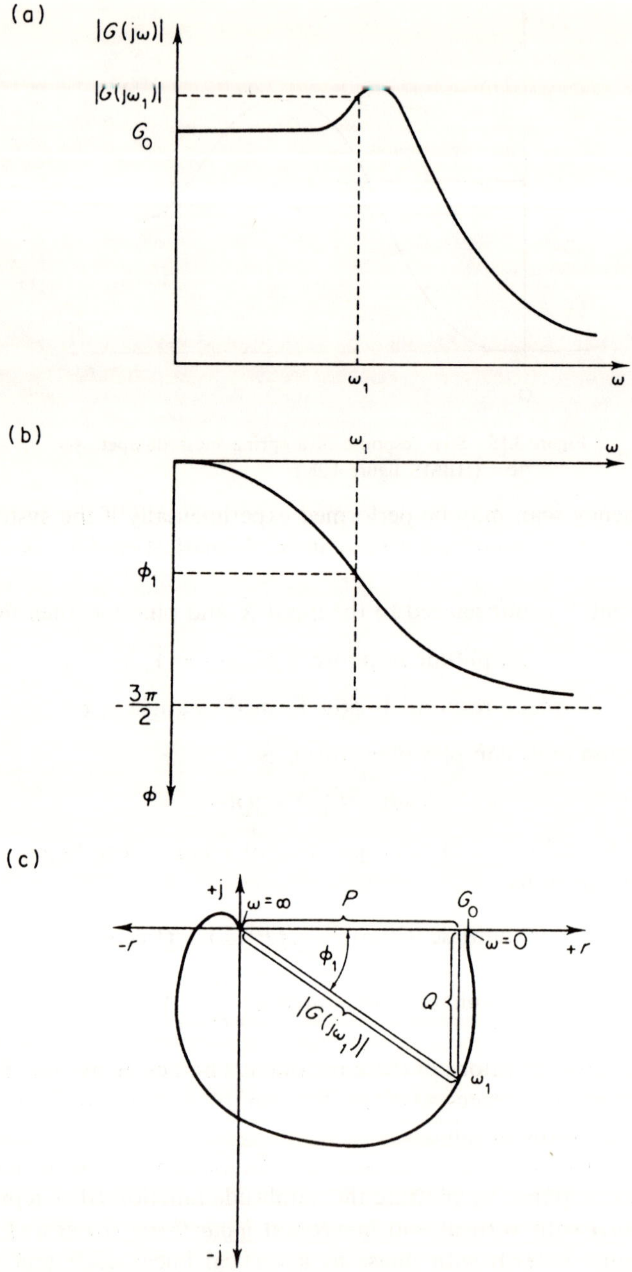

Figure 3.16 Graphical representations of the frequency response. This case would be typical of an extended range audio electronic amplifier: (a) amplitude response; (b) phase response; (c) locus diagram in complex plane showing both amplitude and phase in one plot as seen in vector for w_1. (HBMS, figure 4.31.)

in Figure 3.17. The amplitude of their combined result $|G(j\omega_e)|$ found by simple vertical *addition* of their individual decibel responses.

Another advantage of the Bode plot technique is that a reasonable approximation to the frequency response can be constructed directly from inspection of the transfer function, but that needs experience that must be developed elsewhere to this course.

3.4.4 Laplace transforms

Whilst Fourier-transform modelling of dynamic systems would appear to be extremely useful, it does not always lead to very satisfactory mathematical expressions (for example they commonly involve the sum of an infinite series of terms); nor is it always the most expedient way to analyse sytems. A variation has been developed – the *Laplace transform* – for those cases in which only real (positive) time needs to be considered. (Fourier theory applies for both negative and positive time components).

As the signal now to be transformed is discontinuous (only the positive frequency half is used) the Fourier expressions given in Equations (3.2), (3.3), and (3.4), need modification. Taking the complex frequency form of the Fourier transform; restricting the integration to the positive half of the time domain; substituting s for $\sigma + j\omega$ (p is also used) and denoting $L[f(t)]$ as $F(s)$ leads to the

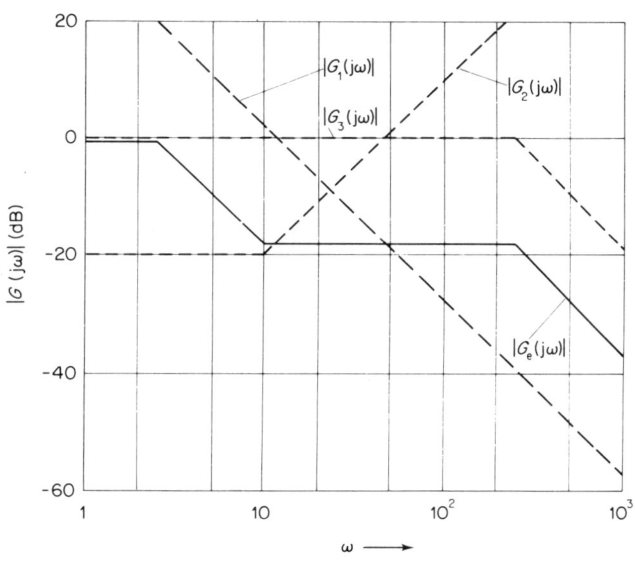

Figure 3.17 Amplitude response for a three-stage system expressed in Bode plot form. (HBMS, figure 4.32.)

Laplace transform for transforming time-domain responses information into the frequency-domain equivalent. Algebraically

$$F(s) = \int_0^\infty f(t)\mathrm{e}^{-st}\,\mathrm{d}t = L\{f(t)\} \tag{3.27}$$

The inverse process to the above, the *Inverse Laplace transform* is given by

$$f(t) = \frac{1}{2\pi \mathrm{j}} \int_{c-\mathrm{j}\infty}^{c+\mathrm{j}\infty} F(s)\mathrm{e}^{st}\,\mathrm{d}s = L^{-1}\{F(s)\} \tag{3.28}$$

Whilst the solution to some unusual system-response problems may need derivation of the Laplace expressions, according to Equations (3.27) and (3.28), many situations allow use of ready-derived *transform tables*. Tables 3.2 gives those of common transfer functions and Table 3.3 gives those for standard test signals.

Table 3.2 Laplace transforms for commonly met transfer functions. (HBMS, table 6.1.)

No.	$F(s)$	$f(t)$
1	$\dfrac{1}{s}$	$1(t)$
2	$\dfrac{1}{s-a}$	e^{at}
3	$\dfrac{1}{s(s-a)}$	$\dfrac{1}{a}(\mathrm{e}^{at}-1)$
4	$\dfrac{1}{s(s+a)}$	$\dfrac{1}{a}(1-\mathrm{e}^{-at})$
5	$\dfrac{1}{1+Ts}$	$\dfrac{1}{T}\mathrm{e}^{-t/T}$
6	$\dfrac{1}{s(1+Ts)}$	$1-\mathrm{e}^{-t/T}$
7	$\dfrac{1}{(1+T_1s)(1+T_2s)}$	$\dfrac{1}{T_1-T_2}(\mathrm{e}^{-t/T_1}-\mathrm{e}^{-t/T_2})$
8	$\dfrac{1}{s(1+T_1s)(1+T_2s)}$	$1-\dfrac{1}{T_1-T_2}(T_1\mathrm{e}^{-t/T_1}-T_2\mathrm{e}^{-t/T_2})$
9	$\dfrac{1+T_2s}{1+T_1s}$	$\dfrac{T_1-T_2}{T_1}\mathrm{e}^{-t/T_1}$
10	$\dfrac{1+T_2s}{s(1+T_1s}$	$1-\dfrac{T_1-T_2}{T_1}\mathrm{e}^{-t/T_1}$
11	$\dfrac{1+T_3s}{(1+T_1s)(1+T_2s)}$	$\dfrac{1}{T_1-T_2}\left(\dfrac{T_1-T_3}{T_1}\mathrm{e}^{-t/T_1}-\dfrac{T_2-T_3}{T_2}\mathrm{e}^{-t/T_2}\right)$
12	$\dfrac{1+T_3s}{s(1+T_1s)(1+T_2s)}$	$1-\dfrac{1}{T_1-T_2}[(T_1-T_3)\mathrm{e}^{-t/T_1}-(T_2-T_3)\mathrm{e}^{-t/T_2}]$

As before our requirement is to obtain the response of a system subjected to a forcing function in the time domain (for example the application of a test signal to a system). Direct solution of the differential equations is often very difficult, if not impossible, and hence the solution procedure is first to convert the time-domain function into the frequency domain using the Laplace transform. The resulting equations (in the 'transform domain') can then often be readily solved and the resultant expression is transformed back into the time domain using the appropriate inverse Laplace transforms. A worked example is given as Example 3.10.

Example 3.10 Finding the time-domain response of a system to a test signal

A sensor system is to be excited with a step signal of amplitude of, say, 2 units. An estimate of the time-domain output response is needed to ensure the test parameters are constructed approximately to capture the result and not overload the various recording equipments.

The sensor system transfer function (typical of a thermometer) is known to be of the form

$$1/1 + s\tau$$

where τ, the sensor time constant of, say 15 s. Explanation of *time constant* is given in Section 4.4.

The transfer function of the step test signal, from Table 3.3, is

$$2/s$$

The response is given by *multiplying* these two expressions together to yield

$$2/s(1 + s\tau)$$

To transform the result back into the time domain it is necessary to recognize if a suitable L^{-1} exists in available tables. (If not then the expression must be rewritten – partial fractions are usually employed – into a form of additive components that can be identified in Laplace tables.) In this case it can be seen that item number 6 in Table 3.2 is of the form of the above equation. Thus the inverse form is

$$L^{-1} = 1 - e^{-t/\tau}$$

As the amplitude of the test signal is 2 units and the time constant $\tau = 15$ s, the response can be plotted as a time graph of the function

$$y = 2(1 - e^{-t/15})$$

as shown in Figure 3.18.

Laplace tranforms have enabled the sensor to be readily modelled before construction. The results enable important parameters to be determined such as

- the length of the record needed to capture the full transient;
- how long the test must run;
- what gain settings should be set on the recorder;

Table 3.3 Laplace transforms for standard test signals

Aperiodic test signal function	Equation	Laplace transform
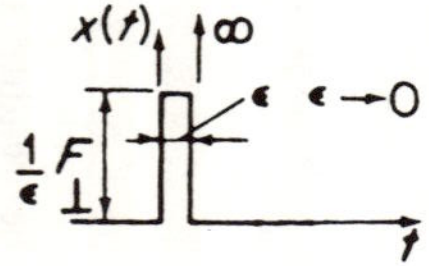	(Dirac) pulse with integral value $F_\perp$ $x(t) = F_\perp \delta(t);\ \delta(t) = \begin{cases} \infty & \text{for } t = 0 \\ 0 & \text{for } t \neq 0 \end{cases}$	$F_\perp(s) = F_\perp$
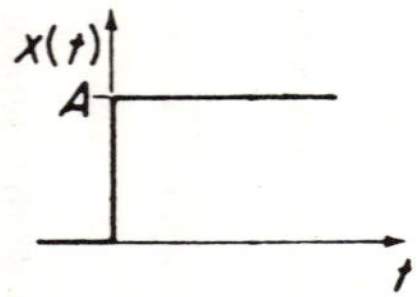	step function $x(t) = Aw(t);\ w(t) = \begin{cases} 0 & \text{for } t < 0 \\ 1 & \text{for } t > 0 \end{cases}$	$F_{\sqsubset}(s) = \dfrac{A}{s}$
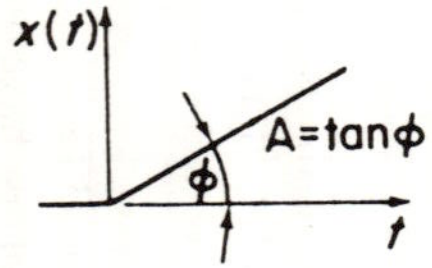	ramp function $x(t) = \begin{cases} 0 & \text{for } t \leqslant 0 \\ At & \text{for } t \geqslant 0 \end{cases}$	$E_{\angle}(s) = \dfrac{A}{s^2}$
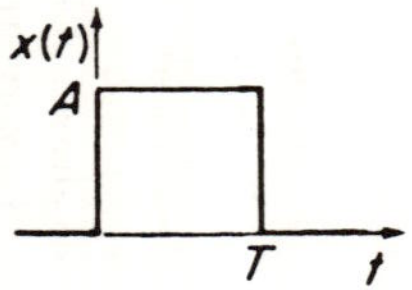	rectangular pulse $x(t) = \begin{cases} 0 & \text{for } t < 0, t > T \\ A & \text{for } 0 < t < T \end{cases}$	$E_{\sqcap}(s) = A\,\dfrac{1 - e^{-sT}}{s}$

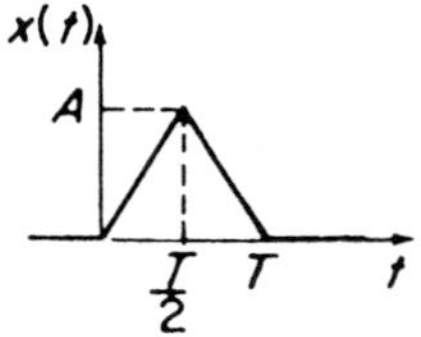

triangle pulse

$$x(t) = \begin{cases} 0 & \text{for } t \leqslant 0, t \geqslant T \\ \dfrac{2A}{T}\,t & \text{for } 0 \leqslant t \leqslant \dfrac{T}{2} \\ 2A\left(1 - \dfrac{t}{T}\right) & \text{for } \dfrac{T}{2} \leqslant t \leqslant T \end{cases}$$

$$E_{\wedge}(s) = \frac{2A}{s}\left(\frac{(1 - \mathrm{e}^{-s(T/2)})}{s}\right)^{2}$$

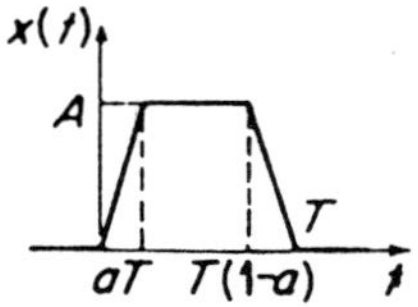

trapez. pulse

$$x(t) = \begin{cases} 0 & \text{for } t \leqslant 0, t \geqslant T \\ \dfrac{A}{aT}\,t & \text{for } 0 \leqslant t \leqslant aT \\ A & \text{for } aT \leqslant t \leqslant (1-a)T \\ \dfrac{A}{a}\left(1 - \dfrac{t}{T}\right) & \text{for } (1-a)T \leqslant t \leqslant T \end{cases}$$

$$E_{\sqcap}(s) = \frac{A}{aT}\,\frac{(1 - \mathrm{e}^{-aT_s})(1 - \mathrm{e}^{-(1-a)T_s})}{s^2}$$

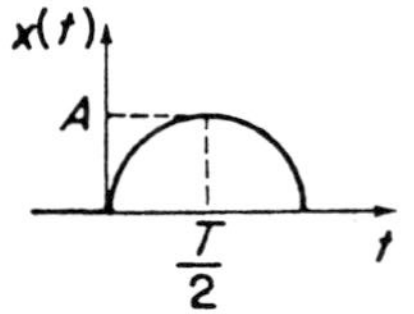

half-sine pulse

$$x(t) = \begin{cases} 0 & \text{for } t \leqslant 0, t \geqslant T \\ A\,\sin\left(\dfrac{\pi}{T}\,t\right) & \text{for } 0 \leqslant t \leqslant T \end{cases}$$

$$E_{\Omega}(s) = \frac{A\pi}{s}\,\frac{(1 - \mathrm{e}^{-sT})}{s^2 + (\pi/s)^2}$$

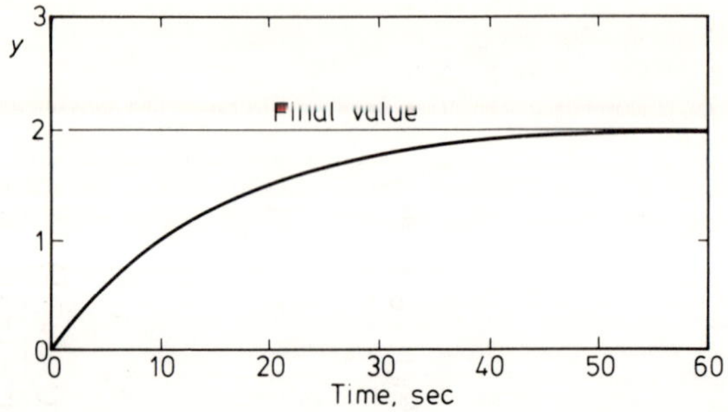

Figure 3.18 Time-domain response of thermometer to step signal

- what the response will look like; and
- a reasonable estimate of the magnitude of the resulting signal.

The analysis was well worth the effort!

3.4.5 Feedback

Before leaving this discussion of system response mention must be made of the situation where *feedback* is employed. This connection arrangement is commonly used in instrumentation and involves the output signal being fed back and added to the input signal. If the addition is as a negative (inverted) signal it is called *negative feedback* and if provided in a positive sense *positive feedback*. These connections can cause a dramatic change in system performance.

Negative feedback is used to bring a system under control by making the output follow a given input reference signal. Examples of instrument use are to force an instrument actuator to respond more quickly, as illustrated in Example 3.11.

Example 3.11 How the linear motor chart recorder operates

The *potentiometric* chart recorder makes a record of variations of a signal by marking a sheet of paper with ink from a pen that is moved relative to the paper according to the level of an input signal.

The pen is mounted to allow it to be driven along a guiding slide by a small, fast motor. In the *linear motor* variant the motor is little more than a coil of wire wound around the steel rod on which it slides. The rod is magnetized by permanent magnets placed at its end and hence when the coil is energized with d.c. current it is propelled along the rod by the electromagnetic fields created.

Also mounted in parallel with the slide is a linear potentiometer, the wiper of which is attached to the motor. Hence when the potentiometer is energized by a d.c.

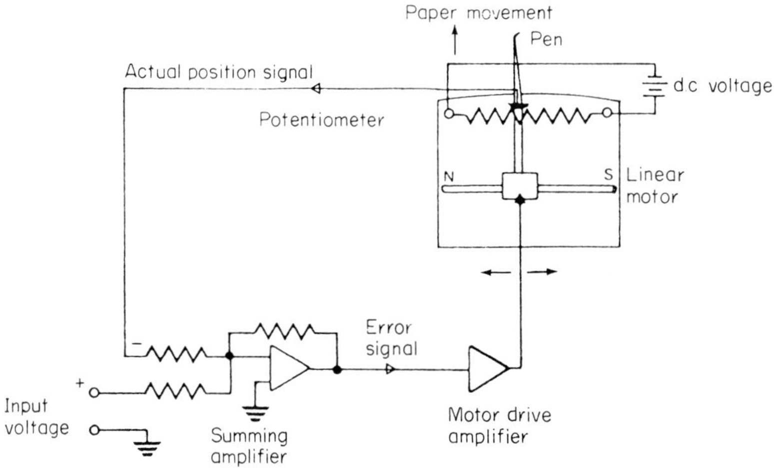

Figure 3.19 Schematic layout of feedback system used in a potentiometric chart recorder

voltage across its length a voltage will be present at the wiper which represents the position of the slide.

It is now possible to control the motor position, and hence the pen position on the paper, by comparing the input voltage with the (fed back) slide position voltage. This comparison occurs at the *summing point* of the motor drive amplifier – see Figure 3.19 – in which the input signal and the *inverted* sensor position signal are added; this is in fact a *subtraction* of the two signals because negative feedback has been used. The difference signal is called an *error signal* and causes the motor to move in a direction that will reduce this error signal to zero. Hence the motor and pen will *follow* the input signal whenever it changes.

Potentiometric recorders of this sort were invented late in the nineteenth century. Their performance did not need much understanding as intuition could help designers obtain adequate performance. Today's chart recorders, however, have greatly superior performance, particularly their ability to accurately follow rapid changes in output voltages. Typically a full traverse of the platten may be made in a fraction of a second without overshoot. This has been achieved by modelling and simulation of the motor, potentiometer and amplifier elements and a thorough understanding of the theory of feedback control.

As Example 3.11 illustrates, negative feedback is used to control and increase the stability of systems. In contrast *positive feedback decreases the stability of systems*. Should the positive feedback be strong enough a system will become *unstable* and the slightest perturbation will cause it to *oscillate*, that is produce a wildly varying output signal for no obvious input signal.

Whilst positive feedback is often undesirable (consider, for example, the acoustic positive feedback between loudspeaker and microphone in a public address or disco system) it can also be employed to advantage. It is applied to *oscillators* to act as a signal source; and it is used to *improve the sensitivity of systems* by taking them almost into their unstable zone of operation. An example of the latter is its use to make a *vertical pendulum* tilt meter so sensitive that it can detect changes from the vertical as small as one part in 10 000 000.

Analysis of feedback systems is well-developed and reported – see, for example, Atkinson (1982) – but it is beyond the scope of this chapter to develop a full working explanation. However, some essential guiding features are covered here.

Firstly the degree of negative feedback applied decides the closeness of the following action. However, if overdone and the system has inappropriate phase relationship in the control loop, then it may become oscillatory – that is, at certain frequencies it becomes a positive feedback system.

Secondly, the design of negative feedback systems aims to obtain an optimum level of following ability without fear of oscillation. Simplistic understanding suggests that this condition is obtained by increasing the feedback gain level to as high a value as practical. Theory, however, shows that use of special *frequency-response compensation* feedback will provide superior response whilst also allowing still higher gain.

If a non-linear element is present in the loop formed by the feedback element and the system stage around which the feedback is applied, then the behaviour becomes complex to analyse. It is preferable to design systems that operate in the linear mode as they are easier to analyse and usually perform better than non-linear systems. However, non-linear systems can be made to perform adequately and are usually less expensive to construct.

Feedback systems can be built using either analogue or digital signals. The classical theory, on which stability assessment is based, was developed in terms of analogue signals, but has been applied to the digital form by extending the theory to cater for the discrete nature of the signals. (A different mathematical formulation is required and this is expressed in terms of z-transforms – see, for example, Oppenheim *et al.* 1983.)

3.5 COMMUNICATION THEORY ASPECTS OF SIGNALS

3.5.1 Communications systems

Obviously as signals are to be sent through a system it is necessary to know something about signal-transmission properties in order to be able to select a suitable system that can send data fast enough and with tolerable error. A considerable body of theory has been developed to assist our understanding of this aspect and whilst the main stimulus is for reasons of telecommunications applications it also applies to instrument systems where connection links are required.

A communication link has the overall system architecture shown in Figure 3.20.

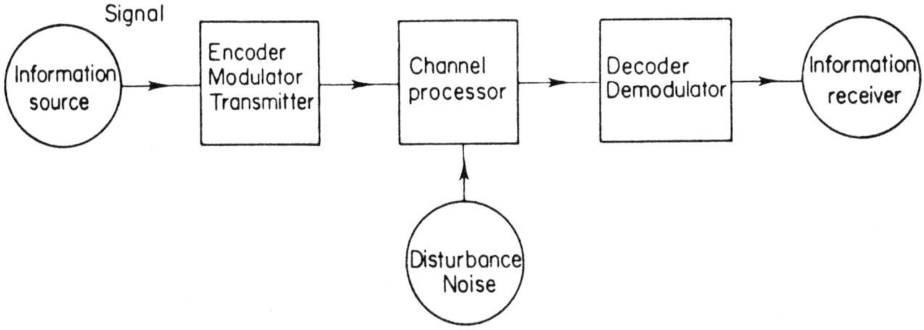

Figure 3.20 Architecture of a typical communication system. (HBMS, figure 4.53.)

The source of the information feeds a suitable signal to a block (the *modulator* for analogue systems or *encoder* for digital) which converts the signal format into one suited for transmission over the channel *carrier* type to be used. Once received the signal is demodulated (or decoded) to retrieve the information sent.

The similarity of Figure 3.20 with a sensing system feeding signals to an actuator is obvious; communications system practice and theory are important to instruments, for they, too, seek to send messages from an input to an output without deterioration. The prime difference is that sensing systems have the additional task to extract information from the physical world, converting it into a form ready to communicate.

Note particularly the presence of a block feeding *noise* into the system of Figure 3.20. Without noise it would be relatively easy to design and establish communications links. In certain cases the effect of noise can be ignored by transmitting, for example, a very strong signal, but in general to get the best performance from use of given resources it is necessary to design for conditions where considerable noise levels can exist.

The forms of physical carrier used in *communications engineering* include copper wires, coaxial cables, optical fibres and radiowave bands. These technologies all have different hardware features but overall they all make use of the same concepts and theoretical considerations. In each case the information is carried on the carrier as some form of variation of the time-varying energy level as explained in Section 3.2.1.

As will be seen below, the bandwidth capability of a carrier is a vital parameter. Often more than one sensing channel of information is required to be sent over a common physical carrier. Use is made of *multiplexing* to send, simultaneously, several different signals along the same carrier.

Time-division multiplexing (*TDM*) uses a switching system, Figure 3.21(a), to bring the several input channels onto the carrier in time sequence. They are separated at the receiving end by a suitably synchronized switch. Instrument systems make extensive use of this form in the *data loggers* that scan across several sensors (placed, say, throughout a process plant, an aircraft or a medical monitoring system). Each sensor's instantaneous value is sampled and stored –

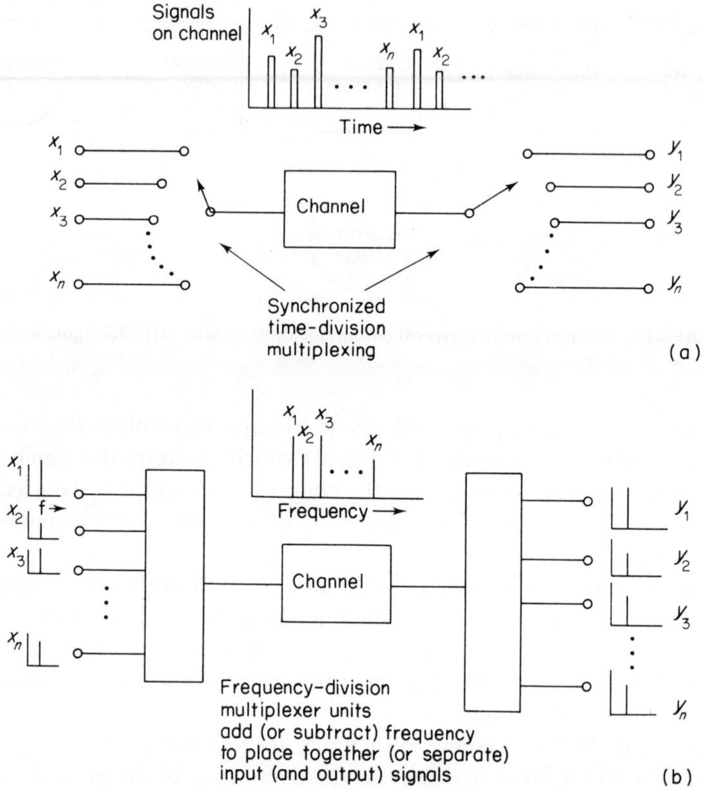

Figure 3.21 Architectures for multiplexing signals on a wide-band carrier: (a) time-division multiplexing; (b) frequency-division multiplexing

in *sample-and-hold* circuit – ready for transmission. To be effective the highest rate of change of the signal must be slower than the rate at which it is sampled; if not information will be lost, as is proved in Section 3.5.2.

The alternative is *frequency-division multiplexing* (*FDM*) in which the frequency of each channel is raised to take a unique place in the higher frequency band, each being placed by the side of the others in the group (see Figure 3.21(b). Upon receipt they are separated and recovered in their original form by the use of filtering and demodulation.

So far we have considered the task of the communications engineer to be one of reproducing as faithfully as possible at the receiver, the current and voltage wave forms from the transmitter. Whilst this may be appropriate for analogue signals (for example, a voltage which represents a temperature) it is not essential for *coded* (digital) signals. In this latter case what is important is the reproduction at the receiver of the *information* contained in the coded message.

This aspect of communication was appreciated only some 40 years ago in the words of C. E. Shannon, the acknowledged 'father of communication theory':

> The fundamental problem of communication is that of reproducing at one point either exactly or approximately a message transmitted from another point.

In his key paper of 1948 he established the mathematical theory of communications, called *information theory*. This is now outlined in the following sections.

This simple premise has been developed into an extensive body of complex theoretical knowledge that enables designers to set up a communications channel that will transmit the required amount of information with a known amount of loss, with the facility to detect errors and even have a degree of automatic error correction. (Details are given in any modern text on communications engineering – for example, Haykin, 1983.)

The concept of information embodied in these ideas is independent of the physical means of communication: it may be applied to any form of communication channel. However, what this theory does not address (Shannon recognized this in his work) is the *meaningful* content of the information. Theory has still to be developed that will enable the measurement engineer to process the meaningful content of a signal as it passes through the signal-processing stages. At present intuition must suffice to ensure this aspect of the measuring system is properly addressed.

3.5.2 Basics of information theory

Two key parameters need to be modelled in a message channel. Firstly, how much information is flowing and, secondly, how much information can a channel carry? We shall develop mathematical expressions for these requirements from the application of information theory to communications engineering. We begin with some definitions.

A *message* is a collection of basic entities, called *symbols*, so arranged that they collectively convey *information* over a communication link. For example, a written message in the English language uses an alphabet of 26 alphabetical characters plus ten numeric characters to form higher-order groups (words and numbers, through sentences to paragraphs, and so on). Other spoken languages use other forms of symbology. Electronic communication systems use signal levels such as voltages and currents to represent letters and numbers.

If the number of different possible symbols that an information source is able to deliver is m, and if there is no *a priori* information about the probabilities of occurrence of these at the receiver before receiving the signals, then all possible symbols, at the receiver, have the same probability of occurrence, which is given by

$$p_1 = p_2 = \cdots p_i = 1/m \tag{3.29}$$

As is usually done, the sum of all possible states is normalized to unity.

$$\sum_{i=1}^{m} p_i = 1 \tag{3.30}$$

The logarithmic measure of information

Experience suggests that the amount of information conveyed in a message must be proportional to the length of the message; that is, for a message of length N units the information I_N may be written

$$I_N = KN \tag{3.31}$$

where K is a constant which will depend on the symbols employed to form each unit of the message.

Now each of the N units of the message may be chosen from over m equiprobable symbols. Hence the number of possible messages that may be sent will be m raised to the power N; that is, m^N. (Consider, for example, if we were to increase N by 1 there would be m times as many different messages that could now be sent.)

Hartley, in 1928, recognized that only a logarithmic measure of information would reconcile this fact with Equation (3.31), that is,

$$I_N = k \log m^N = kN \log m \tag{3.32}$$

in which the constant of proportionality k will be chosen to be unity. Hence the information content per message I will be given by

$$I = I_N/N = \log m \tag{3.33}$$

Note that the base of logarithms is arbitrary: logarithms to different bases are simply related by a constant factor which can be incorporated in our constant k of Equation (3.32), which is defined as unity. Logarithms to the base 2 ('binary logarithms', symbol lb) are universally employed because they are the most convenient for binary signals. The corresponding unit of information is the *bin*ary dig*it* or 'bit'.

Entropy and information

The term *entropy* was first adopted in classical mechanics and physics to describe the state of disorder of a system possessing individual states. This has relevance here, for a communication system also contains degrees of *disorder*. Clearly the more the existence of different decision states in the system, the more the probable level of disorder of the system states. Furthermore, as information is conveyed the disorder may be seen to decrease; and in this sense *information may be considered as negative entropy*.

We can now reconsider the case where our m symbols are no longer equiprobable but have probabilities p_i ranging from p_1 to p_m normalized according to Equation (3.30) above. The information content per message Equation (3.30) should now be written

$$I = \sum_{i=1}^{m} \log (1/p_i) \tag{3.34}$$

The entropy or *average information content per symbol* can be written

$$H = \sum_{i=1}^{m} p_i \log (1/p_i) = -\sum_{i=1}^{m} p_i \log p_i \tag{3.35}$$

(The final form of this equation is the same as that for thermodynamic entropy but with a negative sign, as expected.)

Now the maximum (that is, least negative) value of entropy H_0 is achieved when we have no prior information about the probabilities p_i and must assume that they are all equally likely, as in our initial Equation (3.29). Hence from Equation (3.35)

$$H = \log m = \text{lb bits/symbol} \tag{3.36}$$

and for a single binary symbol the probability of each of the two states is $\frac{1}{2}$, hence $m = 2$, giving $H_0 = 1$ bit/symbol, as expected.

The average information content per symbol, H_0, gives the number of binary decisions needed, on average, to distinguish one state from the set of all possible states (see Example 3.12).

Example 3.12 How many bits are needed to transmit sensor data?

Four different *on–off* protection sensing switches need their information to be sent to a control room by a binary-signal link. They must all be able to operate together if need be. How many signal wire circuits are needed?

At first glance it would appear that four separate circuits can be used, one for each sensor's output. This is one solution but, by the use of coding, it is possible to use only three circuits.

The four binary on–off sensor states (the symbol here is the high/low or 1 / 0 states) provide eight different decision states that are to be conveyed. Using Equation (3.36)

$$H_0 = \text{lb } 8 = 3 \text{ bits}$$

As each wire circuit transmits, for a binary signalling system, one bit, then three circuits are needed, not four. Use of coding has made best use of the signal link in terms of information-carrying capacity.

This result can be seen to be true by assigning decision states to each of the bit positions of the three-bit binary word, as shown in Table 3.4. Complex examples involving hundreds of possible decision states would take too much paper to lay out such charts – Equation (3.36) makes estimation much more straightforward.

Table 3.4

	Switch 1		Switch 2		Switch 3		Switch 4	
Switch state	on	off	on	off	on	off	on	off
Binary number	000	001	010	011	100	101	110	111

So far this explanation has been made for the simple digital 'on–off' signal form. The theory has been extended to cover the case of the continuous probability distribution $w(x)$ for which the entropy $H(x)$ is then given by

$$H(x) = -\int_{-\infty}^{+\infty} w(x)\,\mathrm{lb}[w(x)]\,\mathrm{d}x \tag{3.37}$$

Channel capacity

Channel capacity may be defined as the maximum rate of reliable information transmission over the channel. Shannon's *channel capacity theorem* states that for a signal of entropy H involving r symbols per second the channel capacity C, in bits per second, is required to be

$$C \geqslant rH \tag{3.38}$$

For a noisy channel the entropy (disorder) of a signal will increase (that is, become less negative) as it passes through a channel. Hence the entropy of a transmitted signal $H(x)$ will be changed to that of the received signal $H(y)$ according to

$$H(x) = H(y) + H(x;y) \tag{3.39}$$

where $H(x;y)$ is the conditional entropy or *equivocation* between x and y, and can be regarded as a measure of the average uncertainty of the received sample compared with the transmitted sample. Specifically it can be shown to be the entropy of the additive noise in the channel.

The entropies of the received signal $H(y)$ and the additive noise $H(x;y)$ may be expressed in terms of the signal power P_s and noise power P_n. It can be shown (see, for example, Haykin, 1983) that the average transmitted information or *transinformation* $I(x;y)$ is given by

$$I(x;\,y) = \tfrac{1}{2}\,\mathrm{lb}\,(1 + P_s/P_n) \tag{3.40}$$

where P_s/P_n is the *signal-to-noise ratio* (*SNR*).

Now for a binary data channel of bandwidth B the maximum symbol rate r will be $2B$ samples per second (the 'Nyquist rate') and from Equation (3.40) the channel capacity C is given by

$$C = 2BI(x;y) = B\,\mathrm{lb}(1 + P_1/P_2)\ \text{bits/second} \tag{3.41}$$

This is called the Hartley–Shannon law.

In both Equations (3.40) and (3.41) it has been assumed that the noise within the channel is Gaussian white noise limited to the bandwidth B of the channel, an assumption that is at least approximately true for most physical communication channels.

Two important conclusions follow from the Hartley–Shannon law (Equation (3.41)). Firstly the expression shows that for a desired channel capacity C it is possible to 'trade-off' bandwidth B against signal-to-noise ratio, although not in a linear fashion because of the logarithmic dependence and because the noise power P_n is itself directly proportional to bandwidth.

Secondly, Equation (3.41) describes the optimum performance of a communications channel, although in practice this theoretical optimum cannot be achieved. Sophisticated, fully-utilized, telecommunication channels can be developed to come relatively close to the ideal but measurement systems often fall far short for they often cannot be given the design resources to be optimized in hardware form and system operation. This means that the received signal will contain much redundant information that is not necessarily of any value. Example 3.13 shows how the channel capacity of a measurement system is assessed.

Example 3.13 Setting up a communication link from a sensor output

A given capacitance form of sensor, measuring pressure, produces an output signal from its phase-sensitive detector (PSD) circuit (to be explained in Example 3.15 below) that has to be coupled to a digital recorder. What communication link performance is needed?

The bandwidth of the PSD type of signal output for such an application is taken here to be 100 Hz. (This value could vary, however, according to application, from 0.0001 Hz to 1000 Hz.) The power level of the signal would be typically 10 mW (100 mV signal with an output impedance of 1 Ω). The white-noise level relative to this, in a well-designed PSD stage, would be at least 5000 times less than the signal, typically 2 μW. This means the SNR is 10 000.

Using Equation (3.41)

$$C = B \,\mathrm{lb}(1 + P_s/P_n)$$

we obtain

$$\begin{aligned} \text{Channel capacity} &= 100 \,\mathrm{lb}\, 5000 \\ &= 1229 \text{ bits/s} \end{aligned}$$

This information enables an appropriate communication link technology to be selected. The data rate is at the lowest end of the various options: a simple twisted wire connection (data rate to at least 100 000 bits/s) can be used provided it is shielded to screen the wires from induced external noise. The information is also needed to select following digital stages in which buffer and conversion interfaces are used.

In practice the calculated channel capacity would have to be increased. Practical systems would not be optimum because the noise would not be truly random and the system would not be able to make use of every expected different decision state.

Testing the nature of the signals with a spectrum analyser, noise level meters and other equipment could improve the certainty of the design information but usually the expense of such tests is not worth the effort as bandwidth is relatively cheap to provide when it is this narrow.

3.5.3 Coding and decoding of signals

As shown in Figure 3.20, the first step in setting up a communication link is to modify the signal to suit the carrier system employed. If the signal is digital it may be rendered into a *serial format* (one bit after another over time) and the bit stream organized, often requiring significant transformation, into *code words.*

The most obvious example is the use of *Morse code*, invented by Samuel Morse in the early nineteenth century to enable messages to be sent with the first crude telegraphs. The wire systems of that time could only communicate with on–off signals that were sent, with hand-operated *morse keys*, at around 2 bits per second. The series of dots (short period), dashes (longer period) and spaces (between words) were sent according to preassigned codes for the letters and alphabet parts of the message transmission.

Modern electronic communications systems also make use of codes which have been standardized to allow interconnections to be easily effected on an international basis. In telecommunications links it is the International Telegraph and Telephone Consultative Committee (CCITT) that have set such standards within their overall systems standards. More recently the needs of factory and office hook-ups (which need more kinds of control information to be conveyed) have been standardized by international agreement as the *Manufacturing Automation Protocol/Technical and Office Protocol* (MAP/TOP).

As shown in Example 3.12 one purpose of using codes is to make most effective use of the channel capacity by reducing the level of *redundancy.*

Other reasons exist for coding signals. To make a message secure from interception by unwanted persons it can be *encrypted.* The *scrambler* does this in telephone circuits. Another reason is to reduce the errors in a received message: to detect if errors have occurred and to possibly make corrections. These examples make use of *deliberately introduced redundancy*; the extra bandwidth is deemed to be worth the cost. A prime example of this is the use of advanced codes to ensure data links using satellites do not have errors exceeding one bit in every 10^{10} bits transmitted. In such cases almost as many bits are sent for error detection and correction as are present in the actual message (Lin and Costello, 1983).

The theoretical construction of such codes is beyond the scope of this course; it is sufficient to remember that the level of message reliability can increase to extremely high levels if the bandwidth can be increased accordingly. In practice coding is applied in many instrument systems relatively easily by selecting already developed integrated circuits sold for this purpose. Only an understanding of the overall system needs are required to use these *chip sets.*

In measurement systems certain codes are commonly used in sensor and processing blocks. The physical features of some are now described.

The ubiquitous *binary* signal can be coded in many ways. For a serial transmission of a digital signal between stages the sequence of bit states, marked by a recognizable bit-set state that separates code members, can be of many forms. The Morse code is one such set – but not an efficient method.

Conversion from *binary* to the decimal equivalent can be carried out with several forms of binary input. The *binary coded decimal* (*BCD*) code and *excess-three* codes are shown in Table 3.5.

The 'second tetrad' column in Table 3.5 (binary states for the four bits needed to cope with decimal 9 for the binary-coded decimal state) will be seen to be the expected 'normal' binary sequence for conversion of decimal states from 0 to 9.

The simple arithmetic to be carried out to convert binary to decimal signal forms requires the generation of the *complement* (the binary 'opposite' number–for example the complement of a 101 is 010 of the binary number. The 'normal' binary code sequence is not suited to easy conversion, as the complement, in some states, is not realizable. A solution is to use the *excess-three code* shown in Table 3.5; its mirror image about the 4 and 5 transition is obvious from the table.

A code of particular importance to instrument design in the *Gray Code*, an example of which is set out in Table 3.6. In the four-bit code shown there are sixteen states arranged in an order such that the transition between any two adjacent states involves a change of only one of the bits of the binary word.

The code is designed for applications where the sequence moves smoothly between adjacent states and hence if anything other than a single bit change is detected (by a special, monitoring, binary stage) it is recognized as an error. The

Table 3.5 Comparison of decimal to binary coded signals. (HBMS, table 4.15.)

	Binary-coded Decimal Code		Excess-Three Notation	
Decimal Code	First Tetrad	Second Tetrad $x_3x_2x_1x_0$	First Tetrad	Second Tetrad $y_3y_2y_1y_0$
0	0000	0000	0000	0011
1	0000	0001	0000	0100
2	0000	0010	0000	0101
3	0000	0011	0000	0110
4	0000	0100	0000	0111
5	0000	0101	0000	1000
6	0000	0110	0000	1001
7	0000	0111	0000	1010
8	0000	1000	0000	1011
9	0000	1001	0000	1100
10	0001	0000	0100	0011
11	0001	0001	0100	0100
20	0010	0000	0101	0011
50	0101	0000	1000	0011
51	0101	0001	1000	0100
76	0111	0110	1010	1001
99	1001	1001	1100	1100

Table 3.6 The four-bit Gray code.

DECIMAL	BINARY	GRAY
0	0000	0000
1	0001	0001
2	0010	0011
3	0011	0010
4	0100	0110
5	0101	0111
6	0110	0101
7	0111	0100
8	1000	1100
9	1001	1101
10	1010	1111
11	1011	1110
12	1100	1010
13	1101	1011
14	1110	1001
15	1111	1000

code is not designed for computational purposes and a standard conversion to binary is usually performed before any computation.

The single bit change property includes the first and last codes (decimal 0 and decimal 15 in Table 3.6) so that it is applicable in both 'linear' and 'circular' dimensional measurement situations.

This code is often employed directly in digital transducers. A common example is the *shaft encoder* explained in Example 3.14.

Example 3.14 The digital shaft encoder

The shaft encoder is a 'digital sensor' used to directly transduce the rotational angle of a shaft into a computer or other digital monitoring system. They are used in machines as audio tape deck speed control, in photocopiers, and in aircraft-tracking systems. It is one of a small number of sensing systems that can be said to produce a *direct* digital output of an analogue quantity.

The sensor consists of a small disc attached to the shaft carrying a pattern of concentric rings arranged according to the Gray code. A four-bit version is illustrated in Figure 3.22 and can be compared directly with Table 3.6.

Four position detectors (typically radiation from semiconductor light-emitting diodes passes through the disc to fall upon photodetectors, the whole being mounted inside a lightproof housing) are mounted along a fixed radius such that each ring of the disc will be monitored as the disc rotates. A four-bit digital output signal, representing position, is formed using the output state of each detector to form each bit of the four-bit word. As there are sixteen possible states in this (simple) example, sensor angular position will only be known to within a discrimination of 22.5°. The number of bits is chosen to suit the discrimination needed, with around 22 bits being the upper practical, but very expensive, limit.

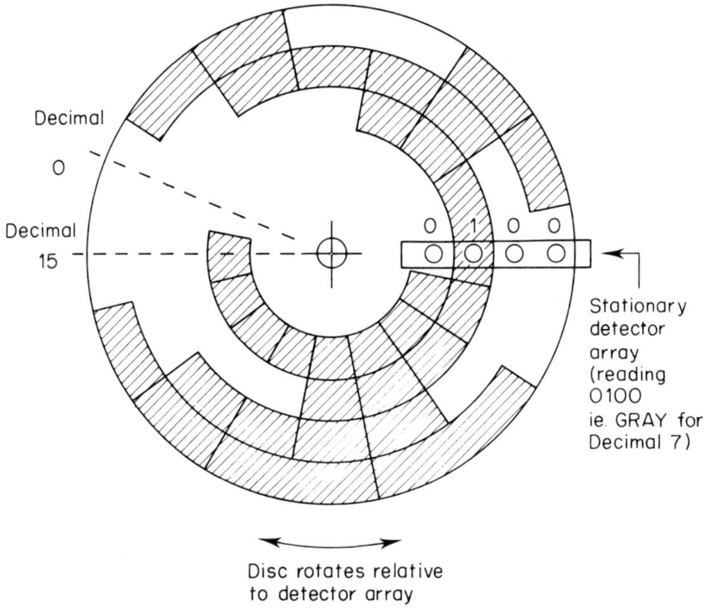

Figure 3.22 A shaft-encoder disc using a four-bit Gray code. (The *leading bit* is innermost.)

As a digital system cannot know which particular state, of two, it is in when it is at the bit-transition position the discrimination is usually quoted as ± 1 bit.

This form of encoder is known as *absolute encoder* because each position has that unique place stored as a value in the disc construction. A restart after system power failure will know where it is. In sharp contrast is the less expensive *incremental encoder*, in which identical radial lines are counted as they pass the detector. In this case counting is made relative to a datum position. The disc incorporated no unique positional data in its construction.

Error detection and correction

Concern about error that may occur in digitally coded information has already been raised in the design of the Gray code. The topic of error detection and its subsequent partial (or, ideally, total) correction is a topic that forms a major part of modern *communications engineering* (see for example, Haykin 1983) and is beyond the scope of this text. However, two simple techniques of error detection are commonly employed in digital information transfer within measurement systems.

A *parity bit* is often added to the end of a short bit stream or 'word'. This bit is chosen to be 1 or 0 in such a way that the sum of all bits in a particular word is always even (or always odd if the other bit option is used). A simple binary circuit receives the complete word and checks that the parity bit is as expected. However, the system is not foolproof - double, and other multiple errors may well cancel out to give a required parity indicating all is correct. It will be obvious that adding more checks, along these lines, could be used to improve error detection – this is the starting point of the considerable body of theory about error detection and correction for merely adding more checks in an obvious manner is not the optimal way to use the extra bits.

For larger blocks of digital data a *checksum* word may be transmitted at the end of the block. This word is similarly the binary sum of all bits in the data block – for example a 256-bit data block will require an 8-bit checksum. The receiving circuit recompiles the binary sum from the data received and compares it with the checksum received. In this system the likelihood of errors going undetected is much less than a simple parity check but still a possibility. In practice error detection can be obtained to very high limits (one bit in 10^{10} has already been mentioned in satellite links) but never taken to perfection.

In both of these simple cases error correction is not possible. The usual strategy is to ensure that the communication is *bidirectional* so that the transmitter can be instructed to *retransmit* the corrupted word of data block when error in transmission has been detected, repeating this operation until the transmission error is no longer present. Clearly there will be measurement situations where this is not possible (the instantaneous value of a process variable, or scientific experiment, often only occurs once) and data will be irretrievably lost. However, at least the data will be recognized as corrupted and disgarded..

3.5.4 Modulation theory

Modulation occurs when an input information-bearing signal varies a property of a *carrier* signal. *Demodulation* is the reverse process used to recover the information-bearing signal at the receiver.

Amplitude modulation (AM) varies the amplitude of the sine-wave carrier. *Frequency modulation* (FM) varies the instantaneous frequency. Other variants are also used. Figure 3.23 shows the time-domain appearance of the wave forms of common analogue methods. For simplicity the input signal is shown as a single frequency but in practice it is usually a band of frequencies.

It is also possible to vary the parameters of digital signals in similar ways to the above. Figure 3.24 shows several schemes which are often referred to as *pulse-analogue* methods. In each case the property of the pulse train (carrier wave) is varied by an analogue information-bearing signal.

Finally the information-bearing signal may be supplied in a digitally coded serial form, and then modulated on the carrier to produce *pulse-code modulation* (PCM).

In amplitude modulation a sinusoidal signal v_a of form

$$v_a = V_a \sin \omega_a t \tag{3.42}$$

where V_a is the signal amplitude, is used to vary the amplitude V_c of a higher frequency sinusiodal carrier v_c of the form

$$v_c = V_c \sin \omega_c t \tag{3.43}$$

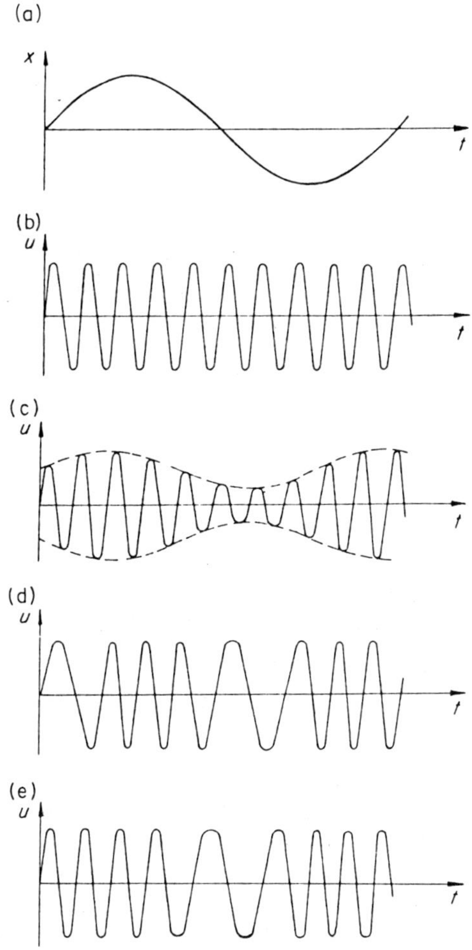

Figure 3.23 Time domain signals from the common analogue modulation methods: (a) modulating sinewave; (b) carrier sinewave; (c) amplitude modulation result; (d) frequency modulation result; (e) phase modulation result. (HBMS, figure 4.62.)

Hence the composite modulated wave v_r is

$$\begin{aligned} v_r &= (V_c + V_a \sin \omega_a t) \sin \omega_c t \\ &= V_c \sin \omega_c t + V_a \sin \omega_a t \sin \omega_c t \end{aligned} \tag{3.44}$$

which is seen to involve a *multiplication* of the two signals. Expanding this expression yields

$$v_r = V_c \sin \omega_c t + mV_c/2 \cos(\omega_c - \omega_c)t - mV_c/2 \cos(\omega_c + \omega_a)t \tag{3.45}$$

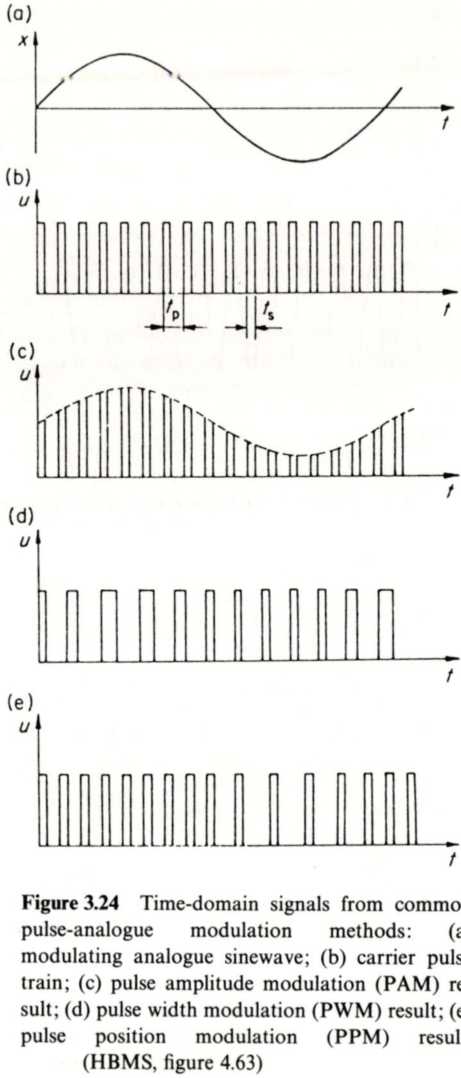

Figure 3.24 Time-domain signals from common pulse-analogue modulation methods: (a) modulating analogue sinewave; (b) carrier pulse train; (c) pulse amplitude modulation (PAM) result; (d) pulse width modulation (PWM) result; (e) pulse position modulation (PPM) result. (HBMS, figure 4.63)

where m is the *modulation index* $= V_a/V_c$: this gives the depth of modulation and is usually expressed as a percentage between 0 and 100%.

Equation (3.45) clearly shows that there are three frequencies now present – the carrier (unchanged) and two sidebands of frequency $\omega_c - \omega_a$ and $\omega_c + \omega_a$ as was shown earlier, in Figure 3.6.

The above procedure of multiplying two signals is used in many different instrument applications. An important variant of this class is *phase-sensitive detection* (*PSD*), also called carrier or synchronous detection. The circuit block performing this function is variously known as a *phase-sensitive detector* (*PSD*), a *lock-in amplifier* or a *synchronous detector*. Example 3.15 explains how this is analysed and applied.

This detection method is used where the measurand of the object to be measured is available either as the amplitude or the phase only (that is, not both together) of an a.c. signal. Examples are reactive electrical transducers which rely on a change in capacitance or of an inductance to convey the measurand information. This technique is also useful where the noise levels are too high to enable a relatively slowly changing signal (typically below 1 kHz) to be recovered. Examples here include electrical resistance bridge circuits where a.c. excitation may be used so that the out-of-balance output signal will also be a.c.; and optical systems in which the sensing light beams are *chopped* (that is, cut to be continuously 'on' and 'off' in rapid succession using a mechanical or electro-optic shutter). In both cases these (and similar) applications the resultant signal will be of the same frequency as, and *synchronous* with, the input excitation frequency.

Example 3.15 Phase-sensitive detection and its application

Figure 3.25 shows the system block diagram for a typical phase-sensitive detection system.

The measurand-sensing component is modulated at a reference signal – this may arise inherently as the result of excitation being already present (as in reactive bridges) or it may be purposely introduced (as in resistance bridges and optical beam situations). The reference signal v_{ref} is of the form

$$v_{ref} = V_{ref} \sin \omega_{ref} t \qquad (3.46)$$

The modulated measurand signal emerging from the sensor output v_m is of the form

$$v_m = V_m(\sin \omega_{ref} t + \phi) \qquad (3.47)$$

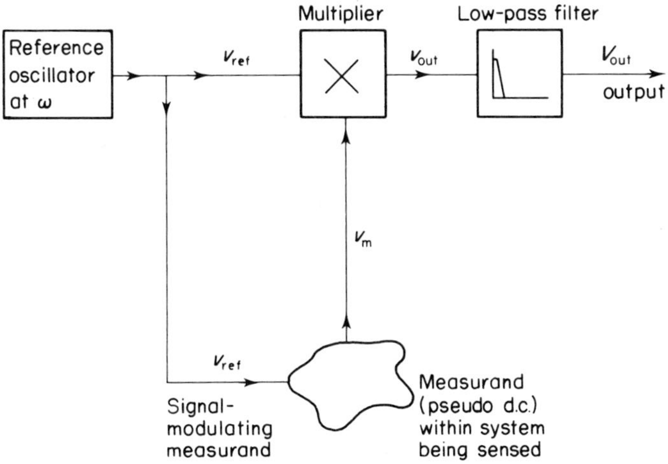

Figure 3.25 Block diagram for phase-sensitive detector system. (From Sydenham P H, 1984, *Transducers in Measurement and Control* (3rd edn). Adam Hilger, Bristol. Reproduced by permission of IOP Publishing Ltd.)

that is, varying at the same frequency ω_{ref} as the excitation frequency but perhaps phase-shifted by an angle ϕ.

These two signals are now multiplied together, as in amplitude modulation, giving

$$V_{out} = (V_m V_{ref}/2)\sin(2\omega_{ref}t + \phi) + (V_m V_{ref}/2)\sin\phi \tag{3.48}$$

The first term is an a.c. signal at twice the reference modulating frequency: this is filtered out with a suitable low-pass filter (the integrator in Figure 3.25).

The second term is a d.c. signal containing the measurand information. It is given by

$$V_{out} = (V_m V_{ref}/2)\sin\phi \tag{3.49}$$

This is the *PSD equation.* It describes the performance of the detection process. It is important to recognize that it shows that the phase-sensitive detection system can be used in two different ways as follows.

(1) *Phase-change detection.* If the measurand is of the type in which phase changes arise in ϕ as the measured variable, such as instruments that use this method to automatically follow interferometer fringes by modulating the instantaneous position of a fringe, then V_{out} will be proportional to the *sine* of the *phase shift*. This is non-linear and periodically passes through sign changes which can cause control instability in closed-loop applications of this form of phase-sensitive detector.

Note that the amplitudes of the modulating signal V_{ref} and output modulated measurand signal V_m must be held constant to obtain the required phase-sensing performance.

(2) *Amplitude change detection.* Another use of the PSD system is for detection of amplitude changes in a measurand. Examples here are the sensing of mechanical strain in resistance strain gauges, displacement in capacitive and inductive sensors and in voltage measurement using bridges where considerable noise levels are present.

In this mode of use the phase angle ϕ is (ideally) arranged to have $\sin\phi$ close to unity and the amplitude of V_{ref} held constant. Equation (3.49) shows that the resulting output is, under such conditions, linearly related to changes in measurand amplitude.

An advantage of the PSD arrangement in all variations of use, is that the modulation signal frequency can vary widely without serious effect – the system will track it as long as the frequency variations do not occur too rapidly.

As a signal recovery method in the presence of noise it works well in situations where the bandwidth can be reduced to lessen the noise energy. However, within that reduced bandwidth, the noise energy must be less than that of the signal, otherwise the signal is also reduced; in this way it quite different to the box-car integrator signal-recovery method dealt with in Example 3.4 in Section 3.2.3.

In practice most PSD circuits make use of square-wave excitation instead of the sine wave used above. As the system is linear in operation the square-wave excitation can be considered as the sum of its sine wave harmonic components with each operating as an independent PSD system. Square-wave excitation is often more convenient from the point of view of construction and gives a marginally improved level of detection.

The PSD method makes use of a multiplier arrangement with an integrator following that operation. Reconsideration of the general correlation process, as set out in Equation (3.8) of Section 3.2.2, shows that phase-sensitive detection is actually a special case of the correlation principle.

TUTORIAL EXAMPLES

3.1 For an instrument of your choice trace the signal path from its input to its output. Draw a system block diagram labelling each node. For each node establish the characteristics of the signal at that point (form, amplitude, phase, code, etc.).

3.2 By graphical and/or computer-programmed methods determine the wave shape of the combination of the following eight sinusoidal signals:

$$
\begin{array}{ll}
3\sin(\omega t + 90^\circ) & 3.2\sin(13\omega t + 120^\circ) \\
5\sin(3\omega t + 130^\circ) & 2.3\cos(16\omega t + 10^\circ) \\
2.5\cos(5\omega t + 20^\circ) & 3.5\sin(17\omega t) \\
1.5\cos(8\omega t) & 1.2\sin(25\omega t + 180^\circ).
\end{array}
$$

It is desired to process this result into a smoother wave shape. What harmonic components would you suggest are needed to achieve reduction of the higher-frequency components?

3.3 Using (as directed) either a spectrum analyser, the relevant theory, or use of standard tables, determine the frequency spectrum of the wave form of the following time-domain signals:

positive half only of a sinewave	heart beat
saw tooth	one toll of a bell
square pulse	ramp

3.4 A box-car integrator is to be set up to recover a 10 nV amplitude waveshape that can be repeated each 10 ms, that lasts for 10 μs and that is buried in 100 mV levels of random noise. The wave-form discrimination is to be around 5% and the desired SNR is to be 10 000.

Devise a suitable system arrangement to recover the signal specifying the sample width and scan rate. What will be the time duration taken to recover the signal?

3.5 A sensor has the transfer function

$$G(s) = 3/(1 + 0.18s)(1 + 0.22s)$$

What will its output time-domain wave form be when it is excited by (a) pulse, (b) step, (c) ramp (of 0.2 s duration) and each being of unit amplitude?

3.6 A correlator-based signal processor system is to be set up to look for changes in the vibration occurring in a gearbox used to drive the final propellor shaft of a large ship. Marked changes in the vibration pattern allow pre-warnings that the mechanical parts are beginning to fail.

Estimate, from sensing and processing considerations at each stage, the communication data rate (order of magnitude only is sufficient) needed to connect:

(a) six vibration sensors to the processor placed on top deck;
(b) the processor output to the Engineer's alarm console in the Bridge area;
(c) the processor output to a recorder of the results of the processing.

(*Hint*: In each case assess the signal information task and the bandwidth involved to obtain that information.)

For each case justify your answers clearly showing the numerical data you have used and how it was employed.

What kind of signal form would be used for each connection?

REFERENCES

Atkinson P, 1982, Closed loop systems. In Sydenham P H (Ed), 1982, *Handbook of Measurement Science* Vol 1, Wiley, Chichester.

Harris R W and Ledwidge T J, 1974, *Introduction to noise analysis*, Pion Ltd, London

Haykin S, 1983, *Communication Systems*, Wiley, Chichester (2nd Ed).

Horiuchi T, 1977, Application of fluctuation to measurement and process diagnosis. In: *Proceedings 'Nature and Scope of Measurement Science'* IMEKO, Budapest.

Lin Shu and Costello R, 1983, *Error control coding: fundamentals and applications*, Prentice Hall, New Jersey.

Oppenheim A V, Willsky A S and Young I T, 1983, *Signals and Systems*, Prentice-Hall, New Jersey.

Sydenham P H, 1982, *Handbook of Measurement Science*, Vol 1, Wiley, Chichester.

Woschni E G, 1982, Signals and systems in the time and frequency domain. In: Sydenham P H (Ed), 1982, *Handbook of Measurement Science*, Vol 1, Wiley, Chichester.

CHAPTER 4

The information machine

4.1 INTRODUCTION

The concept of a measurement system as an information machine has already been introduced in Section 2.4. The nature of signals within measurement systems and their mathematical representation has been explored in Chapter 3. This chapter examines the nature and behaviour of the signal-bearing and signal-processing system that comprise the information machine. Furthermore the operation of these systems under *real conditions* will move us closer toward realization of satisfactory workable instruments.

Concepts presented here concern the form and shape, called the *architecture*, of information machines, the steady-state characteristics of the signal-handling blocks, and how they can alter the information flow.

The sensor forms an interface between the real physical world attribute and (usually) the electronic signal domain. Despite vast numbers of sensors being in use there still does not exist an adequately clear appreciation of what this interface does – several useful models are discussed that help motivate understanding.

Unfortunately all real systems will be influenced by unwanted signals that can look like meaningful information. In measuring systems it is imperative that the errors that can arise due to these unwanted signals are understood so that strategies can be devised to minimize them; this is the subject of the latter part of this chapter.

4.2 THE ARCHITECTURE OF IT MACHINES

4.2.1 Measuring systems are formed of connected blocks

Information-machine systems are formed by joining *blocks* with suitable *connections* the structure of which is referred to as the *architecture*. Given only a few kinds of blocks and a small collection of permissible connections a large, often bewildering, number of alternative system configurations become possible as systems grow in size and complexity. This is the case for all information technology systems.

Fortunately the common measuring kinds of information machine used in measurement systems have relatively simple architectures compared with general forms of systems found in nature and in some other engineering disciplines.

4.2.2 Building blocks

A generalized information machine building block is shown in Figure 4.1. It will be possible to identify many inputs and outputs of energy flow. Note that a distinction must be drawn between energy flows that supply power to operate the block (often not shown) and energy (or mass flows) used to convey meaningful information.

It is clear that any block will have one or more *wanted signal* inputs, one or more wanted signal outputs, various power supply inputs and waste-energy outputs (for example, heat and mechanical vibration).

This level of model does not suffice in measurement systems for there will also be inputs and outputs of *unwanted signals* that must be accounted for. These arise from such effects as the block acting as an antenna picking up signals transmitted by radio stations, as a thermally sensitive device in which its internal behaviour is altered by temperature change, or by the connecting leads allowing unwanted electric currents to flow.

The simplest case is that of just one input and one output and ignoring power supply and unwanted effects. (This is a standard convention in much of systems modelling at a *first level* of understanding.) The block will then be as shown in Figure 4.2. Convention usually shows information flowing from left to right.

An input signal will be processed by the block to yield an output signal and it is usually safe to assume that the effect will be *deterministic*, meaning that the output is predictable if the characteristics of the block are understood. Such systems are said to be *casual* in nature.

4.2.3 The transfer characteristic and the transfer function

The relationship between the input and output signals is termed the *transfer characteristic*. This may be a simple ratio in the case of a *linear* block, where the

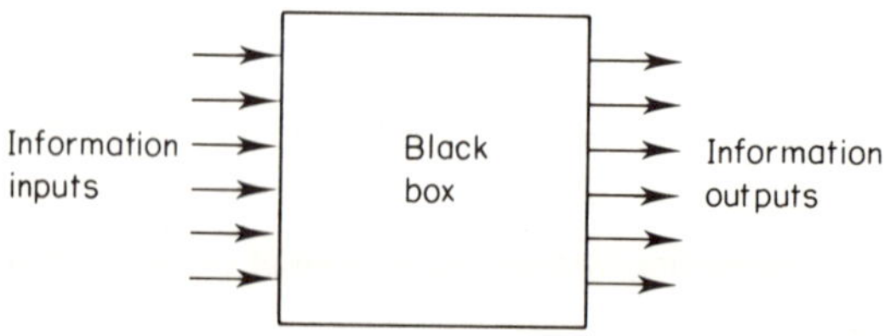

Figure 4.1 Generalized block of an information machine

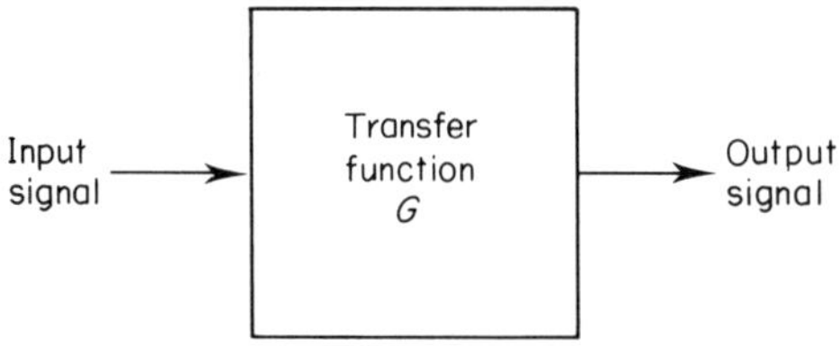

Figure 4.2 Most basic representation of signal-handling building block

output is directly proportional to input; for example, an electric thermometer giving a 15 V output signal swing for a 10°C temperature input swing has a transfer characteristic of 1.5 V/°C (This may also be called the *sensitivity* of the thermometer.) However, if the input/output relationship of the block is *non-linear*, then a more complex mathematical or graphical representation may be required. For example, the thermoelectric voltage V produced at the terminals of a thermocouple when at temperature T is given by the general formula

$$V = a + bT + cT^2$$

where a, b, c are constants determined by calibration of the probe. The quadratic curve has been selected as a 'good fit' for the whole class of thermocouple devices.

Further consideration will soon reveal that the transfer characteristic only represents a steady-state signal situation because real blocks posses *time-varying responses* for which a more complex model must be used to explain its behaviour.

Example 4.1 Static and dynamic response of mercury thermometer

Due to inherent heat capacity it takes a finite time to change the temperature of a bulky temperature-sensing element. A typical mercury-in-glass medical thermometer requires about 2–3 min to come to its steady-state temperature. Its steady-state sensitivity is typically 10 mm of mercury thread length change per degree Celsius. Whilst it is heating up from the ambient to body temperature its indications will not be according to the sensitivity value but will be described by more complex heat flow and glass–mercury expansion relationships, the time-varying nature of the response which is explained later in this chapter.

The *transfer function* provides a mathematical description of both the static and dynamic behaviour. Convention set up in the early days of control theory (ca. 1940–50) is that a non-sensing block uses, for the transfer function, the symbol G. The transfer function is formed by taking the Laplace transform form L of the ratio of input to output signals. Its general form is

$$G = \frac{L\{\text{output signal}\}}{L\{\text{input signal}\}}$$

Any transfer function can be arranged to show that it comprises two distinct elements – the zero-frequency response value and a multiplicative element that shows how the response changes as the frequency of the input signal varies, that is

$$G = (\text{d.c. response term}) \times (\text{frequency-dependent term})$$

Example 4.2 Transfer function of tachogenerator

A tachogenerator is used to measure speed of a shaft. It generates a voltage proportional to angular speed. The *electrical equivalent circuit* of the rotating coil and magnet system is given in Figure 4.3. (An 'equivalent circuit' is a model of a block which *simulates* the same signal behaviour but which is more convenient to use in its study as all components are transformed into a common energy domain; here, that is electrical.)

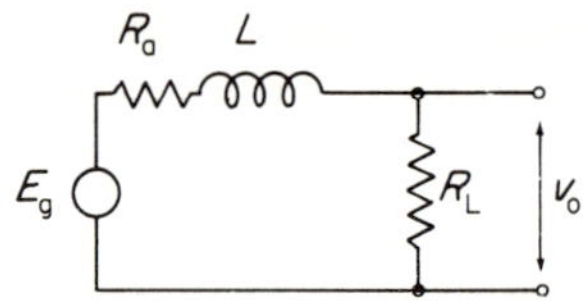

Figure 4.3 Electrical circuit equivalent of generating coil type of tachogenerator. (HBMS, figure 16.1.)

The output voltage generated across the resistance R_L (the output load connected to the generating coil having resistance of R_a) is, in Laplace form,

$$V_0 = \frac{R_L K_g s\theta}{R + sL}$$

where $R = R_L + R_a$, which can be arranged to give the output–input relationship as

$$\frac{\text{volts}}{\text{angular motion}} \frac{V_0}{\theta} = \frac{R_L K_g}{R} \cdot \left(\frac{s}{s + R/L}\right)$$

Thus we see that when the s-dependent right-hand term (the frequency-dependent one) has zero frequency inserted it goes to unity leaving the left-hand side of the expression showing that the generator constant K_g (its transfer characteristic) is modified, when connected to a load, according to the values of the resistance of the generating coil and the output load. The generator resistance, must, therefore, be set up correctly ($R_a \ll R_L$) to obtain the calibration apparently shown in the transfer-function expression.

If we wished to explore the behaviour of the tachogenerator during changes in speed we would study how the output changes for increments of frequency in the expression. A plot of output amplitude versus frequency of input signal will be its frequency response.

Hence our simple dynamic model has predicted the behaviour of the instrument without making it first!

Note that the behaviour of a block is a property of that block, not of the input signal (although use of a feedback connection around it can alter its behaviour). The transfer function is assumed not to change as the input signal varies. This property means that systems of transfer functions can be mathematically combined with ease.

The signal path from a sensor (or sensors) through to the display or actuator is a cascade of interconnected blocks and can, therefore, usually be expressed in terms of an *overall transfer function.* This is simply the *product* of the transfer functions of each of the serial stages concerned.

Such considerations assume that the input and output signal levels for each block are within *allowable limits* of signal level and operation. Where they exceed these levels, the transfer function becomes dependent on the input signal magnitude and the system becomes very complex to model. Fortunately the great majority of instrument design and operation does lie within the limits of linearity: indeed this is a necessity if measurement fidelity is to be preserved as the signal flows through the information machine.

The above concepts have been developed for continuous analogue signals and can be extended, by more complex mathematics, to apply for digital (that is, non-continuous, discrete-time) signal forms – see for example, Oppenheim *et al.* (1983).

4.2.4 Connections

A set of blocks must be *connected* in some way to form a through path for signals. Connections may take several forms as shown in Figure 4.4. This small set of options allows numerous system configurations to be created.

When signals are combined at a *node* a suitable block is usually required to allow a workable connection. Arrows should be added to show the direction of information flow. Connections should not alter the meaningful nature of the information in any unintentional way. Figure 4.5 illustrates the connection of two temperature sensors to give a combined (average) result. A summing amplifier, with a gain of 0.5, is used to derive the average of two temperatures T_1 and T_2 (represented by sensor output voltages V_1 and V_2 respectively.)

Information can be *replicated* as much as is needed using a *fan-out* node. An example is connecting many telephone earpieces to the common line so that several people can hear the same message. Given the correct block to make the join there is no limit to the number of information channels that can fan out.

Conversely, information can be combined at a *fan-in* node, but the information will then be as an information combination. A *multiplexing* (MUX) system joins the information from several different sensors into a single channel in such a manner that they can later be recovered separately by a corresponding *demultiplexing* (DEMUX) system. One common form is *time-division multiplexing* (TDM) in which information values from different (but compatible) channels input are transmitted sequentially over a common link. However, there are many multiplexing schemes available whose appropriateness depends on the

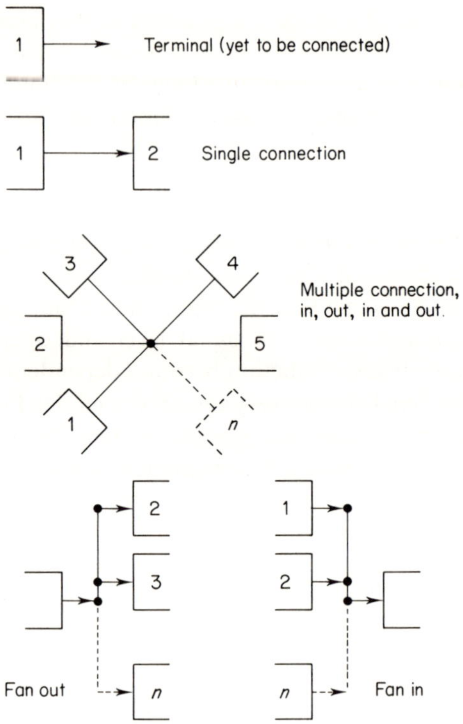

Figure 4.4 Connections used to join blocks in information machines

type, number and characteristics of the input signals. This topic is a field of communications engineering, the theory of which has been considered briefly in Section 3.5.

Another powerful connection type involves the routeing of a signal to an earlier, or later, point in the serial interconnection path. This is labelled *feedback* or *feedforward*, respectively, and the two forms are illustrated in Figure 4.6(a)

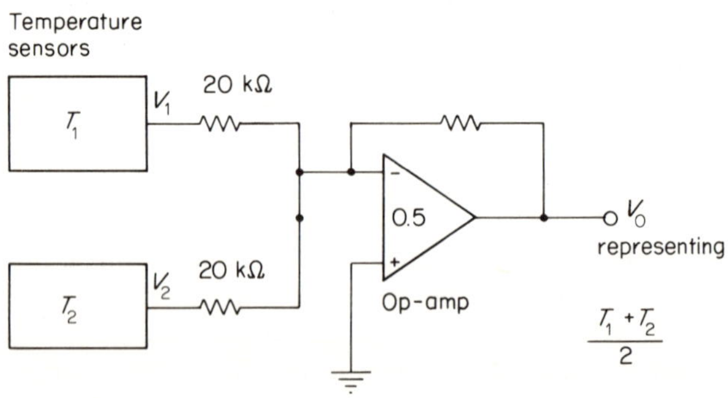

Figure 4.5 Summing amplifier used to average two temperatures

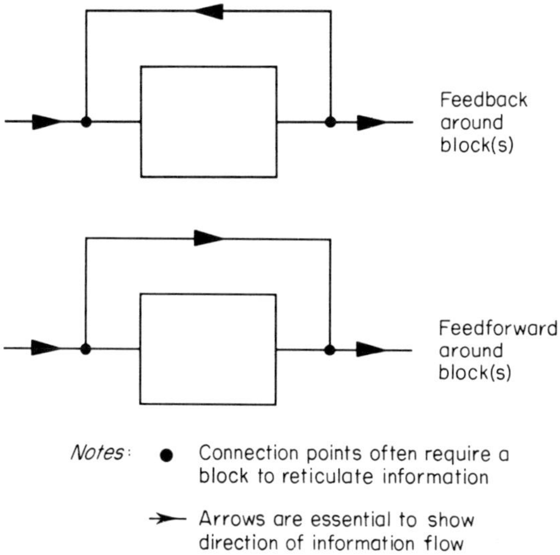

Figure 4.6 Feedback and feedforward connections: (a) feedback; (b) feedforward

and (b). These variants can considerably change the behaviour of a system, which in turn means that they can alter the information content of the blocks that they surround. The signal behaviour within both *positive* and *negative feedback* systems has been presented in Section 3.4.5 – an application is given in Example 4.3.

Example 4.3 An application of feedback – the self-balancing Wheatstone bridge

The dynamic response of an amplifier can be made faster by feeding back output signal to the input. This is carried out in a way that does not alter d.c. steady-state response. One instrument form, the *null-seeking*, self-balancing, Wheatstone measuring bridge uses this connection to keep the bridge set to zero as the input measured variable changes. This gives a more precise operation of the bridge. The output variable then appears as the feedback reference signal used to keep the bridge set at its zero point. At any time the balancing signal is the system output.

By a process of successive reduction from the initial complete circuit diagram it has been possible to break down a system into building blocks and connections. The reverse procedure is used to develop a system; a system representation of this form is called its *schematic block diagram.*

When the information machine is formed of blocks operating in the different energy regimes (i.e. electrical, mechanical, optical, etc.) overall analysis requires that all blocks be described in compatible terms. In most cases it is convenient to express the non-electrical blocks in terms of their electrical equivalent circuit, as was done previously in Example 4.2.

4.3 STATIC BEHAVIOUR

4.3.1 Describing static and dynamic behaviour

Whilst the *static* performance of common, everyday instruments may be simple and straightforward, the *dynamic* performance is often quite complex. This has already been illustrated for the mercury-in-glass thermometer in Example 4.1 of the previous section. Unfortunately for information machine operation these factors are often overlooked and give rise to measuring instruments not operating correctly, yet appearing to do so.

In the interest of simple and convenient performance description a range of apparently simple terms are employed. However, these can often be quite complex; it is essential that they be well understood.

This section concerns the description of characteristics once signals have settled to their steady-state value.

4.3.2 Gain and sensitivity

The change in amplitude at the output resulting from a change in amplitude at the input of a system block is referred to as the *gain* or *sensitivity*. It is usually stated as a simple dimensionless number.

However, the value of gain depends on both the amplitude and the frequency of the signal remaining within specified limits. If the amplitude limit is exceeded the performance will often become *non-linear* and the nature of the output signal may be radically altered. If the frequency limits (either lower or upper) are exceeded the gain will become frequency-dependent and there will also be a frequency-dependent change in the signal phase shift across the block. The gain and phase shift of the operational amplifier used so extensively in electronic systems is typically as shown in Figure 4.7.

Used without feedback the gain of the amplifier, Figure 4.7(a), depends on the frequency of the input signal when it exceeds around 2 Hz. As the strength of the feedback signal is increased around the (closed-loop) block the gain falls but the useful frequency range (the *bandwidth*) is correspondingly increased.

System block sensitivities are often functions of various external physical variables, particularly temperature. These must be considered to prevent error creeping in that limits system measurement fidelity - see Example 4.4

Example 4.4(a) Behaviour changes in a resistance strain gauge weighing system

Consider an electrical resistance element (called a *strain gauge*) mounted onto a metal rod on which hangs an aluminium extraction electrode that is suspended in raw stock material. The purpose of the strain gauge is to monitor small length increases in the support rod. There are related to the increasing weight of the electrode as pure aluminium is deposited onto it, and thus indicate how the electrolysis extraction process is proceeding.

The weighing sensitivity of the system (expressed simply as volts per kilogram (V/kg) is decided by three components of the system.

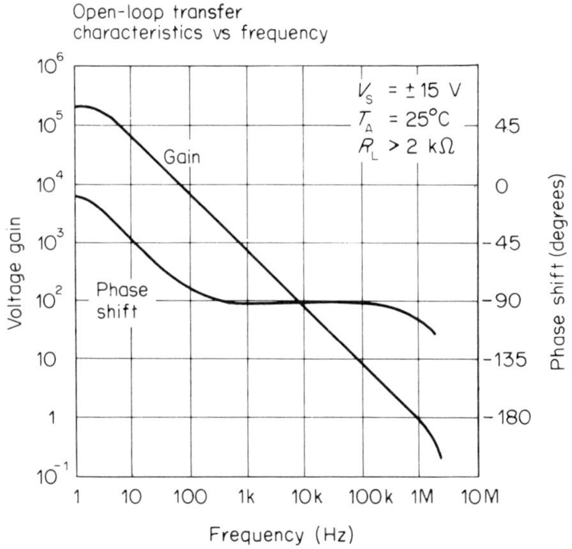

Figure 4.7 Characteristics of common electronic operational amplifier (type 741) (refer to text): (a) amplitude – frequency response; (b) phase-shift – frequency response. (Reproduced by permission of NS Electronics Pty Ltd.)

These are the *gauge factor* of the strain gauge, energizing voltage and component values of the interrogating Wheatstone bridge circuitry and the gain of any following amplification stages.

The value of the gauge factor depends on the temperature of operation. The gain of the amplifier also changes with time, and with temperature. The properties of other circuit components, such as the Wheatstone bridge resistor values, change with temperature. Although some of these changes can be arranged to cancel one another out by good design (see Example 4.12) this is not always possible. Hence if these remaining, although comparatively small, factors have changed since initial calibration, the output signal will be in error by a discernible amount.

4.3.3 Bandwidth

The band of frequencies over which the gain is reasonably constant is termed its *bandwidth.* Unfortunately the gain begins to decrease gradually near the limits of this range so that an arbitrary definition is required. In the operational amplifier example of Figure 4.7(a), the low-frequency response extends unchanged (it is described as 'flat') right down to zero frequency and this is common for many instrumentation building blocks.

The usual method of bandwidth definition is to draw lines across the graph at the maximum value and at the *half-power* point, that is, 3 decibels (dB) below the maximum. (If the signal is electrical and viewed as a voltage this corresponds to 70.7 % of the maximum voltage value.) Figure 4.8 shows how the bandwidth is

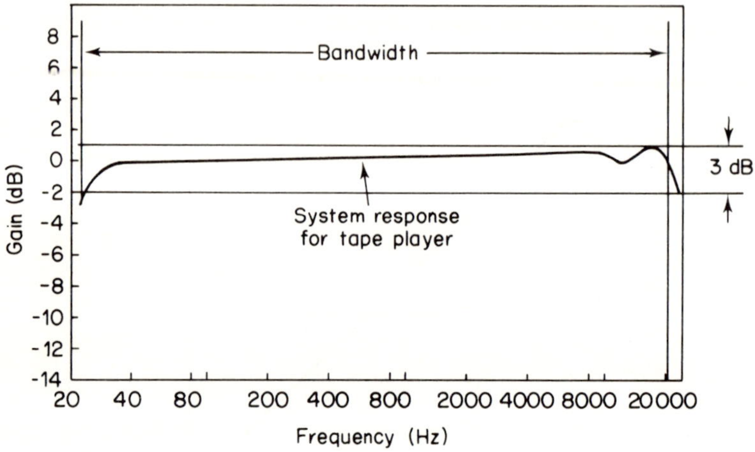

Figure 4.8 Bandwidth of a sound reproduction electronic amplifier

defined for a sound reproduction amplifier – note how the gain varies over the bandwidth. Other conventions are sometimes used, notably the '1 dB down' points, which correspond to 95.1 % of maximum voltage.

4.3.4 Offset

In practice individual manufacturing differences arising in linear devices, such as amplifying stages, lead to a small difference between the input and output signal levels when the input is set to zero – the output would normally be expected to be zero in that condition. This is called *offset*.

Example 4.4(b) Sources of offset in a resistance strain-gauge weighing system

Continuing the application given in Example 4.4(a), the output might not be zero when zero strain occurs, for several reasons.

Differences arise as offset in the strain gauge itself (due to small residual strain in the resistance element that has been captured as the foil was cemented onto the support); in the Wheatstone bridge circuit (due to not being balanced to a truly null condition) and in the following electronic amplifiers (due to manufacturing differences). Each of these is well understood and small adjustments are provided to allow them to be *balanced out* in the original set-up.

Suppose the designer had used a nickel-based material (because it is non-corrosive) as the support element on which the strain gauges were placed. (This assembly is usually known as a *load cell.*)

As large electric currents are flowing in the electrode (1000s amperes) strong magnetic fields exist around the nickel support structure. Nickel, unfortunately, would be a poor choice of material for this particular task for it is highly *magnetostrictive.* The magnetic fields would cause it to change length, giving rise to *apparent* load measurement changes. Offset effects would arise, being of magnitude related to the flowing current.

Once identified, offset effects can usually be kept to adequately small levels.

4.3.5 Drift

In simple terms, *drift* is the rate of change of the signal output with time. Unfortunately, drift is a complex effect usually poorly understood, and seldom occurs at a predictable, fixed rate. It often is a key limiting factor on system discrimination and accuracy and cannot be ignored – see Example 4.4(c) below.

Example 4.4(c) Sources of drift in a resistance strain-gauge weighing system

Drift arises due to change of dimensions and properties of materials. The support stretches and the binding adhesive relaxes the way in which it holds the strain gauge – these mechanical phenomena are referred to as *creep*. The materials from which the electronic components are manufactured also 'age' with time and alter their behaviour slightly in an unpredictable manner.

Through knowledge of the magnitude of these effects (gained much by experiment and experience) it is possible to build workable instruments.

Drift defies generalization but does fall into classes for which convention suggests how to describe it in a simple manner. Figure 4.9 displays several cases that are commonly encountered.

In Example 4.4(c) above, past experience with the material used to form the load cell would be studied in order to estimate its likely future behaviour. Tests would be conducted to observe the change in apparent load as a constant calibrating load is hung on the support. The values recorded would be expected to fall along a reasonably straight line, allowing drift to be quoted as a ratio of strain/time. It would also have to be established that temperature effects were adequately small, otherwise the strain drift observed might be partially formed from length changes caused by thermal expansion. It can be very tedious and time consuming to fully establish the cause and nature of drift for any single instrument.

In electronic amplifiers drift occurs as a function of time and also of its operating temperature. Manufacturers quote these as simple coefficients but in reality the actual drift performance is a non-linear curve. As shown in Figure 4.9 these are placed within *butterfly* zones rendering the quoted numeric value as a conservative value. It is, therefore, commonly observed that an actual amplifier has less drift than the catalogue value – the *worst case* of a batch is quoted by the conscientious manufacturer. *Typical* figures are worthy of careful consideration.

4.3.6 Impedance

When connecting one block to another it is necessary to ensure that the signal is passed on without attenuation due to *loading effects* or, if it is altered by the act of connection, that the magnitude and phase change are accounted for.

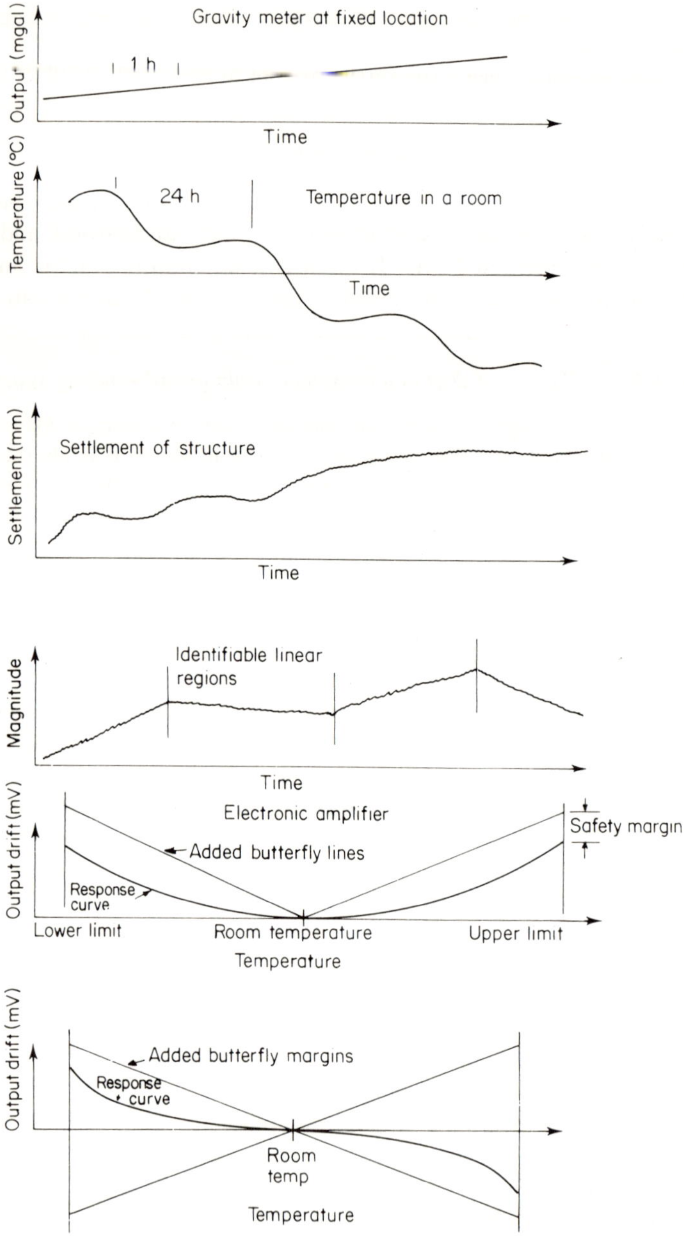

Figure 4.9 Typical drift curves of various classes. (HBMS, figure 16.6.)

A key interconnection parameter is the *impedance* of each of the connecting ports. This allows calculation of the effect of making a connection. If all electrical circuitry used only purely resistive elements, then Ohm's law could be used to work out the effect. In practice many sensor outputs, processing blocks, terminations and actuator inputs are not simply resistive but also possess *reactance X* in those blocks. Consequently *the impedance of a port usually varies with frequency of operation* and may also be dependent on signal level.

To attempt to describe the termination property in a detailed way would require either a complicated mathematical expression or a graphical representation. For expediency the port impedance will often be stated as a number of ohms, which quantifies only the *resistive* component. An 8 Ω loudspeaker, for example, will not be purely resistive at its terminals but the resistive values does assist calculate the power needed to drive it and the appropriate amplifier output port impedance.

In addition to loading effects impedance controls the transfer of energy from the output of one block to the input of the next. The theorem of *power transfer* states that maximum power will be transferred when the output impedance of the generator (in our case, a block) is equal to the impedance of the load (in our case the input impedance of the next block). This is known as *impedance matching* and strictly must involve both resistive and reactive components of impedance. In instrument design impedance matching is particularly important when the energy levels (e.g. from a sensor) are very tiny, or when the frequency of the signals is very high (megahertz and above). Otherwise the usual practice is to ensure that output impedances are low and input impedances are high so that the loading effects are rendered so small as to be negligible.

4.3.7 Dynamic range, span, FSD

Unexpectedly, the *dynamic range* parameter has nothing to do with the time-response capability of an instrument but describes how far it can range from its smallest discrimination level to its largest input value. It is the ratio of the largest to the smallest quantities that can be measured and is expressed as the direct ratio, or in decibels when the range is large.

Consider, for example, the choice of a flow-measuring technique. The inexpensive orifice-plate type of flow meter has a dynamic range of typically 8 (about 9 dB) which restricts it to a very narrow range of flow values. In contrast, the very expensive laser doppler velocimeter flow meter has, potentially, a much wider dynamic range of 10^7 (equal to 140 dB).

The *span* or *full-scale deflection* (FSD or fsd) of an instrument refers only to the maximum permissible value of the input and is quoted in the units of the particular quantity. (The latter term has its historical origins in the 'scale and pointer' display common in older pre-electronic instruments.)

4.4 DYNAMIC BEHAVIOUR

4.4.1 Rise time

As has already been stated the dynamic aspect of systems can be very complex and a full description cannot always be obtained with a few simple terms – pragmatism has, however, developed certain terms that readily convey basic performance characteristics. The frequency dependence on gain and sensitivity has already been stated and the concept of bandwidth introduced.

Two further properties, those of *rise time* and *slew rate*, are discussed below. However, the majority of this section is concerned with the formal classification and mathematical description of information machines used for measurement purposes. Such a description is essential to fully describe dynamic behaviour.

The *rise time* of a system describes the output response when a sudden step change is applied to the input. The input step is assumed to be instantaneous and, in general, the rise time is defined as the time to change from 10% to 90% of the final output value.

There is a common special case, that of the simple exponential rise characteristic displayed by a first-order system (Section 4.4.5), in which the *time constant* τ represents a change from 0 to 63% of the final value. This is related to the rise time t_r by

$$t_r = 2.2\tau$$

4.4.2 Slew rate

The *slew rate* is a similar descriptive term which is applied when the system block can only change its output at a certain maximum rate. The slew-rate figure, expressed for example in volts per microsecond (V/μs) for an electronic amplifier, is the fastest rate at which the amplifier output will change. A square-wave input signal displays this clearly (see Figure 4.10(a)) but the effect can also distort a high-amplitude sine wave as shown in Figure 4.10(b).

It follows that slew rate and bandwidth are related but not in any simple fashion. However, in general it may be noted that if signal amplitudes are small the rate of change of voltage will also be small for a given frequency (that is, of a sine wave) and the slew rate effects will be minimized.

4.4.3 Formal characterization of dynamic response

Measuring instruments convey information on energy carriers. Within systems energy can be:

- added (with power supplies);
- lost (through dissipation in forms such as heat in electrical resistors or friction of a mechanical brake pad)
- extracted (deliberately taken out with energy *sinks* such as electronic active devices);
- stored (charge in an electrical capacitor, magnetic energy in an electrical conductor, or compression of a mechanical spring).

If a signal path within the system is without significant storage effects, then the energy will flow from place to place virtually instantaneously. Signal response will be without time delay or modification of its frequency content. There is,

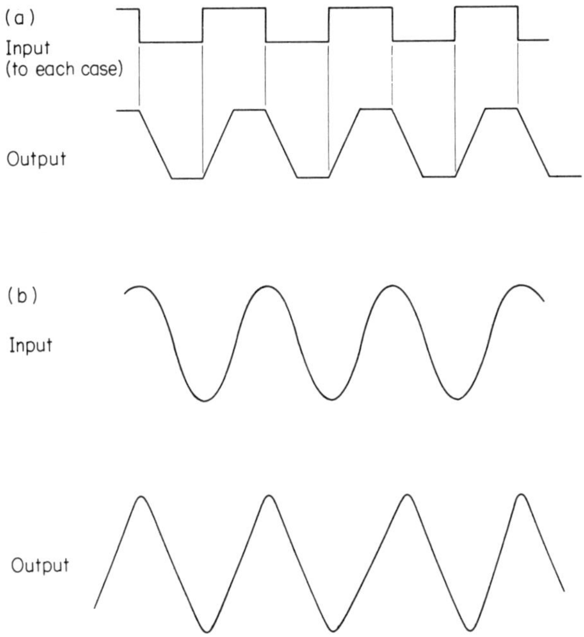

Figure 4.10 Slew rate limitations on signals: (a) ramps, instead of vertical rises, in square wave operation; (b) sine waves become triangular

however, no such thing as an instrumentation system without some degree of energy storage, time delay and bandwidth limitation. In mechanical systems, for example, there are only close approximations to truly frictionless bearings and inertialess linkages. Similarly a few picofarads of stray capacitance in an electronic system is seldom significant to signals of a few kilohertz (although it would be for signals of high megahertz or gigahertz).

If storage effects are significant, then signal-processing effects are *time-modified* according to the type and relative magnitude of the storage elements present. Amplitude and phase change effects can occur, these varying with the frequency content of the signal. The nature of such changes is explained below.

When systems of energy sources, sinks and stores operate in a linear manner they can be readily analysed to understand their behaviour in a quantitative manner. If the behaviour is not linear then, once again, mathematical modelling treatment becomes very much more difficult. Whilst a mathematical description of non-linear instrument systems is possible the inherently low cost value of an individual sensor system seldom allows expenditure of enough effort to carry out an in-depth analysis.

Fortunately linear description in highly applicable to instrument systems; it finds extensive use. This is now outlined.

The differential equations describing energy relationships of a linear energy system take the general form:

$$a_n \frac{d^n y}{dt^n} + a_{n-1} \frac{d^{n-1} y}{dt^{n-1}} + \cdots + a_0 y = x(t) \tag{4.1}$$

where x is the input variable and y the output variable.

The coefficients $a_n, a_{n-1}, \ldots, a_0$ are constants of the system. The equation obtained by equating the left-hand side of (4.1) to zero is called the *characteristic equation*; this is a feature of the system alone. The input to the system (on the right-hand side) is usually a function of time $x(t)$ and is known as the *forcing function.*

In principle at least, any equation of this form may be solved to give the output response $y(t)$ for a given forcing function $x(t)$; it is only necessary to know the system coefficients $a_n, a_{n-1}, \ldots, a_0$.

The general expression, Equation (4.1), shows that an ascending order sequence exists as higher-order derivatives are added. This family is as follows:

Zero order:
$$a_0 y = x(t) \tag{4.2}$$

First order:
$$a_1 \frac{dy}{dt} + a_0 y = x(t) \tag{4.3}$$

Second order:
$$a_2 \frac{d^2 y}{dt^2} + a_1 \frac{dy}{dt} + a_0 y = x(t) \tag{4.4}$$

In practice only these three orders are usually considered because higher orders are difficult to handle mathematically and most instrument situations can be understood adequately without going beyond the second order.

In Laplace form these become:

$$\text{Zero order} = \frac{Y(s)}{X(s)} = 1 \tag{4.5}$$

$$\text{First order} = \frac{Y(s)}{X(s)} = \frac{1}{\tau_s + 1} \tag{4.6}$$

$$\text{Second order} = \frac{Y(s)}{X(s)} = \frac{1}{(\tau_1 s + 1)(\tau_2 s + 1)} \tag{4.7}$$

in which the τ coefficients are termed *time constants*. (The formulation and properties of Laplace transforms were introduced in Section 3.4.4.)

4.4.4 Zero-order responses

This order is mathematically simple for, as seen above, the transfer function, Equation (4.5), has no frequency-dependent term, meaning that the input signal is preserved without its phase or frequency content being changed. Hence, the output signal frequency spectrum is identical to that of the input. Only the amplitude may be modified and this will occur uniformly across the spectrum of possible frequencies.

In many instrumentation circumstances a zero-order system response is desired but this is not always the case; often the properties of higher-order systems are needed because of their frequency-modifying properties, which enable information to be combined, split and modified.

Example 4.5 Measurement system with zero-order response

Consider the measurement of shaft position using an electrical, rotary, potentiometer mounted rigidly to the end of a shaft that is oscillating over a 100° angular excursion, as shown in Figure 4.11.

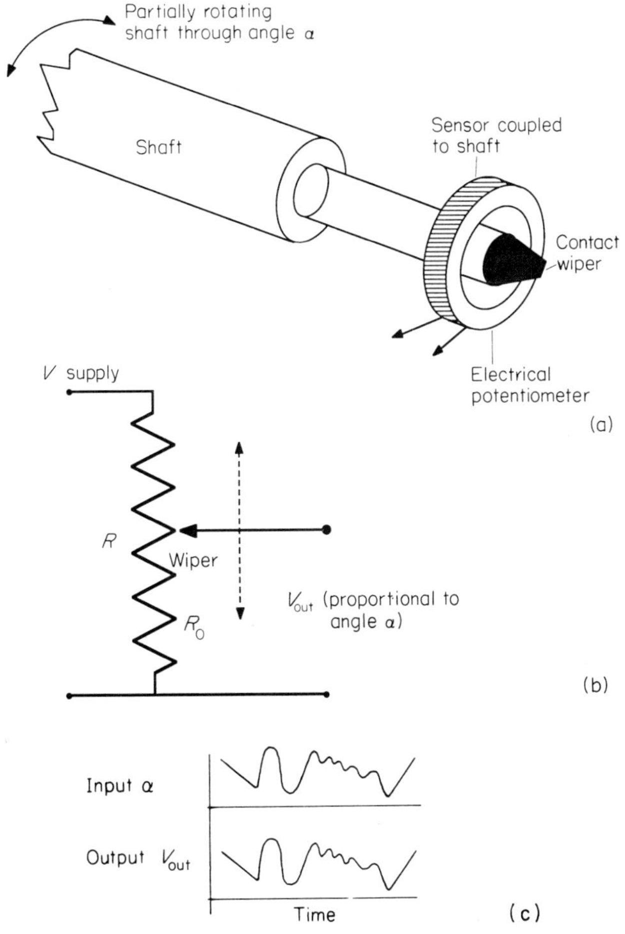

Figure 4.11 Measurement of shaft angle with mechanically driven electrical potentiometer: (a) mechanical setup; (b) electrical circuitry; (c) input/output – time response graphs

We must first inspect the various *organic* (the term used by hardware system designers for the actual units that can be observed) parts of the system to see if there are any relevant energy storage elements.

The shaft is of large diameter and thus cannot twist enough to store significant energy in its spring compliance. The low rotary inertia of the wiper and shaft of the potentiometer does not present significant inertial loading on the drive shaft to cause shaft torsion. The output wiper is, therefore, caused to oscillate identically with the mechanical input movements, that is, with the same relative amplitude, with the same frequency and with zero phase shift.

The electrical circuit is solely resistive with no apparent inductive or capacitive storage effects arising at the low frequency of oscillation. Hence the output signal will be a faithful representation of shaft motion related simply by a numerical *transfer characteristic* converting input angular rotation to electrical output voltage.

However, if the shaft connection were too compliant, as say with the use of a low compliance flexible coupling to drive the potentiometer, or if the potentiometer wiper assembly were of large mass, then a quite different dynamic response would arise. The shaft would twist during dynamic operation causing the wiper to lag, or lead, its correct position. This is because an energy storage element has become significant – in this case the flexibility of the coupling shaft. (The response would then normally be second order, as discussed below.)

Clearly whether this, and any, system is zero order or higher is decided by the relative magnitude of system storage elements, building-block types, signal amplitudes and speeds of signal changes. Whilst experience can often allow the expert to rapidly assess the situation, the more certain method is to mathematically model the system with practical material constants and dimensions inserted. In many measuring instruments quite small effects can be most significant.

In this example note also that energy can only be dissipated (from bearing and wiper friction) and that this example is of an *irreversible* system – driving electric current into the potentiometer produces significant heat but does not produce noticeable shaft motion. In contrast, many transducers are *reversible*. In the piezoelectric crystal, for example, an electrical input will cause a mechanical size change, and a mechanical input (such as an applied force) will cause an electrical output – this is the principle of piezo gas lighters.)

4.4.5 First-order response

If one store of energy becomes significant in capacity and is reversible, that is it can also discharge energy back into the system, the time and frequency responses are very different from those of the zero-order system.

The differential equation for the first-order system, Equation (4.3), shows a single time-variant term (that with s) and hence when this is Laplace transformed, Equation (4.6), there is a single time constant τ. This is the coefficient of the frequency dependent term and means that the output varies as a function of frequency. This type of response *is the same for all first-order systems.*

To derive the system response for any given input signal the transfer function is multiplied by the Laplace transform of the input driving function. The result is then transformed back into the time domain using inverse Laplace transforms (see Example 3.10). Two useful driving functions are (i) the single step change to the input, and (ii) the repetitive sine wave. These allow a good appreciation of how a system will respond even if the input is not exactly either of these cases.

The time-response equation for a *step* applied to a first-order system is given by

$$y(t) = AKe^{-t/\tau} \tag{4.8}$$

where A is the amplitude of the input step, K the transfer coefficient of the first-order system and τ is the *time constant*. Graphs of this function, for a rising and falling step, are given in Figure 4.12. The form of the response is always the same, the time constant dictating the particular exponential response envelope.

Thus it is possible to concisely describe the dynamic behaviour of a first-order system as

'first order with time constant τ of x seconds and gain of K'.

In many cases the input signal is a complex time-varying function instead of the simple step change. To analyse the *frequency response* the transfer function is similarly multiplied by the Laplace transform of a generalized sine wave (of frequency f and angular frequency $\omega = 2\pi f$ in the standard notation) and then transformed back into the time domain. The resulting equation is

$$Y(t) = \frac{AK\tau\omega e^{-t/\tau}}{\tau^2\omega^2 + 1} + \frac{AK}{\sqrt{(\tau^2\omega^2 + 1)}}\sin(\omega t + \phi) \tag{4.9}$$

where $\phi = \tan^{-1}(-\omega\tau)$ and from which it can clearly be seen that it is frequency dependent.

This expression comprises two parts. The first component of the right-hand side is a *short-lived transient* which decays with time and usually becomes insignificant in a few time constants after application of the sine-wave input. The

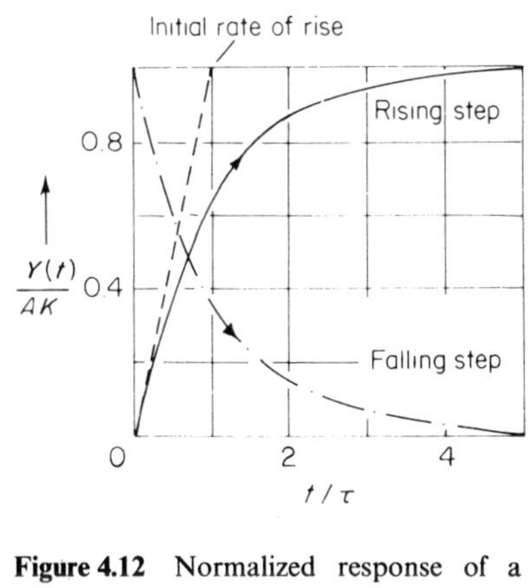

Figure 4.12 Normalized response of a first-order system to a step input function. (HBMS, figure 17.4.)

second component is the *steady-state response* which does not change with time and is usually the component that is of interest.

The steady-state response may be decomposed into two parts: the amplitude response

$$|G(\omega)| = \frac{K}{\sqrt{[(\omega\tau)^2 + 1]}} \tag{4.10}$$

and the phase response

$$\arg G(\omega) = -\tan^{-1}(\omega\tau) \tag{4.11}$$

Both are frequency dependent and are plotted in Figure 4.13.

When plotted in the Bode plot form (see Section 3.4.3) in which straight lines are drawn across the response curves when they are plotted with logarithmic axes, it will be found that the curve of Figure 4.13(a) approximates to two lines intersecting at a *corner* or *cut-off* frequency at

$$\omega_c = 2\pi f_c = 1/\tau \tag{4.12}$$

which lies at the '3 dB down' point. This point is often, therefore, referred to as the upper limit of the *bandwidth* of the system response. At frequencies above this point the slope, often referred to as the *roll-off*, always has the value 6 dB/octave (meaning a halving of the amplitude for every doubling of the frequency) which is equal to 20 dB/decade (a tenth of the amplitude at ten times the frequency).

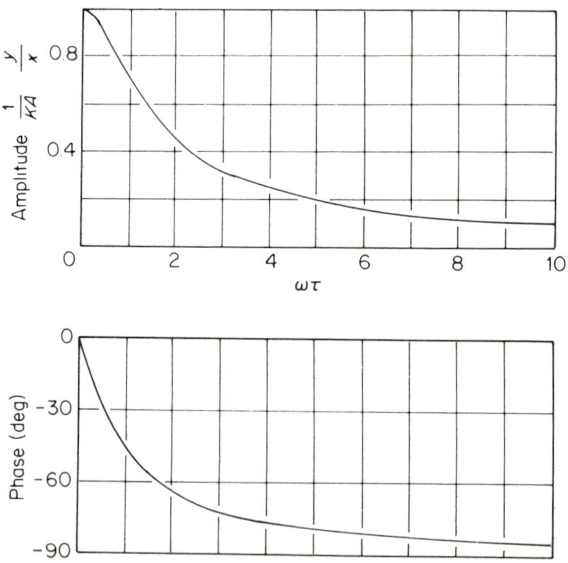

Figure 4.13 Normalized response of a first-order system to sinusoidal signals of varying frequency: (a) amplitude; (b) phase. (HBMS, figure 17.6.)

The phase response, Figure 4.13(b), shows a gradual development of phase lag rising from zero at low frequencies to a maximum of 90°. This means that any zero-order controller block (such as a low-frequency op-amp application) connected to a zero-order load (such as an electrical resistance heater) and being driven by a sensor of this first order (such as a thermometer) cannot ever break into oscillation due to the occurrence of positive feedback (180° feedback).

Both the 6 dB/octave slope and the 90° phase shift are hallmarks of a first-order system and any system displaying these responses would be identified as first order.

Consider now a complex signal having a range of frequency components. If some of these components exceed or even approach the corner frequency ω_c they will be modified in amplitude and phase such that the output signal may be dramatically different from the input. From the information machine perspective this means that the meaningful content of the signal passing through the block may be changed.

In measuring-instrument systems it is first essential to recognize, quantify and design for these effects that may occur, for example, in sensors and transducers. This is illustrated in Example 4.6. In addition these effects may be deliberately introduced into the transducer design or signal-processing stages of an instrument.

In both cases the effects described above are properly described as *first-order, low-pass filtering* effects, and signal-processing blocks would often be labelled as *filters*.

The general treatment of first-order systems also includes those in which the output amplitude can *increase* with frequency (at 6dB/octave) and output phase *lead* the input (by 90°). Commonly met physical systems used in measurement systems seldom display these properties but these ideas are often employed in the design of electronic signal-processing blocks.

Example 4.6 Response of a first-order measurement system

Consider a platinum resistance thermometer installed in a bath of water in order to measure and record temperature changes taking place. An adequate model of the system is as given in Figure 4.14. The response of the electrical part of the sensor will be assumed here to be zero order and therefore insignificant (an assumption which is often true for thermometers.)

Heat reaching the sensor from the water changes the electrical resistance of the thermometer monitored by incorporating the sensor in a Wheatstone bridge, the electrical output then being used to represent the input temperature changes.

Under dynamic conditions, that is when the temperature is changing, some heat has to flow into, or out of, the sensor assembly. This heat flow is impeded by the thermal resistance of the protective metal *thermowell*, by the glass surrounding the platinum element and also by the *boundary layer* of fluid adhering to the outside of the thermowell. This boundary-layer effect is likely to be quite significant and be the dominant thermal resistivity factor.

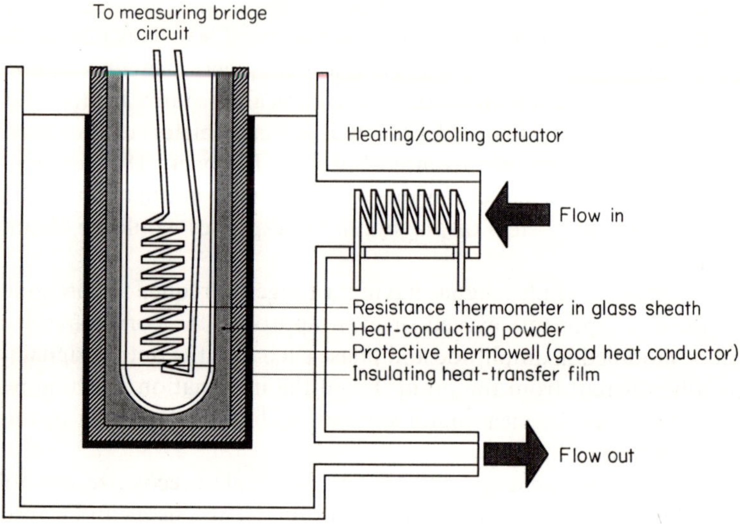

Figure 4.14 Diagram of platinum resistance thermometer used to measure temperature changes in a water bath

The mass of the metal thermowell, glass envelope around the sensor, the platinum element and its mounts provide storage of heat according to the specific heat and mass of heat-storage components. The thermowell mass is almost always the dominant storage element. Together the energy storage of the thermowell and the thermal resistance of the heat transfer film form a system that is first order in nature. Its behaviour can now be considered. The time constant will be given by

$$\tau = \frac{\text{heat storage in thermowell}}{\text{resistance to heat flow decided by area of heat-transfer coefficient}}$$

(For a typical process control application this time would be expected to be in the range 20–100 s.)

If a step change to the temperature occurs by releasing hot water into the flow the sensor will not give an exactly similar step rise in the measured output. Instead the sensor would indicate an exponential rise according to Figure 4.12. The final value (within 98 %) of the new temperature level will be reached after five time constants have elapsed.

Now assume the heater and cooler system shown in Figure 4.14 is used to produce varying frequency, cyclic cooling and heating between prescribed maximum and minimum temperatures, for example when the sensor is used as part of a closed-loop controller.

The relevant response will be the frequency response shown in Figure 4.13. For slowly changing cycles the response of the observed and recorded output from the sensor system will correspond closely to temperature changes occurring in the water bath. However, as the frequency rises the amplitude will fall off due to the *smoothing* or *averaging* effect inherent in the sensor construction. This effect can be so large that quite incorrect values may be observed.

Note also that as the frequency of the temperature cycle rises the sensor signal phase falls behind the driving input temperature changes. For a first-order system the phase shift can never exceed 90° lagging.

How can we ensure that the averaging and lagging effects of the first-order response be minimized or avoided? In general there are two possible approaches. Firstly the time constant of the sensor may be made sufficiently small by redesign of the thermowell assembly to reduce the heat storage. Significant averaging and lagging effects will then only be observed at higher frequencies of temperature oscillation. However, there may be practical limitations such that significant heat storage in the sensor assembly is unavoidable. In this case *compensation* can be applied to the sensor output signal to counteract the gain reduction and phase shift at higher frequencies. Compensation is achieved by means of non-linear signal processing and is designed from the mathematical model of the sensor response expressed in the form of Equation (4.9).

In both approaches, reduction of the system time constant or the addition of frequency compensation, the result is to seek to render the overall system *effectively zero order over the design frequency range.*

This *low-pass filtering* effect arises in many instruments. In flat-bed chart recorder use, for example, the plotter may not have adequate response – the record can be reduced in amplitude and delayed in phase. Similarly first-order compensation techniques may be used to improve the response over a given frequency range.

Cascaded first-order systems

Where several energy stores can be recognized the immediate thought would be that the system will be of higher than first order. This is not always so. If the stores are each associated with their own energy dissipation element and are of the same reactive kind (for example, different thermal storages within a thermometer/sensor assembly of Example 4.6) then the system is a cascade of first-order systems.

However, there do exist reactive elements of different and opposite 'kinds', for example electrical inductance and capacitance, mechanical compliance (of a spring) and friction damping (of a damper), which when acting together produce phase changes of up to 180° and display second-order response. These are considered formally in Section 4.4.6.

The overall response of a cascade of first-order stages depends on the relative magnitudes of the time constants involved. If one stage is dominant, that is, it has a time constant larger than any other by, say, a factor of ten or more, then the dynamic effect of the rest can be ignored. In the temperature sensor of Example 4.6, the thermal mass and resistance of the glass mounting form another first-order stage in series with the dominant heat transfer (boundary layer) film and mass of the sheath system. The time constant of the glass part is, however, much smaller (being about 5 s) than for the heat-transfer film.

However, if all stages of the cascade of first-order stages have similar time constants the overall system response is modified by introducing an initial delay period before the signal begins to rise and by providing an overall longer time constant than for one stage alone. As the number of stages increases the overall response approaches the *Gaussian response*. This response has some special properties and is considered formally in Section 4.4.8.

4.4.6 Second-order response

Where two energy stores exist that are of opposite kind, for example the capacitor and inductor of a tuned electrical circuit or the mechanical spring and mass, the energy can then shuttle back and forth between the stores giving an oscillatory response. This is quite different from responses arising in zero and first-order systems. Second-order systems obey the laws of *simple harmonic motion* and are described by *linear wave equations.*

Damping

If the system is disturbed or *perturbed* by a sudden single input (such as a step function or delta function) the oscillation of energy between the stores will occur at the *natural frequency* $\omega_n = 2\pi f_n$ of the system. The amplitude of this oscillation will decay with time according to the relative magnitude of the energy-dissipating elements (for example, electrical resistance and mechanical dashpot or friction dampers) in the system. If only a little is dissipated each time the energy moves between stores the system will oscillate for several (or perhaps many) cycles and is then said to be *underdamped*: if the energy dissipates rapidly in the first transfer between stores no oscillatory behaviour is directly observed and the system is said to be *overdamped.*

The *step response* is obtained, as before, by multiplying the transfer function of the general second-order system in Laplace form (Equation (4.7)) by the transfer function of a step input, then taking the inverse Laplace transform of the result. The step response is then given by the quadratic equation

$$Y(s) = \frac{1}{a_0 s} \cdot \frac{1}{[(a_2/a_0)s^2 + (a_1/a_0)s + 1]} \tag{4.13}$$

General solution of this equation involved use of the complex algebra - it is more convenient to consider three forms of solution that arise depending on the values of the two roots of the quadratic equation. If we define the *damping factor* ζ to be given by

$$\zeta = \frac{a_1}{2\sqrt{(a_0 a_2)}} \tag{4.14}$$

in terms of the coefficients of Equation (4.13) the three forms will be as set out in Table 4.1 and illustrated in Figure 4.15.

Note that for any system there is a particular rate of energy dissipation for which $\zeta = 1$ and which is labelled *critical damping.* In many cases when second-order systems are designed this will be the optimum damping factor

Table 4.1

Roots of Equation (4.13)	Damping Factor	Description
Complex	$\zeta < 1$	Under-damped
Real and equal	$\zeta = 1$	Critically-damped
Real but unequal	$\zeta > 1$	Over-damped

which will maximize the responsiveness of the system to changing inputs while minimizing the excessive system oscillation.

Because Equation (4.13) is quadratic and has two roots, two parameters are required to characterize performance of a second-order system. Those used are usually the natural frequency ω_n (or f_n) and the damping factor ζ. The complete performance may then be expressed in a statement of the form:

> 'second-order with natural frequency f_n Hz, damping factor ζ, and passband gain of K'

This may be compared with the first-order system which is fully characterized by the time constant τ and the gain K alone.

Resonance

As shown above a lightly damped second-order system can be caused to oscillate at its natural frequency. However, if such a system is driven at a

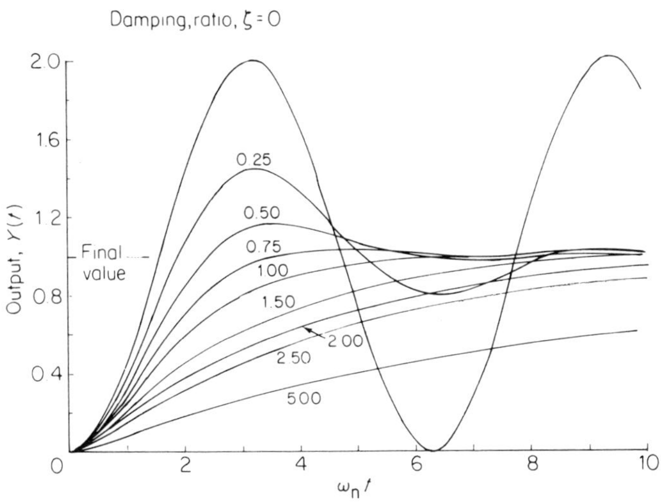

Figure 4.15 Normalized response of a second-order system to a unity step input change. (HBMS, figure 17.7.)

frequency close to the natural frequency these oscillations can become very large and potentially destructive. Such a system is said to be in *resonance*. The *resonant frequency* ω_{max} is the frequency at which the maximum amplitude is obtained and is close to, but not always exactly equal to, the natural frequency. The amplitude at resonance is controlled by the damping factor as well as the amplitude of the input driving frequency, and will increase with time until a balance is reached between the energy supplied at the input and the energy dissipated in the damping elements of the system.

The phenomenon is formally analysed by considering the frequency response of a second-order system in the same manner as the first-order system (Section 4.4.5). The transfer function in Laplace form, Equation (4.7), is multiplied by the Laplace transform of the input sine wave, and the product transformed back into the time domain. The result is

$$Y(t) = \frac{AK}{\sqrt{\{[1 - (\omega/\omega_n)^2]^2 + (2\zeta\omega/\omega_n)^2\}}} \cdot \sin(\omega t + \phi) \tag{4.15}$$

which gives the *frequency response* of the system as the input frequency is varied. Once again this can be decomposed into two parts; the amplitude response given by

$$|G(j\omega)| = \frac{K}{\sqrt{[(1 - \omega^2/\omega_n^2)]^2 + 4\zeta(\omega/\omega_n)^2}} \tag{4.16}$$

and the phase response given by

$$\arg G(j\omega) = -\tan^{-1}\left[\frac{2\zeta\omega/\omega_n}{(1 - \omega/\omega_n)^2}\right] \tag{4.17}$$

Equations (4.16) and (4.17) are presented in Figure 4.16 in normalized form. These may be directly compared with those for a first-order system shown in Figure 4.13.

The most notable feature of Figure 4.16 is that as the damping ratio is reduced the resonance (at the natural frequency) becomes more pronounced, rising in amplitude to be theoretically infinite when no damping exists. Note also that the resonant frequency ω_{max} shifts slightly as the damping changes particularly when the damping is light and resonance becomes pronounced. For $\zeta < 0.7$ the resonant and natural frequencies are related by the equation

$$\omega = \omega_n\sqrt{(1 - 2\zeta^2)} \tag{4.18}$$

Hence although the terms resonant and natural frequency are often used interchangeably this is not strictly correct.

For systems with relatively light damping the sharpness of the resonance peak of Figure 4.16(a) is commonly described in terms of a *quality factor* (Q) usually defined as

$$Q \simeq 1/2\zeta \tag{4.19}$$

Whether a second-order system is described in terms of damping factor or Q-factor is entirely a matter of choice as both are determined entirely by the

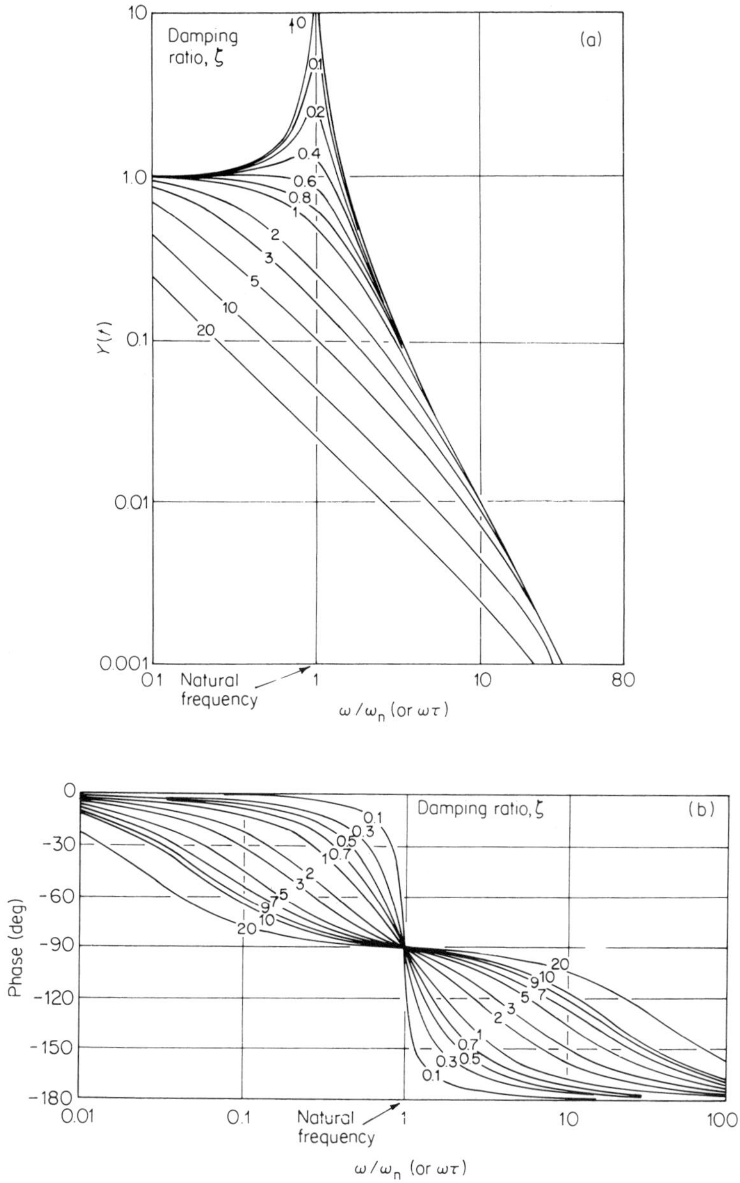

Figure 4.16 Normalized response of second-order system to sinusoidal excitation: (a) amplitude; (b) phase. (HBMS, figure 17.8.)

relative magnitudes of the energy storage and energy dissipation. It can be shown that an equivalent definition of Q-factor is

$$Q = 2\pi \text{ energy stored / energy dissipated per cycle} \tag{4.20}$$

(In an L–R–C electrical circuit, for example, Equation (4.20) gives $Q = \omega L/R$.)

When plotted in Bode form, Figure 4.16(a) shows that there is a *high-frequency roll-off* which tends toward −12 dB/octave (−40 dB/decade) as frequency increases well above the natural frequency. Note that this roll-off rate is independent of damping factor. Also if the two straight-line regions are extrapolated they intersect once again at the natural frequency in every case. However, the response at the natural frequency does depend on the damping factor, as shown in Figure 4.16(a). In particular with $\zeta = 1.0$ it gives −6 dB (50%) response and with $\zeta = 0.707$ it gives a −3 dB (70.7%) response. These may be compared with the corresponding figures for the first-order roll-off, namely −6 dB/octave (−20 dB/decade) and a single roll-off characteristic with a −3 dB (70.7%) response at the corner frequency. Note particularly that while the corner frequency is equal to the bandwidth for the first-order system this is *not* true in general for second-order (and higher) systems. As previously stated, bandwidth is defined between −3 dB points and hence for second-order systems bandwidth equals the natural frequency only when the damping factor $\zeta = 0.707$.

Inspection of the phase response, Figure 4.16(b), shows that the phase shift of a second-order system is also a function of the damping factor and frequency; however, regardless of the damping level it will always be 90° lagging at the natural frequency and always approach 180° at high frequencies. The corresponding figures for first-order systems are 45° and 90°. Use of these systems with any other, including zero-order blocks, in a control loop can lead to oscillation at high frequencies as positive feedback can occur (180° shift).

The response of Figure 4.16 may be properly described as a *second-order low-pass filter* response, although the bandwidth of such a filter generally differs slightly from the natural frequency as explained above. The design of second- (and higher-) order filters is a common requirement in measurement systems and unfortunately uses a theoretical description different from the 'systems' description given here. Filters are designed with respect to a desired *cut-off frequency* ω_c defined as the −3 dB point relative to the *peak* amplitude response. This leads to a different definition of damping factor and different presentation of amplitude and phase responses. Detailed consideration of filter design is beyond the scope of this book and the reader is referred to an electronic circuit design text, for example Millman and Grabel (1987).

The effect of this filtering action is particularly important in measurement systems because it can radically alter the shape and, therefore, information content, of any non-sinusoidal signal. This arises because *any* non-sinusoidal waveform can be decomposed by *Fourier analysis* (see Section 3.2.2.) into a number of sine-wave components of which each will be affected by the filtering to different degrees in amplitude and phase – see Example 4.7.

Example 4.7 Signal modification due to second-order filtering

Consider two signals of 10 Hz and 30 Hz shown in Figure 4.17(a) and Figure 4.17(b) respectively. With amplitudes in the ratio 2:1 and relative phase as shown, the

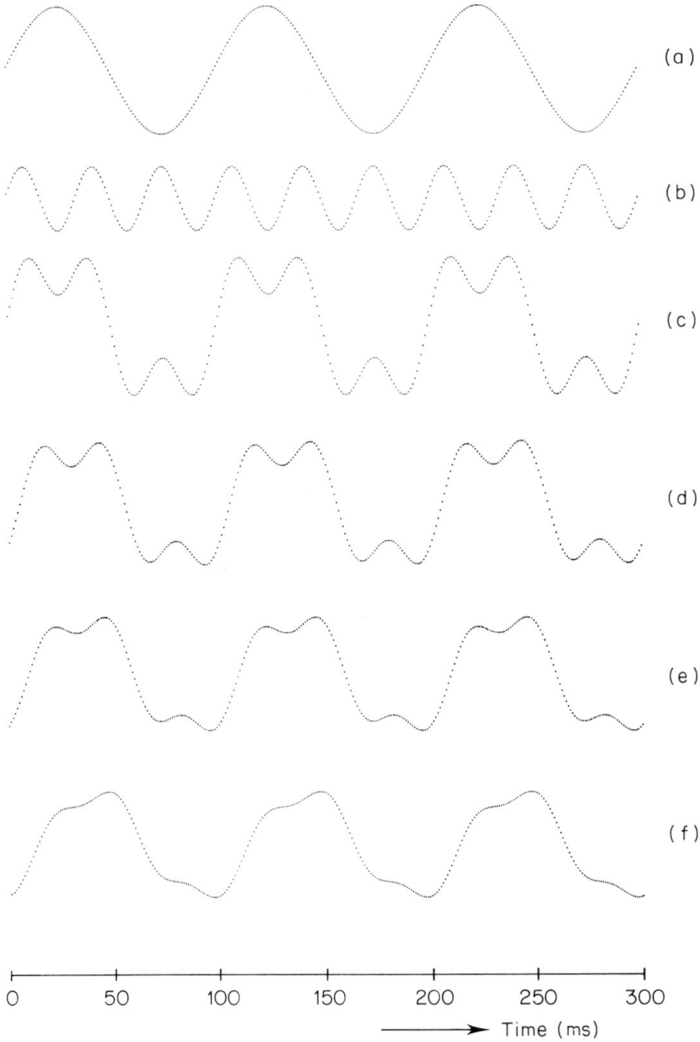

Figure 4.17 Signal modification due to second-order filtering – Part 1. A composite signal (c) comprises two components at 10 Hz (a) and 30 Hz (b). Output waveforms due to second-order filter having: (d) natural frequency 50 Hz; damping factor 1.0 (e) natural frequency 30 Hz; damping factor 1.0 (f) natural frequency 20 Hz; damping factor 1.0

composite signal has the symmetrical shape shown in Figure 4.17(c) and begins to approximate a square wave.

Figures 4.17(d)–(f) show the result of processing this signal through a second-order filter of natural frequency 50 Hz, 30 Hz and 20 Hz respectively. In each case both component sine waves are attenuated according to Equation (4.16) plotted in Figure 4.16(a) and phase shifted according to Equation (4.17) plotted in Figure 4.16(b). The attenuation and phase shift (lag) is greatest for the higher-frequency

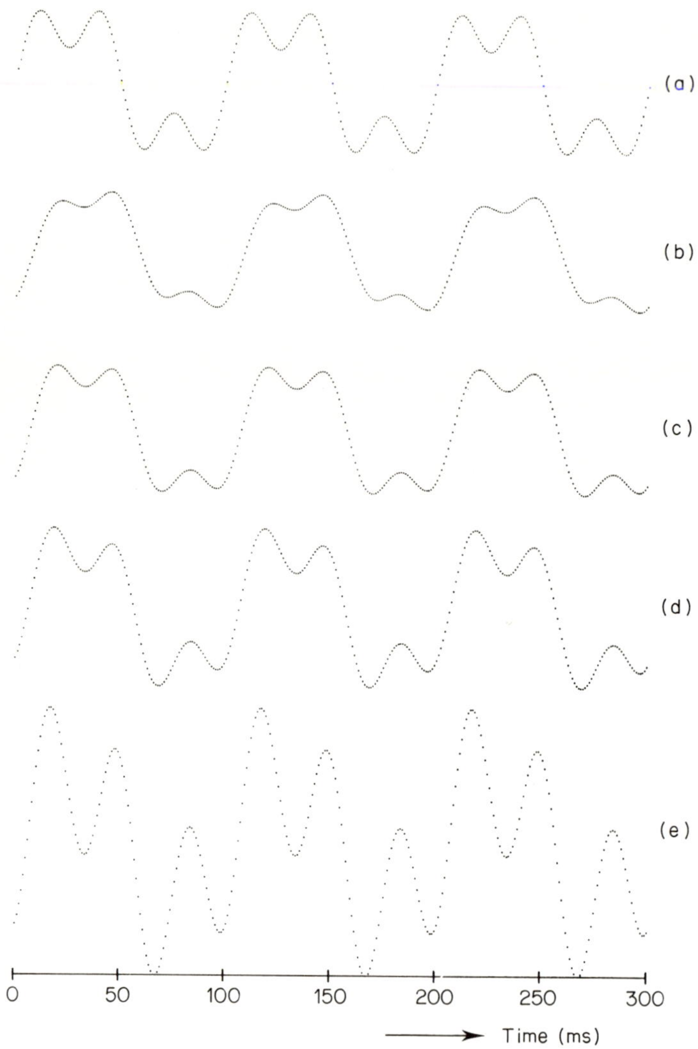

Figure 4.18 Signal modification due to second-order filtering – Part 2. The input signal (a) comprises two components at 10 Hz and 30 Hz. Output waveforms due to second-order filter having: (b) natural frequency 30 Hz; damping factor 1.0 (c) natural frequency 30 Hz; damping factor 0.7 (d) natural frequency 30 Hz; damping factor 0.5 (e) natural frequency 30 Hz: damping factor 0.2

component so that the symmetrical shape is lost. As the effect becomes more pronounced toward Figure 4.17(f) it may be said that the 30 Hz component is being 'filtered out'. However, it is important to note that even when both components are well within the system natural frequency, for example Figure 4.17(d), there is significant modification of the wave-form shape.

In each of the cases of Figure 4.17 we have considered 'well-behaved' second-order filtering with a damping factor $\zeta = 1.0$. Figure 4.18 shows the effect of reducing the damping factor.

The same composite input signal is used and the natural frequency is kept constant at 30 Hz; but as the damping factor is reduced the symmetry of the output waveform is changed again. In particular the amplitude of the high frequency (30 Hz) component is amplified rather than attenuated as the *resonance* effect becomes more apparent at the lower damping factors.

The final case, Figure 4.18(e), is an extreme example of a highly underdamped system which will selectively amplify frequencies close to its natural frequency. Note that in this case the output signal is only just recognizable as having been derived from the input signal.

Some second-order design considerations

The filtering action, illustrated in Example 4.7, is not as catastrophic as it might appear to be because the process is *reversible*. The signal modification is *non-destructive* of the information content, in the sense that the original signal can always be reconstructed. Signal reconstruction requires processing blocks which perform the *inverse* modifications on the amplitude and phase of each of the sine-wave components and in most cases this is readily implemented in electronic circuits. Detail of how this is done is beyond the scope of this text but clearly the details of the original filtering must be fully known before the inverse can be designed.

In general second-order systems present some particular design problems primarily because of two related reasons. Firstly they can be caused to resonate, either deliberately or accidentally, a phenomenon which is strongly frequency dependent and can lead to major signal changes as demonstrated in Example 4.7. Secondly in high gain and even some moderate gain systems the 180° phase shift can lead to self-oscillation.

Many sensor and transducer arrangements are inherently second order as illustrated below in Example 4.8. Control loops built around such sensors can become unstable at high frequencies due to phase shifts across the sensor. *Frequency compensation* must be provided, usually in the form of signal processing of the transducer output, to ensure this does not occur.

Similarly, observations made with second-order sensor systems can give misleading results if not conditioned correctly. Adequate damping is usually needed to control resonances but must not be too large unless slow response times are acceptable – choice of damping factor is always a compromise between response time, response overshoot (Figure 4.15) and the risk of resonance (Figure 4.16). The resonant frequency is usually set to be at least ten times higher than the highest frequency of interest in the measurand. Example 4.8 explores these relationships. However, sophisticated techniques, seeking to extract the most out of limited sensors do sometimes make use of systems with lowest damping and operating at their resonance point.)

Example 4.8 Response of a highly damped galvanometer – a second-order instrument

A common, second-order response instrument is the moving-coil galvanometer used in the high-speed ultra-violet and laser-based chart recorders. In these a beam of radiation is reflected from a small mirror onto a moving paper chart which records movement of the mirror by a photographic process. (The same kind of actuation system is used, but here in closed-loop operation, in compact disc (CD) players to keep the laser reading head following the optical track.)

A lightweight mirror (see Figure 4.19) is mounted on the electromagnetic coil assembly of the galvanometer and suspended on a vertical ribbon between the pole-pieces of a permanent 'horseshoe' magnet such that the mirror rotates as electric energy is fed to the coil through the ends of the suspension. The ribbon acts as a torsional spring and the operation of galvanometers of this design (often called 'mirror galvanometers') is not different from the moving coil meter movement used for some electrical indicating instruments. However, the small size (match size and shape), lightweight construction and relatively high compliance of the ribbon spring, lead to a much higher natural frequency and operational bandwidth.

The restoring spring of the mirror system, the moving mass of the mirror (with its support) and electrical inductance are all energy stores: the air friction, in-built

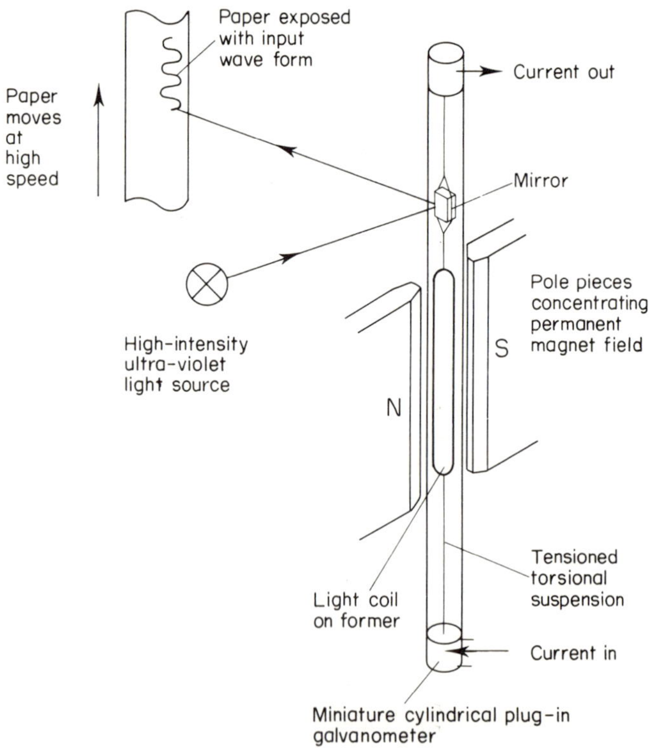

Figure 4.19 Construction of a mirror galvanometer used in high-speed chart recorders

damper of the mirror mount and electrical resistance of the coil along with the driving electronic amplifier are dissipative damping elements. This means that energy can be shuttled back and forth potentially causing oscillations that can be damped by the loss elements. It is, therefore, a second-order damped system and will not provide truly meaningful records of input electrical energy variations unless carefully conditioned in operation.

Commercial instruments usually provide the operator with a choice of plug-in mirror drive assemblies (simply called *galvanometers*) each having different natural frequencies up to around 10 kHz. The higher natural frequencies are achieved at the expense of sensitivity as spring compliance is increased and the number of turns on the coil, and hence moving mass, is reduced. In consequence the input impedance varies between galvanometers and for accurate measurements each needs use of specific driving-amplifier input impedance.

Calculations for setting up these recorders are made according to the theory of damped, second-order systems. These are made by first reducing the mixed energy system (mechanical and electrical forms are dominant – the optical path of a beam being deflected from the mirror to the paper surface is certainly of zero order) to an electrical equivalent circuit. Given the required damping leads to the correct amplifier impedance to be connected and to the effective overall sensitivity. To streamline this essential need recorder manufacturers usually supply easy-to-use application information.

Recognition of energy stores and energy dissipators, which can often be done, in the first instance, by a casual inspection of a given measuring instrument situation and its application, allows rapid assessment of the kind of response problems that might arise. More detailed modelling is then undertaken to confirm and refine the initial analysis and to optimize performance.

However, practical cases do arise when the filtering action of a transducer, or more commonly a complex sensing system, cannot be determined by disassembly and inspection. In these cases the system response may be determined directly by using a range of *test signals* such as step functions, sine waves and others, as described in Section 3.3.3. (Experimental identification of system modelling parameters is taken up again in Section 5.3.3.) The transfer function of the system $G(\omega)$ can then be derived, in principle at least, from the relation

$$G(\omega) = Y(\omega)/X(\omega) \tag{4.21}$$

where $X(\omega)$ is the Fourier transform of the input (test) signal and $Y(\omega)$ is the Fourier transform of the output signal.

Appropriate parameters for the system, such as the order, natural frequency and damping factor, may then be derived from the transfer function. The overall procedure is known as *process identification.*

Other examples of second-order systems are vibration measuring accelerometers and related seismometers; weighing systems using load cells including chemical balances and gravimeters; pendulum based devices such as tilt-meters, pendulum and mass-and-spring clocks, gravimeters and strain measuring systems where the base is made of noticeably compliant (springy) material. If any of these instruments or measurement techniques are to be employed for dynamic measurement a second-order, damped, system analysis is essential to demonstrate that they can measure an input measurand in an adequately, error-reduced, manner.

4.4.7 Higher-order systems

Whilst the methodology for analysing higher than second-order systems exists the mathematics becomes progressively more complex. For example, it will be recognized that third-order systems lead to cubic equations and, hence, three roots in the solution and three parameters for the characterization of the frequency response.

The effort of higher-order analysis is rarely worthwhile in instrument applications, for much can be learned about performance by working with the dominant effects up to second-order. Exceptions are cases where such factors as mass-production savings, extreme safety requirements and defence application can allow more resource to be devoted to the modelling exercise.

In a weighbridge, for example, there will certainly exist resonances other than the fundamental produced by the measuring load cell (a spring) and the mass of the platform and load. They are, however, usually much higher than the fundamental and so have little effect on measuring performance under normal operating conditions. However, for exceptional dynamic applications (for example vehicles being weighed whilst in motion across a weighbridge) a full analysis of the system dynamic performance would have to be undertaken.

Non-standard dynamic performance may also arise when a system can be seen to act in a non-linear manner such as those with *backlash effects* and other discontinuous actions. Analysis of such a system is usually both practically and mathematically difficult. Some special, but limited, analytical methods have been developed and these involve digital computational methods that are the subject of more advanced course levels than is covered by this text.

4.4.8 Gaussian response

The Gaussian frequency response does not arise in simple physical systems. However, the response of many complex systems, including much electronic instrumentation, such as oscilloscopes and spectrum analyser filters, does approximate the Gaussian response or may be deliberately and carefully designed to have such a response.

The Gaussian response is so-called because the roll-off with frequency follows the equation of the Gaussian probability distribution with a mean of zero. The Gaussian curve was introduced in Section 3.3. (Equation 3.10) and the resulting frequency response is reproduced in Figure 4.20. It is one half of the familiar 'bell-shaped' curve. Note that in contrast to first and second-order responses this roll-off is plotted on *linear axes* in terms of percentage response. The cut-off

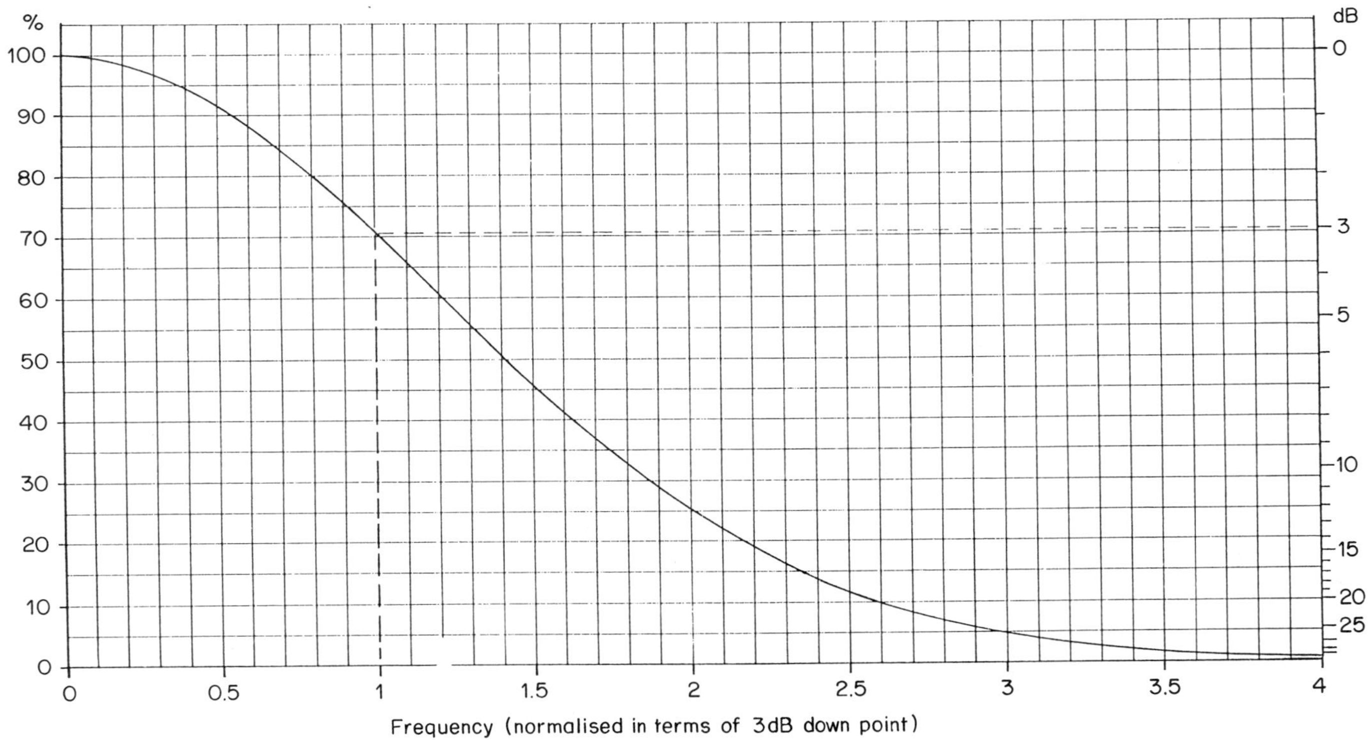

Figure 4.20 The Gaussian roll-off curve, in which the frequency axis has been normalized with respect to the 3 dB-down (70.7% response) point. (Reproduced by permission of Tektronix Inc.)

frequency f_c is defined as the -3 dB (70.7%) response point as usual so that this frequency is the system bandwidth.

The corresponding step function response is shown in Figure 4.21 with a rise time t_r defined from 10% to 90%, as usual.

Note that the curve is symmetrical and has no overshoot, a property which is partially desirable for instruments providing time-domain display of a signal such as the oscilloscope - see Example 4.9. Furthermore for any Gaussian response system the product of rise time and bandwidth is constant, that is

$$t_r f_c = 0.35 \tag{4.22}$$

This property is exploited in the interpretation of oscilloscope displays as discussed in Example 4.9.

For a complex system of cascaded stages, each of Gaussian response, the overall system response will also be Gaussian. For n stages, each of bandwidth f_i and rise time t_i the overall system bandwidth f_r will be given by

$$f_r = 1 \Bigg/ \left[\sum_{i=1}^{n} t_i^2 \right]^{1/2} \tag{4.23}$$

From either equation it is clear that one low bandwidth, or one long risetime, will dominate the overall response and degrade the system performance. Hence for optimum response systems are usually designed, as far as practicable, with all f_i (and t_i) approximately equal.

Equations (4.22) and (4.23) are also of value when component responses are not all Gaussian although the result is then only approximate. However, the closeness of the approximation increases as the number n of subsystems increases and, for example, first-order subsystems (for which $t_r = 2.2\tau$) may be incorporated without serious error providing they do not dominate the summation.

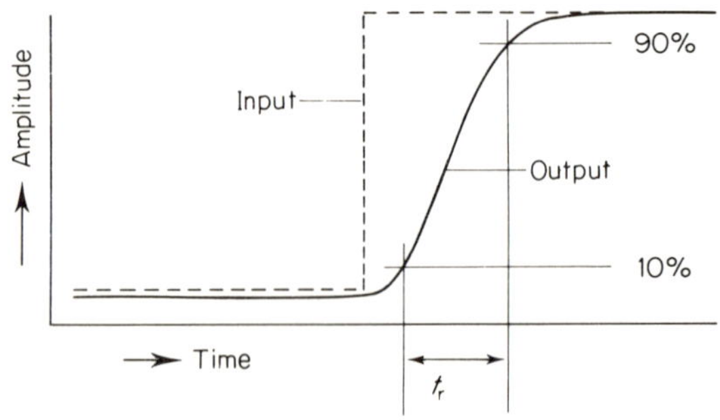

Figure 4.21 The time-domain response of a Gaussian roll-off system to a step input

Example 4.9 Use of an oscilloscope for accurate amplitude and rise-time measurements

The oscilloscope is a common tool for the inspection and measurement of electrical signals. It is essentially a voltmeter; however, between the probe tip where the signal is acquired and the display there exists a chain of signal-processing stages which includes the probe assembly, input attenuator, preamplifier, delay line, vertical deflection amplifier and cathode ray tube (CRT) display. Each stage will have a high frequency roll-off and good commercial oscilloscopes are designed such that the overall response is Gaussian. (The probe is usually a passive, first-order, stage and must be chosen to be of adequate bandwidth such that it does not degrade the overall system performance). The oscilloscope will be labelled with a bandwidth, for example, 60 MHz for a medium performance laboratory model.

Inspection of the Gaussian frequency response, Figure 4.20, leads to two important conclusions critical to signal amplitude measurement. Firstly, *the ampli-*

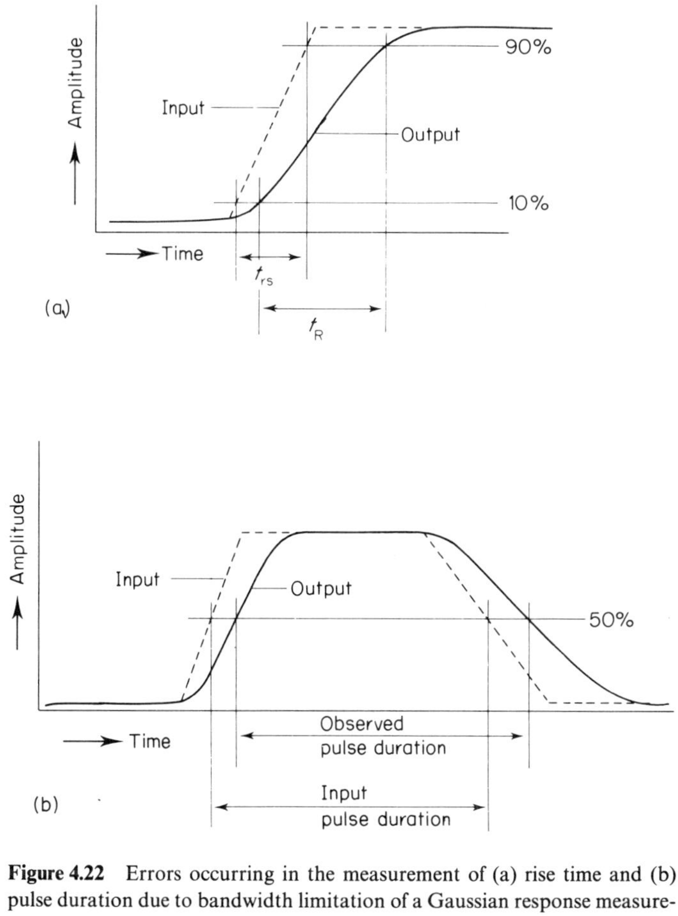

Figure 4.22 Errors occurring in the measurement of (a) rise time and (b) pulse duration due to bandwidth limitation of a Gaussian response measurement system

tude display will only be 100% *at frequencies well below the stated bandwidth.* For our 60 MHz unit a 30 MHz signal will only produce 92% of the real signal amplitude. (At 10 MHz it is still not fully responsive, giving 98%.) However, with the Gaussian curve to hand a correction factor can easily be applied. Secondly, the oscilloscope can produce useful displays at frequencies *above* the stated bandwidth. Our 60 MHz oscilloscope, if accurately Gaussian, will produce, for example, a 25% display at 120 MHz.

Oscilloscopes are also commonly used to measure the rise times of signals, for example, the step edge of a pulse. Consideration of Figure 4.21 immediately suggests that there is doubt as to whether the displayed rise time is due to the signal or due to the rise time of the oscilloscope itself. The rise time of the oscilloscope can be calculated from Equation (4.22) – in our case $t_r = 0.35/60\text{ MHz} = 5.8$ ns. However, even if the rise time indicated is longer than 5.8 ns the displayed rise time is t_R, the combination of the signal rise time t_{rs} and the oscilloscope rise time. Hence we have

$$t_R = [t_{rs}^2 + t_r^2]^{1/2}$$

according to Equation (4.23), and the erroneous display is illustrated in Figure 4.22(a). However, the required signal rise time t_{Rs} may be readily calculated as

$$t_{Rs} = [t_{rs}^2 - t_r^2]^{1/2}$$

Figure 4.22(b) shows that pulse duration measurement is also affected when measuring between two edges which do not have similar rise and fall times. However, once again the result may be readily corrected by compensation for oscilloscope rise time.

4.5 THE SENSING INTERFACE

4.5.1 The popular definition model

In foregoing sections of this chapter we have explored measuring-system architecture and response of such systems to stimuli. It is now necessary to return to discussion of the sensory interface and consider its role in the information machine system.

The sensor is that part of the information machine that interfaces with that part or aspect of the physical world that we wish to observe. It is, therefore, the 'input stage' of our machine and its adequate specification and design is crucial if the machine (the instrument) is going to function satisfactorily. To understand the sensor (and often there will be several sensors in the complete instrument), modelling, particularly the mathematical kind, must be undertaken.

Whilst measurements have been made since the earliest time of man the sensory interface has attracted little philosophical thought with regard to its modelling. In consequence the sensor is, in many ways, the least understood part of the information machine.

Models do exist but they have only been used to 'motivate the mind'; they have yet to play a significant role in formulating or assisting design algorithms. By contrast, the algebra of Boole provides the methodology for digital processing; in process control stability theory allows simulation and performance prediction; in communications engineering Shannon's sampling theorem enables maximum data rate to be estimated. At present sensor interfacing is awaiting such counterparts to be developed.

Despite there apparently not yet being a model that is recognized as being as foundational as those mentioned above, there do exist three types of abstract model of the sensory interface. Each helps understanding of what is occurring and which factors must be addressed in setting up such an interface.

The first and, best-known, model is best described as the *popular definition* of what occurs when a measurement is made. It can be defined as:

> *A measurement is the comparison of an unknown magnitude of a quantity with an agreed standard declared as the unit, the measure taking the form of the difference expressed in numerical form.*

Figure 4.23 portrays this process as set up to perform an (apparently) very simple measurement task, that of determining the length of a bar.

This model highlights the need in a sensory interface for:

- an *agreed unit* for each particular measurand;
- a *scale* for interpolating between integers of a unit;
- *physical standards* for setting up agreed units in a uniform manner between different users of the same unit, and
- a *practical method* with which to make the comparison.

In simple measurement systems this comparison is usually made by the observer, often visually, but this is not the case in more complex systems or automatic instruments.

Whilst providing some basic help in understanding sensor design and operation this model is far from being a complete enough description. In particular it does not explain important requirements such as the energy relationships that exist as information is transferred.

4.5.2 The energy/information model

Modern sensor technology, which often does not use human observation to make the comparison, requires use of another model that covers the energy interactions taking place during sensing.

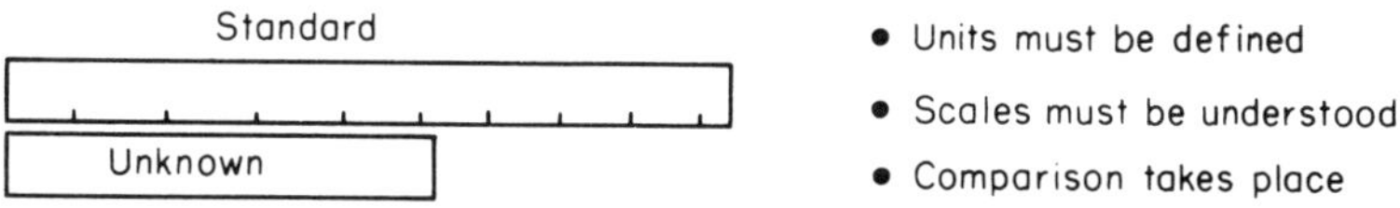

Figure 4.23 Popular understanding of the measurement process. (From Sydenham P H, 1984, Sensors and transducers for physical variables, *Proc. PROMECON*, R-4 paper, The Institute of Measurement and Control, London. Reproduced by permission.)

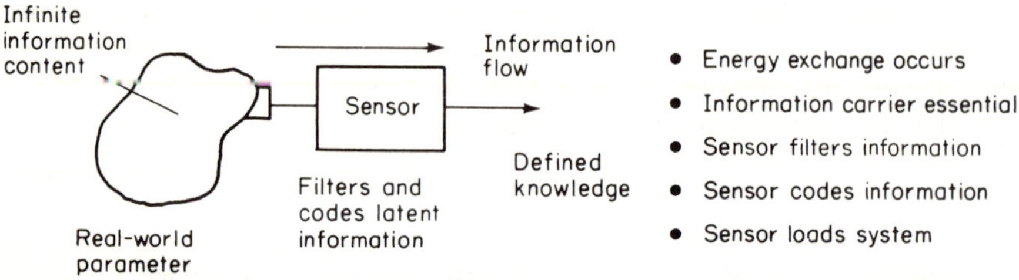

Figure 4.24 Energy/information model of the sensory interface. (From Sydenham P H, 1984, Sensors and transducers for physical variables, *Proc. PROMECON*, R-4 paper, The Institute of Measurement and Control, London. Reproduced by permission.)

The sensor is interfacing the appropriate energy modulations existing in the observed region, transmitting these to the subsequent signal-processing stage, as shown in Figure 4.24. Within any real observed subject there exists an *infinite* amount of information. The sensor, therefore, must select just that needed from this large available quantity, conveying it as suitable signal without distortion (that is, without error) of its meaningful content.

The energy/information model illustrates that:

- *energy* is always involved in sensing;
- *energy modulations* represent the desired information;
- there is an *infinite amount* of information residing within the energy of the observed system; and
- the sensor is an *information filter.*

Any other energy perturbations that are present may give misleading, or *misinformation* signals. These unwanted signals are called *noise.* It is important to note that noise is only regarded as such when it is not wanted. In a different measurement context this same noise signal may contain required information usually related to a different measurand (as has been already explained in Example 3.8).

The energy/information model also makes it clear that an energy connection between the sensor and the observed object must take place and in doing this it may upset the energy relationships in the latter leading to error. Such unavoidable connections are said to *load* the observed object. The *loading effects* must always be considered in sensor application and design, as has already been discussed in the context of impedance (Section 4.3). Example 4.10 considers loading effects in the context of energy flow.

Example 4.10 Loading error of a thermometer

Consider the differential expansion form of thermometer illustrated in Figure 1.5(g). (Temperature is sensed by the changing rates of expansion that occur between the central aluminium, *actuation* rod, and the steel support.)

This design is, by its nature, bulky to implement and thus differential expansion thermometers usually possess considerable *thermal mass.*

When the thermometer is placed to observe a process temperature the thermal *regime* (the distribution of heat and how it flows) that existed prior to the thermometer being put into place will not be *exactly* the same as after placement. The presence of the thermometer creates (from basic laws of physics) an automatic redistribution of the thermal regime.

If the heat storage content and heat conductivity of the thermometer are significant compared with the magnitudes of the counterparts of the surrounding region in the process it will be clear that adding the thermometer changes the temperature of the now combined region. It will, therefore, register what the temperature is with it in place, not that without it.

Adding a pencil-size thermometer to the side of a steel-making blast furnace is unlikely to *load* that process as its thermal mass and conductivity will not extract much heat energy compared with that available in the greatly larger furnace. However, projecting the same thermometer into the gases passing up a small boiler flue pipe can introduce considerable loading as it may thermally conduct significant heat out of the flowing stream. Making *truly accurate* temperature measurements can be deceptively difficult because of loading errors.

Loading is also commonly encountered when connecting electronic and mechanical stages together.

At the limit are some, subatomic, energy physics measurements that cannot yet be made because the interception of the process by any yet devised observing interface introduces loading to such a degree that the original process cannot be observed – the energy available to link into is just to small too sense.

The usual practice for dealing with loading is to design the interface such that loading effects are adequately small. That, however, requires some mathematical-modelling considerations to provide the necessary quantitative information. An alternative, not always possible, is to observe changes in the output of the actual process when the sensor is added. For example, adding a voltmeter to measure the output of a microphone stage in a public address system will reduce the amplified output in the speaker (or better, observed as a voltage) if significant loading is taking place.

4.5.3 The set-theoretical model

Whilst the energy-flow model introduced above is valuable and indeed essential to practical sensor design it does not encompass the ideas inherent in the popular definition model. What is sought is a more fundamental model that incorporates the features of both of the above models – and more. No such model has yet been devised.

One other model, however, that appears to be approaching this need from a suitable direction (but needing more consideration) is the *set-theoretical* model. This has been evolving over the past fifty years but is yet to find routine application in mathematical modelling of sensors. (Such delays in scientific and technological development are quite common – Boolean algebra formed the

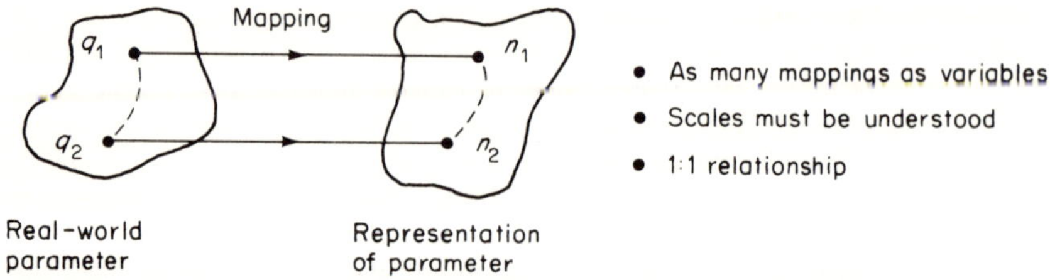

Figure 4.25 Set-theoretical model of the sensory interface. (From Sydenham P H, 1984, Sensors and transducers for physical variables, *Proc. PROMECON*, R-4 paper, The Institute of Measurement and Control, London. Reproduced by permission.)

basis of digital computing some 40 years after its publication – its own development can also be traced back many decades before that!)

The set-theoretical model is based on the premise that the sensor is *mapping* the real world measurand into an *equivalent representational* system. Figure 4.25 is the diagram that portrays this model.

Study of the conceptual level of this model reveals that:

- the *observed measurand* has an *associated scale*;
- the *output signal* has its *associated scale*;
- the *scales of each may be different*;
- there must be a *known mapping* between the two scales;
- there must be a *separately defined mapping* (channel) for each measurand to be observed;
- the relationships can be *expressed mathematically*.

This model can be conveniently expressed in terms of *set theory* symbolism, which is very concise. However this introductory text cannot encompass the mathematics and symbolism required in the space available. The reader is referred to Finkelstein (1982).

4.6 ERROR AND UNCERTAINTY WITHIN THE INFORMATION MACHINE

4.6.1 Classification of errors

In previous chapters we have seen how it is that the measuring kind of information machine has the task of extracting information that has associated with it a meaning that can be used to glean knowledge.

The knowledge gained by the process of using the measuring instrument is expressed in terms of relationships and numerical values. Unwanted signals give rise to incorrect interpretion of information because error is introduced into the system data set. Errors, therefore, subsequently limit the usefulness of an instrument as a knowledge-gathering tool.

To quote P.K. Stein

> all measuring instruments should be regarded as guilty until proven innocent.

The performance of instruments is decided by the magnitude of their errors. Two thermometers may use the same physical principle to map the measurand into signals but design differences are decided by their need for different error magnitudes.

The design of an instrument and its effective application is a matter of first realizing a suitable transduction arrangement and then seeing how this is affected by system parameters that introduce errors of operation. The cost of error reduction is a prime limiting factor on what instrument performance can be offered for a given price level.

Presented here is an introduction to the error regime of an information machine – the full treatment is mathematically complex and needs considerable time to develop. A less brief overview is given in Hofmann (1982) from which some of this material is extracted.

A first step in controlling errors is to appreciate their sources and nature. Three kinds, based on the predictive nature of each, can be identified.

Systematic error

Consider an 'electronic voltmeter' using electronic preamplification to enable it to measure low voltages without introducing circuit-loading effects. The gain of amplifier and the indicating voltmeter are both sources of error – the amplifier gain and the voltmeter's voltage/indicated scale value must be set correctly by comparison with the accepted standard – see Chapter 2.

Physical considerations of how these parameters might vary show that once they have been *calibrated* (that is, determined) they can be ascribed an *error correction* that can, in future, be used to obtain the correct value (provided the calibration is time invariant or its characteristics with time are known) for any nominated future value.

Errors that can be predicted on this *value-by-value* basis are called *systematic*. The causal nature of these errors allows the mathematics of deterministic systems to be used to describe the error behaviour.

Error Δx is defined as

> the difference between the standardized correct value x_S and the observed incorrect value x_I

that is,

$$\Delta x = x_I - x_S. \tag{4.24}$$

Where these errors are time dependent the formulation, using a notation to denote those actually arising in the output of the instrument, is then given by

$$\Delta x(t) = x_{\text{aI}}(t) - x_{\text{aS}}(t) \tag{4.25}$$

The *error function* Δx can take the form of a simple correction factor (such as a multiplier to correct gain); be an additive value (or set of values) to correct a point (or set of points); be a more complex model using linear and cubic expressions that are fitted to experimental results (such as the correction curves for thermocouples and platinum resistance thermometers); or be highly complicated system equations obtained by *system identification* procedures.

To obtain the *error transfer function* $G_E(s)$ of the instrument stage we take the ratio of the *actual* transfer function $G_I(s)$ and the *error-free* transfer function G_S, from which the error contribution is extracted by subtracting unity.

A typical form for systematic error is

$$G_E(s) = \Delta x_a(s)/x_{sS}(s) \tag{4.26}$$

The availability of the error transfer function enables the behaviour of errors to be studied when the system is fed (excited) different input signals - by the method explained in the study of the signal response of linear systems given in Section 4.4.3.

Random error

Now consider another measuring instrument example, one that uses a flexible wire to convert *distance* into *rotation* of a precision drum. The wire is fastened at one end to the drum and held wrapped around it by spring tension. The wire is pulled off the drum from a datum point to the other end of a length interval; drum rotation is then a measure of distance.

It is found, experimentally, that the values of the individual measurements (of a set of length measurement made of a standard test length) vary about a mean value, each having a value that varies *randomly* about the mean by a small amount.

The precise magnitude of any individual future value (to within the available discrimination of the instrument) cannot be predicted in the same way as can the future results for the systematic process existing in the voltmeter example, given above. It is, however, possible to make predictions about various properties of a *group* of future measurements.

This class of measurement error is generally known as *random error*. Experience suggests that random error arises as the result of several different sources that collectively manifest themselves by arising at differing times with differing amplitudes. The net result is randomness. Whilst some sources may well be truly random, experience also suggests that most sources can be rendered systematic if enough effort could be devoted to discover, and model, the various causes. It is, however, often more expedient to use *non-deterministic*, mathematical description for modelling purposes. The phenomena being described (in our case random or pseudo-random errors) are said to be *stochastic*, this is, governed solely by the laws of probability.

The relevant theoretical considerations of stochastic signals (at least the starting point) have been introduced in Section 3.3.1. The *normal* or *Gaussian probability density function* $w(x)$ (Equation 3.12) was introduced. This basic theory has been extended to allow for the modelling of both analogue and digital signals.

We shall now re-examine the use of Gaussian distribution theory for experimental measurements made with an instrument. Once again there is not the space in this text to provide a full derivation of the formulae presented and the reader is directed to any elementary applied statistics text (for example, Chatfield, 1970).

In practice an instrument is used to make n repetitions of a single measurement of a quantity x. It is *assumed*.

- that the quantity x does not change during the sequence of measurements;
- that the properties of the instrument do not change during the measurement (for example, influence variables such as ambient temperature);
- that any human observer/operator involved does not become tired or bored or careless during the sequence;
- that the errors involved are *entirely* random; and
- that the random errors for this instrument are indeed 'normally distributed', that is, follow the Gaussian distribution function.

If these assumptions are reasonable then the best estimate of the true value of x will be given by the mean x,

$$\bar{x} = \frac{1}{n} \sum_{i=1}^{n} x_i \qquad (4.27)$$

The *standard deviation* σ of our *sample* of n measurements is then given by

$$\sigma = \left[\frac{1}{n-1} \sum_{i=1}^{n} (x - \bar{x})^2 \right]^{1/2} \qquad (4.28)$$

(This is not quite equivalent to equation 3.13 presented in Chapter 3 because our σ value here is only an estimate of the standard deviation of the random error in our measurement of x. We would need to take an infinite number of measurements before we could determine the true value of the standard deviation.)

From the instrument user's point of view the most important requirement is to estimate how close the calculated mean $\bar{x}$ is to the true value of x - clearly if the user continued to take more measurements x_i $\bar{x}$ would become a better estimate of the true x. The required estimate of the quality of $\bar{x}$ is the *standard deviation of the sample mean* σ_m, also known as the *standard error*, which is given by

$$\sigma_m = \left[\frac{1}{n(n-1)} \sum_{i=1}^{n} (x - \bar{x})^2\right]^{1/2} \tag{4.29}$$

and it follows that

$$\sigma_m = \frac{\sigma}{\sqrt{n}} \tag{4.30}$$

Hence the result of this repeated measurement exercise would be to quote the measured value of x as

$$\bar{x} \pm \sigma_m$$

It is worth repeating that the standard error σ_m must be used with care. When the number of readings n is made very large σ_m may become deceptively small. The five assumptions listed above must be reconsidered to see if they justify the high accuracy implied by the quoting of a very small σ_m value.

Human error

A human observer is often part of an instrument chain. In such cases it has been well proven that errors can arise that are clearly ascribed to the performance of the human operator, not to the instrument. These are referred to as *human error.*

In 'seeing' forms of observation (for example, a pointer against a scale), errors arise from drift of definition of observed 'coincidence' as the eye tires and lighting conditions change and from bias of the 'apparent' concidence between two fiducial lines, existing between different observers (parallax error assessment is different from person to person).

In 'hearing' observations human errors arise from variation between observer ability to discriminate when an interval is believed to have begun or ended; when sound levels and tones are thought to be equal and when characteristics of complex sounds appear different to different observers. These factors also vary with the tiredness of an observer.

Additional to these human sensing errors there also exist *transcription errors* that arise when making a record (such as incorrect values, transposed numbers and decimal point errors). Furthermore, 'outlier', 'rogue' or 'gross' (wildly wrong and extreme) values appear without explanation. These are usually eliminated where it is *clearly reasonable* to do so. (It is dangerous to automatically remove them – history shows new discoveries may come from these, see Example 1.4.)

These three kinds of human error can arise simultaneously in a single measurement situation, as explained in Example 4.11.

Example 4.11 Error sources in a heat-measuring bolometer

A bolometer (heat) detector is formed with a very fine grid of electrical resistance wire (of diameter typically a few micrometers). It is used to sense the heat changes caused in it by impinging focused, radiation from, say, an operational electronic component for which the temperature rise is being determined.

Errors arising due to ambient temperature effects on the mounting of the bolometer are systematic in nature — a mathematical model could be constructed or a calibration test conducted that would allow the operator to decide what the error will be, given knowledge of the ambient temperature.

The error caused by stray radiative heat entering the system from a fluctuating background would be of random form (unless a systematic reason could be identified such as a known steady heat source). By monitoring this stochastic effect, the error of a temperature determination made with the bolometer can be assigned a probable level of error — but the actual 'next value' could not be given in any other than probabilistic terms.

In using the instrument to measure a component the operator can incorrectly point the bolometer, thereby measuring the temperature of the wrong area, misread the output values and erroneously record the values read. These are human errors.

4.6.2 Origins of error

In determining the performance of an instrument it is necessary to appreciate how errors arise in practical instruments and use this information to develop an *error budget*. This is developed by systematic assessment of the magnitudes of the many sources involved. When these have been identified they are tabulated and then combined, see Section 4.6.3, to decide the system *uncertainty*.

Every application must be studied on its own merits, tempering the exercise with considerable judgement lest it become too time consuming. The experienced instrument engineer will be able to identify potential sources of such error from *intuition*; it is, however, preferable to make use of a check list developed for the particular application area — Section 8.2.2. The importance of identifying the error of an instrument is so great that Chapter 8 addresses the selection and testing aspect.

This section discusses the various *sources*, as opposed to kinds, of error. Section 4.6.3 develops how they propagate through an instrument. Section 4.6.4 explains how they can be reduced at, or close to, their source.

Three main sources of error in the performance of an instrument can be identified:

- *imperfections in components* changing their performance;
- *external influence variables* changing the system's performance;
- *inherent physical fundamental laws* of nature that generate unwanted noise within the system.

Measurement error can arise, in a *direct* manner, within the hardware of instruments by virtue of changes in the values of components that form part of the information chain. These changes take place over time and with stress level. They are also caused by such effects as temperature and voltage-level changes. Such changes can cause shifts in the gain and offset of stages and alter the mathematical relationships built into analogue systems.

Examples are a small change in gain in an operational amplifier stage caused by drift in the magnitude of either of the two resistors forming the gain ratio; expansion of a metal measuring scale as mechanical stresses of manufacture relax; and drift in sensitivity of a silicon optical detector as the 'holes' and 'junctions' drift as the chemistry changes.

In many cases these have a *direct*, significant, effect in that changes in system performance are directly proportional to such shifts without reduction.

Other sources of internal errors are those of *indirect* nature where the component value change does not influence the system so strongly because it only influence another less dominant parameter. For example, variations in the power supply of an electronic amplifier will alter the amplifier gain by a small amount due to deliberately in-built reduction mechanisms.

Another example is where the rapid motion of an inspection machine slideway causes short-term heating effects that temporarily alter the reading accuracy (as the scale expands and contracts). Yet another example is how dynamic stress, arising during rapid automatic positioning of a zoom lens system, gives a temporary variation of the optical focusing system dimensions.

These effects can be estimated and allowed for by good design. Component suppliers provide data on many of these effects in the form of graphs, coefficients and higher-order identified equations.

Error reduction usually carries an increasing level of cost penalty as the errors are made smaller; they are normally only reduced as much as is necessary for the *overall* accuracy level sought. Error budget information shows which design parameters must be improved to yield the greatest reduction in overall error.

Influence variable effects

If measured finely enough it will be found that all materials, and thus components, exhibit some variation of their physical properties when subjected to one or more of many existing external effects.

The most common effect is response to *environmental temperature*. Such external effects are called *influence variable effects*. For example, steel expands with temperature change; the resistance of a resistor changes with temperature; and the electrical leakage resistance on a circuit board changes with relative humidity.

Data is widely published for many of the effects that arise in instruments but as the error level is reduced it becomes more difficult, if not often impossible, to obtain precise enough correction data. At the limit it is necessary to test each individual instrument to obtain this information.

The commonly met *influence variables* are:

- temperature (which is almost universal)
- pressure
- relative humidity
- vibration or acceleration
- electromagnetic interference (EMI)
- tampering by humans, animals and insects.

The instrument designer works to reduce these effects to stated practical limits within the allowable price for the instrument manufacture. Application of a measuring instrument to an environment where influence variable effects exceed the design values can be expected to give rise to errors greater than given in the performance specification.

Whilst all influence variable effects rarely occur at their maximum values simultaneously (the *worst-case condition*) there usually can be no certain way of deciding the combination in which they might occur. Specification of overall uncertainty often, therefore, assumes the worst-case condition. Example 4.12 discusses how the temperature influence variable affects operation of a strain gauge system used in a pressure gauge. Some other effects are discussed in Example 4.4.

4.12 Influence variable effects in a strain gauge system

A common form of pressure gauge uses electrical resistance foil strain gauges to monitor strains of a stiff diaphragm that deflects as the pressure exerted on one side changes. Example 1.2 introduced the principle. Figure 4.26 provides more definitive engineering design data and typical assembly arrangements are shown later in Figure 6.7.

The *gauge factor* of the metal foils commonly used is given by

$$G = \frac{\Delta R/R}{\Delta L/L} \tag{4.31}$$

where R is the nominal electrical resistance of the gauge element, L the length of the active part of the grid and Δ values the changes in nominal values. Standard values are $R = 120$, 350 and 1000 Ω. G is close to 2.1 for metal foils.

A change in temperature of the foil, mounted (Figure 4.26(a)) on a substrate of different thermal coefficient, gives rise to an *apparent strain* – one that appears to be present but that is not caused by the input measurand.

As the temperature of fluids to be measured and of the ambient of the sensor are rarely constant the temperature coefficient (*tempco*) of the foil must be carefully chosen. This is done by making the foil from material with the same thermal coefficient of expansion as the material on which it is mounted. A perfect match

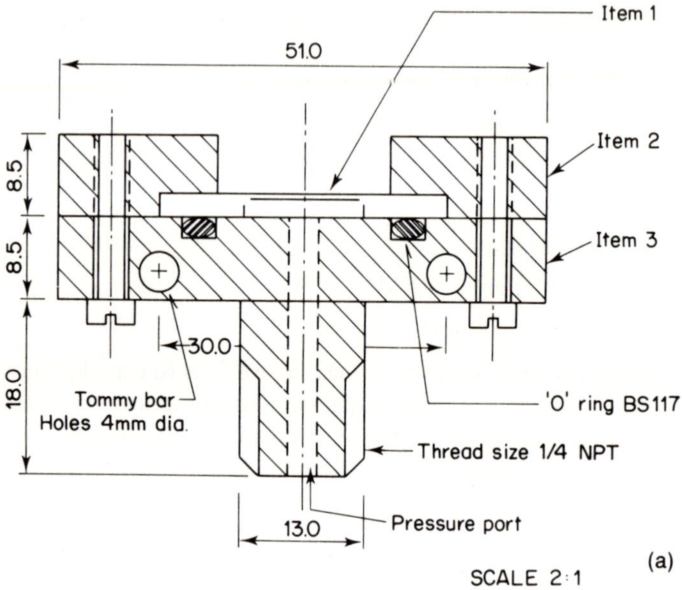

Figure 4.26 Physical structure of a strain-gauge-based pressure gauge. (a) Assembly drawing (b) strain-gauge foil and circuitry

cannot, however, be obtained, so some influence variable error still arises from this thermal differential expansion effect.

The electrical resistance of the gauge is also noticeably temperature dependent (this cannot be avoided) giving rise to another source of apparent strain that must be allowed for. An obvious way to compensate for this source is to use a second, identical, foil that is mounted to experience the same temperature effects but without input strain change from pressure variations. (This also further reduces the thermal expansion effect mentioned above.)

The two foils are connected into opposite arms of the Wheatstone Bridge detector such that temperature effects largely cancel out by *common-mode* differential action. Whereas this method does, indeed, greatly reduce the effects, some residual imbalance still remains. Thus is formed a second source of influence variable error.

An extension of this method is that shown in Figure 4.26(b) where four gauges form a complete bridge circuit and, in doing so, can be used to sense strain whilst also providing temperature compensation – achieved in this case by using a strain-gauge assembly that measures, simultaneously, both compressive and tensile strains averaged across the diaphragm.

In addition to these effects the gauge factor is also slightly temperature dependent and designs needing a wide range of ambient temperature of operation (−20 °C to 120 °C for example) need still more compensation and this is usually provided by measuring the temperature with a sensor that alters the bridge sensitivity. This effect varies with temperature level, material and actual device, so that it is difficult to obtain complete compensation in practice.

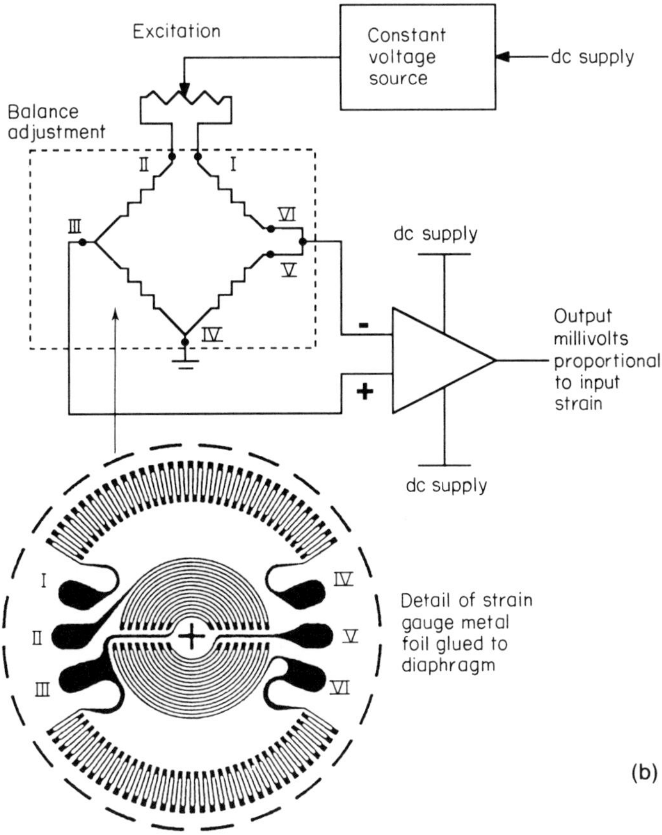

Figure 4.26 (*continued*)

Using all of the above methods of compensation a basic gauge concept that can only measure to around 10% uncertainly when operating in an environment of $\pm 10\,^{\circ}C$ can be made into a reliable sensing system that can operate to 0.01% uncertainty in an environment from $-20\,^{\circ}C$ to $120\,^{\circ}C$.

Practical use of sensing effects need considerable design effort to give them enough 'immunity' to influence variable effects.

Coping with the usual six or so influence variables adds considerably to the complexity and cost of the task. Many sensing methods also have to work with principles and materials for which influence variable effects are interdependent.

Where costs justify it, another approach to error reduction is to control the influence variable magnitudes by providing a *controlled environment*, such as a temperature-controlled laboratory – but this still requires the availability of capable sensors to control the environment. The advantages and disadvantages of such an approach are illustrated in Example 4.13. Internal compensation is illustrated in Example 4.14.

Example 4.13 Accuracy and stability in laboratory frequency counters

Standard laboratory frequency counters, often labelled as 'Universal' counter timers, measure frequency, period, intervals and other time-related quantities by counting (totalizing) events over an interval known as the *gate time*. The gate time is derived from a reference clock, called the *time base* and the accuracy of the frequency (or other) measurement displayed depends directly on the accuracy and stability of the time-base oscillator.

The accuracy of this oscillator, and hence of the whole instrument, is determined by the manufacture of the reference quartz crystal, and is usually specified to the nearest 1 part in 10^6 (1 ppm). However, the frequency of this oscillator varies by a few parts per million according to the age, temperature and circuitry in which the crystal is located. Hence typical stability specifications for the time base of a laboratory frequency counter are:

- aging — less than 3 in 10^7 per month
- temperature — less than ± 1 in 10^5 over 0 °C to 50 °C
- power supplies — less than ± 1 in 10^7 for $\pm 10\%$ variation

For more exacting measurements the reference oscillator circuit is usually temperature controlled and the specifications then become (typically):

- aging — less than 1 in 10^7 per month
- temperature — less than ± 1 in 10^6 over 0 °C to 50 °C
- power supplies — less than ± 1 in 10^8 for $\pm 10\%$ variation

Even more careful temperature and voltage control of the crystal oscillator circuit may be provided (at greater cost) to give, typically, a further tenfold increase in stability

Fundamental physical law errors

A major class of error sources that also arise are those known as *fundamental physical law* errors. These arise as the result of physical processes – laws of nature – that take place internally in materials and devices. They are set apart from the errors explained above in that they are immutable effects of nature. They cannot be *designed out* but their magnitudes can be *reduced* by appropriate methods based upon our understanding of their physical laws. Those fundamental effects commonly met in instruments are now reviewed.

The resistive, dissipative, component of an electrical energy system has associated with it *Johnson noise* – also called *thermal* or *white* noise. (This has already been introduced in Example 3.8 as an example of a random process.) It arises as the result of the statistical variation in dissipated energy within material structure and hence noise energy is directly proportional to absolute temperature. The practical effect of these minute individual energy variations is a bulk change to the instantaneous signal level as given by Equation (3.14),

$$\overline{v^2} = 4kTRB$$

where $\overline{v^2}$ is the mean square voltage generated, k the Boltzmann constant, T the temperature (in kelvin) of the resistance R (in ohms) of the component and B the frequency bandwidth (in hertz) of the observing system.

Typical thermal noise voltages will be in the order of nanovolts to microvolts. Although this may seem too small to be of consequence electronic sensor practice commonly works with primary sensor signals in this region. It is the most frequenctly met instrument fundamental limit.

Wherever electric current charges flow discrete packets of energy move. When working with very small signal levels (single electrons are sometimes involved) in electronic sensors and amplifiers the output signal will have a discrete randomness rather than be a macro-average of discrete events. This effect is called *shot noise*. It was first experienced in vacuum tube devices; it is now commonly observed in semiconductor practice.

Shot noise, expressed as a current i_s, is characterized by the expression

$$\overline{i_s^2} = 2eIB \tag{4.32}$$

where e is the electron charge constant, I is the average current level flowing, and B is the frequency bandwidth of the observing stage.

A third important fundamental effect is called *flicker*, $1/f$ or *hyperbolic* noise. Despite much theoretical study its exact nature has not been quantified by a general theory.

Observation shows that as the frequency of the signal rises the flicker noise falls off as an inverse function and hence the name 1/f-noise. This is a logical expectation – consideration of how thermal influence variable effects become increasingly difficult to isolate (average out) with reducing thermal cyclic frequency lead to the expectation that many noise effects may have this characteristic because many are thermally based. Flicker noise has an infinite value at zero frequency. It is usually neglible, compared with other noise sources, from about 400 Hz and above. It is very troublesome when signal *periods* rise to longer than minutes duration, this is, frequencies below 10 mHz.

Figure 4.27 shows the nature, but not relative magnitude, of these various sources as the frequency rises from zero.

As well as the fundamental sources of noise there also exist several commonly encountered potentially controllable sources. *Pink* noise, also shown in Figure 4.27, is a defined noise type that is easy to generate for testing purposes. *Grey* noise is the term given to a noise spectrum that does not follow any of the above given forms. The *narrow band* noise arises from such effects as electromagnetic radiation from electrical mains supplies, fluorescent lights and combustion engine ignition systems, and is so called because the noise is largely concentrated at a single frequency.

The thermodynamic cause of electronic thermal noise also manifests itself in the mechanical regime of materials. Variations of size arise due to the thermal agitation of the molecules. This is called *Brownian motion*. It rarely poses a noise source problem in instruments as it is very small in magnitude (a block of brass

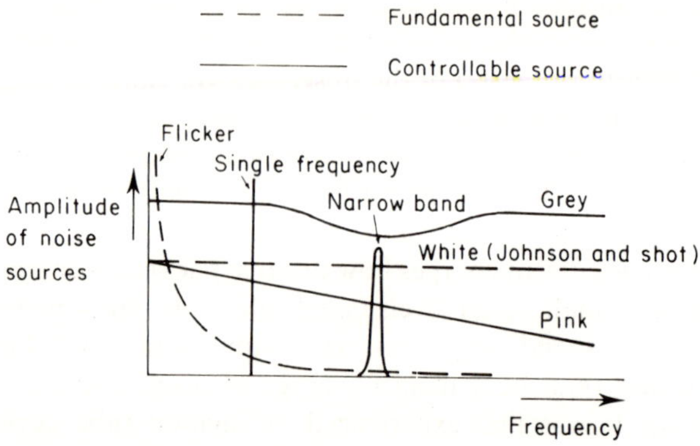

Figure 4.27 Frequency spectra of fundamental and other electronic noise sources. (HBMS, figure 22.14.)

at room temperature and measured with a 1 Hz bandwidth changes size by only, 1 part in 100 000 000.

Brownian motion error became important in the 1930s when the most sensitive electrical quantity detectors were electromagnetic, moving-coil, galvanometers. When set up with a constant input current they showed continuous movements that limited their resolution. The arrival of electronic amplifiers circumvented the problem because they could provide lower detection levels. Today Brownian motion error has to be allowed for in very sensitive instruments, such as seismometers, gravity wave-detection experiments and ultra-sensitive weighing balances used in chemical analysis.

A counterpart of electronic shot noise exists in optical systems as the discrete *photon* noise caused by the corpuscular nature of photons (energy packets in transmission). Photon noise and the ingress of *stray radiation* into the desired field of view are the two main fundamental noise sources to be dealt with in optical (radiation) regimes.

To extract the meaningful content of a signal the level of noise, always present to some degree, must be controlled. The *signal-to-noise ratio* (SNR) is a simplistic expression of the situation existing. The *noise equivalent 'measurand variable'* is another way of defining a sensor's detection capability. For example, a thermal detector with 'NET = 0.01 K' (noise equivalent temperature of 0.01 K) has a noise level equal to that of a resistor at temperature of one hundredth of a degree absolute to 0.01 K. As optical detectors will often have differing geometry, electrical impedance, field of view and wavelength sensitivity it is now common practice to provide a *normalized* detectivity statement that enables them all to be compared on a uniform basis.

4.6.3 Error propagation and sensitivity analysis

Error arises from many sources and collectively it produces a level of *uncertainty* in the measurement result made with an instrument. After the errors for each part of the system have been identified they need to be combined – the error of a whole system is not simply the sum of all errors. This section provides an introduction to how this process is carried out; further detail and a worked example are available in Hofmann (1982).

Propagation of systematic error

The output of the complete measuring system, the *measurement function* y, will usually be formed as a function of n other, indirect, measurements, an example being the measurement of the quality of a loaf, given in Example 2.5, where several measurands are combined to map a set of measurements into 'quality'. This can be expressed as

$$y = f(x_1, x_2, \ldots, x_n) \tag{4.33}$$

where x_i are the various indirect measurement values (not the errors) used to form the output.

The combined effect of *errors*, Δx_i, arising in each x_i, can be determined using the *linear error-propagation law*. It can be shown (see Hofmann, 1982) that, using Taylor series expansion and removal of all second and higher-order terms, the error in the combined output Δy is given by

$$\Delta y = \sum_{i=1}^{n} \partial y/\partial x_i \, \Delta x_i \tag{4.34}$$

Expressed in terms of *relative error* (that is, as $\Delta y/y$) this is

$$\Delta y/y = \sum_{i=1}^{n} \partial y/\partial x_i \, \Delta x_i/y \tag{4.35}$$

This function can be used to show, mathematically, how the errors of various combinations will propagate into the final value y. Table 4.2 gives relationships for combining error when the transfer function and system architecture are known.

When the form of measurement function y is known, the result of combination of the various x_i errors can be determined using Table 4.3.

As errors are usually small compared with their associated nominal value the formulation given in Table 4.3 can often be simplified using the various approximation expressions (for example, the cosine of small angles is close to unity). Hofmann (1982) gives a table of such approximations.

Propagation of random error

In this case if the measured quantities x_i are independent of each other, then the approach leading to Equation (4.33), given above, is also applicable. However, from the theory of the Gaussian distribution the *variance* of each component ($= \partial^2$) is added, each suitably weighted by the derivative of the function with

Table 4.2 Relevant measurement errors of typical measurement circuits

No.	Circuit	Structure	Transfer function	Relative measurement error*
1	series	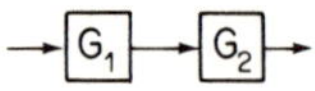	$G_1\,G_2$	$\frac{\Delta x_1}{x_1}+\frac{\Delta x_2}{x_2}$
2	parallel	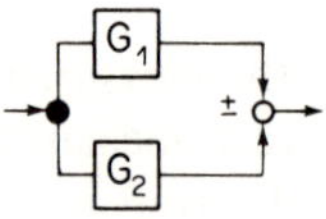	$G_1 \pm G_2$	$\frac{G_1}{G_1 \pm G_2}\frac{\Delta x_1}{x_1} \pm \frac{G_2}{G_1 \pm G_2}\frac{\Delta x_2}{x_2}$
3	feedback (positive feedback)	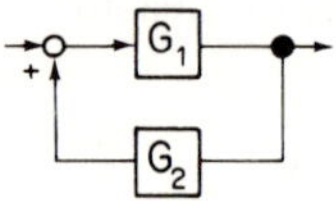	$\frac{G_1}{1-G_1G_2}$	$\frac{1}{1-G_1G_2}\frac{\Delta x_1}{x_1}+\frac{1}{(1/G_1G_2)-1)}\frac{\Delta x_2}{x_2}$
4	feedback (negative feedback)	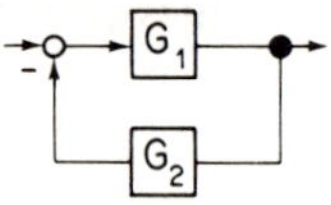	$\frac{G_1}{1+G_1G_2}$	$\frac{1}{1+G_1G_2}\frac{\Delta x_1}{x_1}-\frac{1}{(1/G_1G_2)+1}\frac{\Delta x_2}{x_2}$

* The measurement errors $\Delta x_i/x_i$ are the relative measurement errors of the blocks i before their interconnection.

respect to that variable as before. This is referred to as the *squared error propagation law* and is given by

$$\Delta y = \pm\left[\sum_{i=1}^{n}\left(\frac{\partial y}{\partial x_i}\Delta x_i\right)^2\right]^{1/2} \tag{4.36}$$

in which the ∂ quantities may be any consistent measure of error including standard deviation ∂, and standard deviation of the mean, ∂_m.

Sensitivity analysis

As measurement performance depends upon the errors of an instrument which, in turn, depend on how the various system components are affected by internal and external parameters it is worthwhile, in an instrument study, developing an appreciation of the *sensitivity* of these factors to the various influencing effects. Such specific studies are known as *sensitivity analysis.*

Sensitivity of each design parameter may be obtained by varying each (in turn and with all others held constant) observing the effect on the output. This can be carried out

- *practically*, using an actual instrument in a test situation, or
- *mathematically*, by using suitable mathematical models.

For example, the power supply voltage for an operational amplifier can be varied, by a few percent, to see what effect that has on the stage gain. In the

Table 4.3 Combination of measurement errors when the form of the measurement function is known. (HBMS, table 6.4.)

	FUNCTION	ABSOLUTE MEASUREMENT ERROR	RELATIVE MEASUREMENT ERROR
No.	$y = f(x_1, \ldots, x_n)$	Δy	$\Delta y/y$
1	$x_1 + x_2 + x_3 + \cdots$	$\Delta x_1 + \Delta x_2 + \Delta x_3 + \cdots$	$\frac{\Delta x_1 + \Delta x_2 + \Delta x_3 + \cdots}{x_1 + x_2 + x_3 + \cdots}$
2	$x_1 - x_2 - x_3 - \cdots$	$\Delta x_1 + \Delta x_2 + \Delta x_3 + \cdots$	$\frac{\Delta x_1 + \Delta x_2 + \Delta x_3 + \cdots}{x_1 + x_2 - x_3 - \cdots}$
3	$x_1 \cdot x_2 \cdot x_3 \cdots$	$\Delta x_1 x_2 x_3 + x_1 \Delta x_2 x_3 + x_1 x_2 \Delta x_3 + \cdots$	$\frac{\Delta x_1}{x_1} + \frac{\Delta x_2}{x_2} + \frac{\Delta x_3}{x_3} + \cdots$
4	$\frac{x_1 x_2 x_3 \cdots}{x_a x_b x_c}$	$\frac{x_2 x_3 \cdots}{x_a x_b x_c \cdots} \Delta x_1 + \frac{x_1 x_3 \cdots}{x_a x_b x_c \cdots} \Delta x_2 + \cdots$ $- \frac{x_1 x_2 x_3 \cdots}{x_a^2 x_b x_c \cdots} \Delta x_a + \frac{x_1 x_2 x_3 \cdots}{x_a x_b^2 x_c \cdots} \Delta x_b$	$\frac{\Delta x_1}{x_1} + \frac{\Delta x_2}{x_2} + \cdots - \frac{\Delta x_a}{x_a} - \frac{\Delta x_b}{x_b} \cdots$
5	ax	$a \Delta x$	$\frac{\Delta x}{x}$
6	x^n	$n x^{n-1} \Delta x$	$\frac{n}{x} \Delta x$
7	$\sin x$	$\cos x \Delta x$	$\cot x \Delta x$
8	$\cos x$	$-\sin x \Delta x$	$-\tan x \Delta x$
9	$\tan x$	$\frac{1}{\cos^2 x} \Delta x$	$\frac{2}{\sin(2x)} \Delta x$
10	$\cot x$	$-\frac{1}{\sin^2 x} \Delta x$	$-\frac{2}{\sin(2x)} \Delta x$

theoretical approach the mathematical expression relating stage gain to power-supply voltage may be investigated by differentiating it with respect to each parameter of interest.

In either case the resultant family of curves (plotted from the results; Figure 4.28 is an example, being one of a family for different wire diameter) gives a clear understanding of how the various parameters impinge on the design decisions and effectiveness of the system. An example of the mathematical procedure is given in Example 5.3, where modelling is taken up in more depth.

A combination of the experimental and mathematical methods is recommended as this is usually the most economic path to a successful instrument. Sensitivity analysis should always be carried out on a design but often the relatively low level of resource devoted to a particular instrument design prevents in-depth studies in many cases. Applicable techniques may be adopted from other well-established engineering areas, notably electronic and manufactured mechanical systems.

4.6.4 Error-reduction strategies

Control of error magnitudes is an essential part of the design of a measuring instrument. Several approaches can be used.

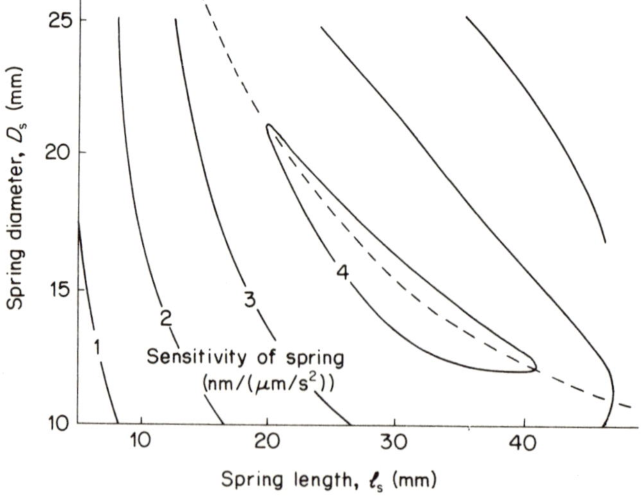

Figure 4.28 Family of curves showing how spring sensitivity varies with spring-wire and wound coil diameter. (From Hugill, 1984.)

Design for minimum error

The safest, but not necessarily the least cost, approach to error reduction is to ensure that each sub-unit of the system has its hardware performance designed with an error level well below that required of it. The usual 'rule of thumb' is to make such errors around one tenth of the final error of the system.

This simplistic methodology, however, can be misleading because of the way that error magnitudes propagate through several stages – it may not actually lead to the optimum result. Furthermore, the strategy can impose unnecessarily high cost penalties as hardware is made more complicated and formed of more expensive modules and materials. In many cases the available technology for implementing hardware may not be available to provide a solution.

Given that modelling and testing could derive a satisfactory *minimized error system* of hardware there is still some doubt as to whether this is the optimum design approach.

An alternative method for controlling the magnitude of error is to *sense* the magnitude of each (or a group of) errors by supplementary means using these extra signals to compensate for error through *hardware performance modification*. This methodology is illustrated in Example 4.14.

Example 4.14 Compensation of errors in laser-based length measurement

Lengths from 0 to 10 m can be measured to 0.1 μm discrimination with the *laser interferometer* instrument, the component layout of the simpler form being as shown in Figure 4.29. Error arises from several sources when using this arrangement.

Forming the interfered beams would be very difficult to set up and maintain in operation if *plane* mirrors were used, as by Michelson. This is easily overcome by the use of a more complicated *corner cube* reflector that will return a beam parallel to the incoming beam to close limits, regardless of wide angular variation of angle of the reflector to the beam axis.

The measuring accuracy of the system depends on the *optical path lengths* of each arm of system. These 'lengths' are partially determined by the optical refractive index of the medium through which the radiation passes. This, in turn, is decided by the pressure, relative humidity and temperature of the normally used ambient air paths. To reduce the system measurement error to less than the around 1 part per million possible in the normal ambient environment it is required that some means be devised to reduce the magnitude of error arising from optical path length changes.

One possibility for overcoming these influence variable errors is to create a special environment where the optical length remains constant, that is, make use of evacuated tubes in which the optical path length variations are controlled to allowable limits. Such additions are expensive to build, for *high-vacuum* enclosures are needed. These are expensive to operate because a vacuum pump (300 W at least) must run continuously. It can also take several hours to 'pump the system down' ready for use — and then access is no longer possible. Such measures, however, can control errors to a degree that the interferometer can be made to operate with an error of less than 1 part in 100 000 000.

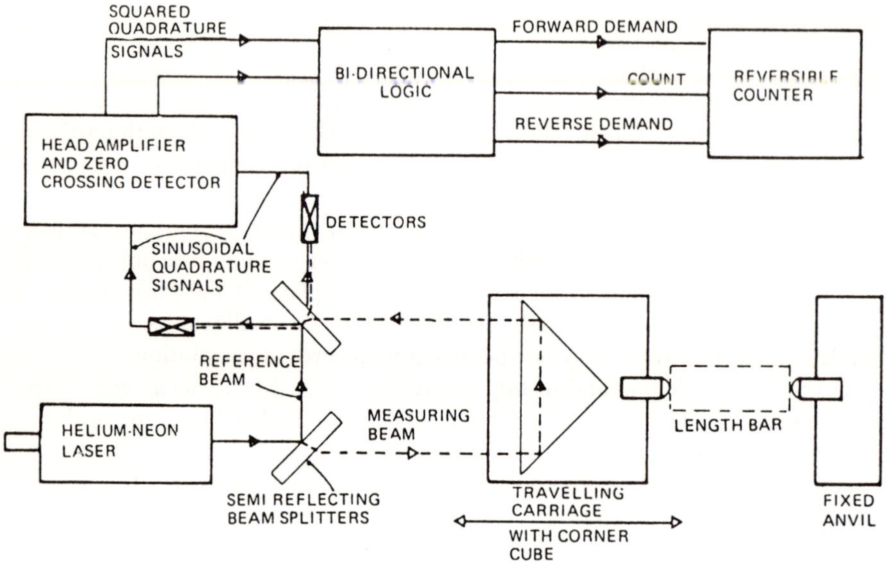

Figure 4.29 Layout of a simple form of laser interferometer. (From Sydenham P H, 1984, *Transducers in Measurement and Control* (3rd edn). Adam Hilger, Bristol. Reproduced by permission of IOP Publishing Ltd.)

The alternative methodology, that used in industrial practice where evacuated enclosures are not appropriate, is to measure the ambient influence variable levels and apply electronic corrections to the output signals. Top-of-the-line systems use several pressure, temperature and relative humidity sensors that are placed alongside the paths of the beams. This *active compensation* can be done effectively because the influence variable effects are systematic and well understood. Combination of the several influence variable sensor signals, however, into a single correction factor at any instant of time, makes use of statistical methods because individual values at each point in space vary randomly.

Another source of error is size change caused in the structure by ambient temperature variations. Interferometers measure with a discrimination of fractional micrometres by comparing the length of the measuring arm with that of the reference arm. It is obvious that the length of the reference arm (in Figure 4.29 this is the small distance - typically 50 mm - needed between the two flat, semi-reflecting beam-splitting mirrors) must be held constant to within the required level of dimensional stability.

Steel structural metals change length at 12 μm per metre per °C. A steel reference arm of 50 mm length, therefore, alters length for a 20 °C ambient change by a (relatively massive) 12 μm - this is 1200 times the discrimination and stability sought!

This error source can be reduced by several means. A material with a lower thermal expansion coefficient, such as quartz (1 μm per metre per °C or better) can be used to construct the mount that holds the mirrors apart. The mount could also be temperature controlled to about 0.1 °C. Another way is to design a *re-entrant* mechanical mounting in which expansion outwards of the mounting arm is compensated by expansion of another arm that moves the mirror in the opposite direction (as shown later in Figure 6.11); reduction by a factor of ten is possible.

If all of the above methods were combined the overall result would (potentially) be able to reduce this error source to 1/20 000 of the original magnitude.

It has long been recognized that where a measuring instrument's ability to repeat measurement values is precise over time (that is, its *reproducibility* is known) then its accuracy can be increased by the process of traceable calibration. (These terms were introduced in Section 2.5.) In this way a data set is generated that can be used to slightly alter, and correct, the actual measured values as they occur. Such devices as calibration graphs, tables and mathematically 'fitted' equations are commonly used by manual means to reduce the error of a measurement. This process is appropriately termed *post-measurement correction*.

In recent years a common approach to post-measurement correction is to make use of a computer to store the magnitudes of the various errors, correcting (in real time) the output values by the appropriate amount. Some so-called 'smart' or 'intelligent' instruments go as far as to incorporate *automatic calibration* to develop and maintain the up-to-date correction information.

Post-measurement correction usually requires the system design to be reasonably close to the desired error level for several reasons. (Only rarely can it be used to bring about massive corrections.) Practical operational requirements dictate this because gross errors often cause subsequent stages of the instrument chain to go into non-linear operation. For example the *input offset* of an integrated circuit operational amplifier of, say, 100 mV, will cause an output offset, with a stage gain of 200, of 20 V, and this would drive the following stage into saturation because the operating power supply is usually 15 V. Selection of adequately small error components, coupled with system design considerations, can lead to a satisfactory system in which small errors are best reduced by post-error compensation.

In practice various combinations of the above error-reduction alternatives are used to obtain the desired level of accuracy and uncertainty.

Noise energy reduction

The ideal sensing interface extracts just that information needed to gain the knowledge sought. Practical reasons, however, often require a sensor to extract more information than this ideal and also to cope with high levels of noise intrusion into the sensor and the observed area itself. It is, therefore, often necessary to process the sensor signal to further *filter the information content*– this is highly relevant to system error reduction.

The discipline of signal processing and signal recovery, when it is buried in noise, is extensive and cannot be detailed in a first course such as this. An outline of the main methods for improving the signal-to-noise ratio (SNR) is now given to assist appreciation of what is possible and how it is carried out.

The SNR ratio can be improved by reducing the noise-energy level or increasing the signal magnitude. Given knowledge of the nature of the noise it can be reduced by several different strategies.

The thermal noise case study of the mathematical model, given in Equation (3.14), shows that reduction of the bandwidth B of observation, resistance R of the sensing component, and component temperature T will all reduce the noise energy. Sensitive sensors commonly make use of each of these options. For example, the thermal detector used in airborne *remote sensing scanners* will usually be cooled with liquid nitrogen, be constructed to have low electrical output impedance and be used with as narrow a bandwidth as is allowable.

Shot noise, (defined in Equation (4.32)) is reduced by reduction in bandwidth and operating current level. For this reason electronic sensors make use of low signal levels.

Flicker noise is reduced by devising a system that operates with a measurand frequency that is adequately high – above 400 Hz is normal. If, in its original form, the measurand response is less than this (for example, slowly changing daily temperature) then *chopping* is commonly added – see *phase-sensitive detection* below – to take the low-frequency signal into a higher region.

Noise pick-up from the mains and radio sources is reduced by improved earthing arrangements and the use of effective shielding.

Increasing the energy of the signal is not always possible because amplification of a signal also amplifies the noise content and it always adds in other noise components in the process.

Bandwidth reduction

Consider a signal with a noise component present. The combined, pre-processed, frequency spectrum often comprises a comparatively narrow band of wanted signal placed within a wide band of white noise.

Limiting the bandwidth and centring the response over the signal band greatly reduces the noise energy without changing the signal energy – thus the SNR is improved. Many methods make use of this concept to extract signals.

The obvious method for doing this is to use appropriate electronic filters that simply cut the bandwidth down to just wider than the signal. When the bandwidth is small this becomes difficult because the filter centre frequency drifts and so does that of the signal.

A more sophisticated approach is *phase-sensitive detection* – also called *synchronous detection* and *lock-in amplifier*. The system architecture and mathematical model have been explained in Example 3.15.

The observed measurand is deliberately modulated with a reference signal of frequency about ten times higher than the highest signal frequency of interest. The output of the now modulated measurand is multiplied by the reference signal, possibly with some phase adjustment, and the resulting signal integrated with a long-time-constant integrator. The application of this method to interrogate electrical reactance type sensors (an inductive unit in this case) is given in Example 4.15.

Example 4.15 Phase-sensitive detection applied to the LVDT displacement transducer

The linear variable differential transformer (LVDT) is used to sense displacement. As shown in Figure 4.30, a central coil is driven with a.c. excitation so as to magnetically induce signals into two other coils. Excitation by a.c. is essential for reactive sensors for there the rate of change of energy is too slow for satisfactory operation. An armature, connected to the motion to be sensed, moves inside the coils changing the inductive coupling to each outer coil.

When the armature is in the central position each outer coil has induced in it the same a.c. magnitudes displaced by 180°. If moves to one direction, one coil output rises and the other falls. If the two outputs are connected into a bridge arrangement the output will then be an a.c. signal that varies in both amplitude and phase as armature position changes.

Application of the theory given in Section 3.5.4, coupled with system operational considerations, shows that in this case of phase sensitive detection the system needs to operate as a detector of amplitude changes in the voltages induced in the differentially connected coils. This requires an adequately constant oscillator voltage and constant phase shift between the modulating and modulated signals – which means that the inductance of the sensing-coil assembly must remain largely constant.

Simple rectification of the a.c. output will yield displacement magnitude information as a pseudo-d.c. signal, but cannot indicate direction. It will also rectify noise energy induced into the leads and in the coil system adding this to the signal as output and increasing uncertainty in the displacement measured.

Use of phase-sensitive detection not only overcomes the above deficiences but also adds extra performance, providing excellent SNR outputs for comparatively little extra complexity or expense. It reduces noise through highly effective bandwidth

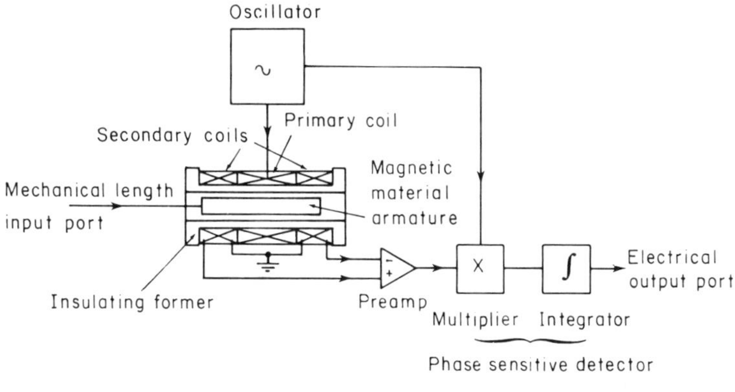

Figure 4.30 Block diagram of LVDT displacement sensor using phase-sensitive detection. (HBMS, figure 24.5(a).)

reduction; it operates the measurand above noticeable flicker noise frequencies; and it tracks the narrow band of detection to follow any drift in the excitation frequency. It is so effective at reducing stray noise pick-up that it allows twisted wire, unshielded, connection leads to be used in many applications.

Phase-sensitive detection methods are very much better than use of a narrow-band filter alone – in simple forms these can provide about 40 dB SNR improvement. Addition of extra auxiliary filters and use of the best performance blocks makes SNR improvements of 120 dB possible.

Bandwidth reduction methods only work as long as there is more signal than noise energy in the signal bandwidth. If the noise energy is greater than the signal energy, then other methods can be used, but with restrictions.

Box-car integration – an averaging method

If the noise energy exceeds the signal energy in the signal bandwidth, then an averaging process can be used *provided several samples of the signal and noise are available.*

The *box-car integrator* (so-called because oscilloscope waveforms look like a set of railway box-cars) averages the values of several events. The architecture and methodology of this method have been explained in Example 3.4. Each point of the signal wave form, being the same each time, builds up in amplitude as the successive samples are added. The noise level at that point, if it is random in nature and uncorrelated from sample to sample, averages toward zero as the number of samples increases. The signal is subsequently recovered point by point by reducing the noise energy and increasing the signal energy. The price paid for this increase in sensitivity is, of course, reduced response time of the instrument. The measurement is not available to the user (or next block in the information machine architecture) until enough samples have been totalled to give the required improvement to the SNR.

This limitation is often referred to as a 'sensitivity/response time trade-off'. In principle sensitivity (or system resolution) can be increased indefinitely as integration times are lengthened. In practice there is a limit to the process set by the reducing effectiveness of averaging the noise component.

Correlation

As we have already considered formally in Chapter 3, the above methods are special cases of the general class of *correlation techniques.* The basic concept here is that two similar patterns will yield the largest *correlation coefficient* when they are identical in form and position. Consider, for example, looking through two identical line grids drawn on two transparent films. Whey they are aligned and overlapping the transmission through the correlated pair is the greatest. Correlation has been explained in Section 3.2.3, and Examples 3.5 and 3.6 explain how it is applied to the recovery of signals buried in noise.

TUTORIAL EXAMPLES

4.1 For a complex instrument and circuitry, of your choice, identify the building blocks and connections. Prepare a table of building blocks, list their functions and identify the types of connections existing.

4.2 Using data sheets for the 741 operational amplifier, plot the gain-frequency response for closed-loop gains of 1, 10 and 100. In each case determine the slew rate. Input and output impedances are quoted; study how these vary with gain and frequency. What is the drift in terms of temperature and time? Are those dependent on any circuit factors, and if so how are they related?

4.3 Identify a first-order sensor system that is available for inspection. Draw the organic system layout, labelling the various building blocks. By identifying the storage and dissipative elements develop a quantitative first-order model. Plot the estimated output responses for a step input change. Plot the estimated frequency response over the range of frequencies that are likely to arise in the input signal.

(The exercise can also be carried out in a qualitative sense by making estimates of the time constant and gain where measured coefficients cannot be obtained. Where measurements can be made the fully developed mathematical model is to be preferred.)

4.4 Making use of the 'application notes' provided by a manufacturer of a high-speed galvanometric recorder, calculate and specify the galvanometer and amplifier circuit resistances needed to record a complex signal having a highest frequency of 8.5 kHz and amplitude of 5.5 V. The accuracy requirement is to record the signal to within ± 0.1 V. Comment on the transient accuracy and slew rate.

4.5 A length-measuring machine of high industrial accuracy (0.3 μm in 300 mm uncertainty) is being purchased to measure the diameter of nominally 200 mm diameter aluminium pistons. A glass measuring scale is used on the cast-iron machine.

(a) Specify, with quantitative reasoning, the ambient temperature conditions needed for correct operation of the machine and how long it would be expected to take to stabilize after being placed in such an environment after delivery from the factory.

(b) Suggest the measurement procedure needed to ensure that ambient temperature effects are all uniform and minimal before beginning to make valid measurements.

(c) Estimate how long it would take for the piston to come to the same temperature as the measuring scale after it is taken from the grinding machine used to finish the diameter. How would you know when it had stabilized?

4.6 Observations are to be made of the heat radiated from the Moon (first performed by Callendar around AD 1900). Suggest, with justification and systems diagrams, a suitable detection system including the form of electronic signal processing.

REFERENCES

Chatfield C, 1970, *Statistics for Technology*, Penguin Books, Harmondsworth.

Finkelstein L, 1982, Theory and philosophy of measurement. In: Sydenham P H (Ed), 1982, *Handbook of Measurement Science*, Vol 1, Wiley, Chichester.

Hofmann D, 1982, Measurement error, probability and information theory. In: Sydenham, P H (Ed), 1982, *Handbook of Measurement Science*, Vol 1, Wiley, Chichester.

Hugill A, 1984, A gravimeter with automatic readout. PhD Thesis, Flinders University of South Australia.

Millman J and Grabel A, 1987, *Microelectronics*, McGraw-Hill, New York

Oppenheim A V, Willsky A S and Young I T, 1983, *Signals and Systems*, Prentice-Hall, Englewood Cliffs, NJ.

CHAPTER 5

Modelling the measurement system

5.1 INTRODUCTORY REMARKS

5.1.1 Importance of modelling

Throughout the previous chapters mention has frequently been made to the use of models of systems and of parts of systems. These models allow consideration of some aspect of the system *before it is built*, thus saving money and time in establishing expected performance and in setting up an optimal system. An example of this is the use of a computer-aided-design draughting tool that allows a mechanical system to be *simulated* as pictures on the computer screen. Another well-known example of a system model is an aircraft flight simulator used to train pilots without their actually taking to the air.

Models also enable existing systems to be studied in modes of operation in which they may be called to provide, yet do not arise in normal practice. Here an example is the model incorporated into nuclear power station control systems that is used to explore likely performance of the reactor.

Models also find application as an integral part of a control system. Mo automatic sheep-shearing plants being developed incorporate a size and sha model of the *average* sheep. This model enables the clippers to be directed mo rapidly to the *expected* position where wool should be removed: close-ran sensors then take control to set the final positioning.

The use of models to assist design and performance of systems is exten *Models are a means to motor ingenuity.* Their application in measure systems is equally as beneficial as it is in other disciplines.

It has been reported (Abdullah, 1985):

> Until recent times sensor and transducer design had been largely a 'black art' often rel on a lifetime's experience of a single designer with a company. Rules of thumb and c experimental 'cut and try' methods were generally employed. According to one com

> the design of a (pressure sensor) capsule for a new instrument could easily take on year of effort and involve considerable tooling and testing costs . . . Using CAD met . . . the design of nesting capsules now takes a matter of weeks. One manufacturer has b able to develop a whole range of pressure sensors . . . with smaller diameter capsu . . . without sacrificing accuracy of performance.

This chapter provides an introduction to the way in which measureme systems, in particular, can be modelled. It also aims to develop appreciation o when and where models can, and should be used.

5.1.2 Types of model

Several types of model exist. Each has its own relevance and characteristics. It is important to recognize immediately that none is an exact and complete representation – only the final real system is that. All models provide appreciation of an *aspect* of the whole. Whilst models *assist* understanding they cannot replace the need to make and test the final system.

Linguistic models

The first level of abstraction of a topic is its description in the spoken language. A set of words allows definition of organic components of a system and records and describes their qualities and interactive behaviour. This is called a *linguistic* representation.

Major shortcomings of this model form are the possible inexactitude of the description due to *semantics* of language, comparatively slow speed of communication of technical concepts and ideas and space-consuming nature of the record. They are generally descriptive, lacking ability to communicate quantitative knowledge.

Notwithstanding these disadvantages, linguistic representations are always essential, complementary, components of a complete modelling operation and they are always needed to begin such an exercise.

An example is a *technical specification* in which words describe the required performance of an instrument. A section of such a document is given in Figure 5.1.

Iconic models

An *icon* (ikon) is an *image, figure or other pictorial representation of a concept.* Engineering and science have always made extensive use of icons to capture and expose ideas. Figure 5.2 shows an assembly of some different types of icons used in instrument engineering.

Some icons, such as block diagrams, model organic and conceptual characteristics of systems. Others, graphs being examples, show quantitative relationships.

3.2.4 Maintainability

The thermometer does not need to be designed to be repaired. In the event of malfunction, the instrument is to be replaced with a new one.

3.2.5 Environmental Conditions

The environment the instrument is to withstand is divided into operating and non-operating environments. The thermometer shall meet the performance characteristics specified in section 3.2.1 when operated within any of the combinations of environmental conditions defined in the operating environment. It is not necessary for the thermometer to be able to meet its specified performance whilst being subjected to non-operating environment extremes that exceed those for the operating environment. The instrument shall survive the storage environment for 1 year without deterioration.

3.2.5.1 Operating Environment

3.2.5.1.1 Operating Temperature Range

Figure 5.1 Portion of specification of sensor - example of linguistic representation

Icons can convey considerable information in a meaningful manner and are less subject to misinterpretation than linguistic descriptions. They rapidly motivate the user.

Manufactured scale models

It is often appropriate to construct physical models of the process in order to expose characteristics. The most obvious form is a full-size (scales are also used) model just like the final item - this is generally called the *prototype.* It is constructed and used in testing.

In many situations, however, it is more cost effective to make models of only aspects of a task, not of the whole. When a complex and expensive instrument is being developed wooden and plastic mock-ups are usually made, examples being scale models of giant astronomical telescopes and of process plants. These static models are not always operational but do show relationships between the subsystem parts. They assist development of the organic structure - and help explanation of operation.

Many physical scale models are made to be operational in some manner. Theoretical understanding of the physical behaviour of systems, as they are scaled, shows that meaningful observations of experiments will only be obtained if features of the system are scaled according to their physical relationships. To obtain similar working relationships in the model, as would arise in the full-size system, often requires quite different physical parameters for the fluids, materials and relative dimensions used in the model. Figure 5.3 shows full-size models built to test dimensional stability of mounts in rock over long periods of time. Only the important parameters of interest are made in their correct and final form.

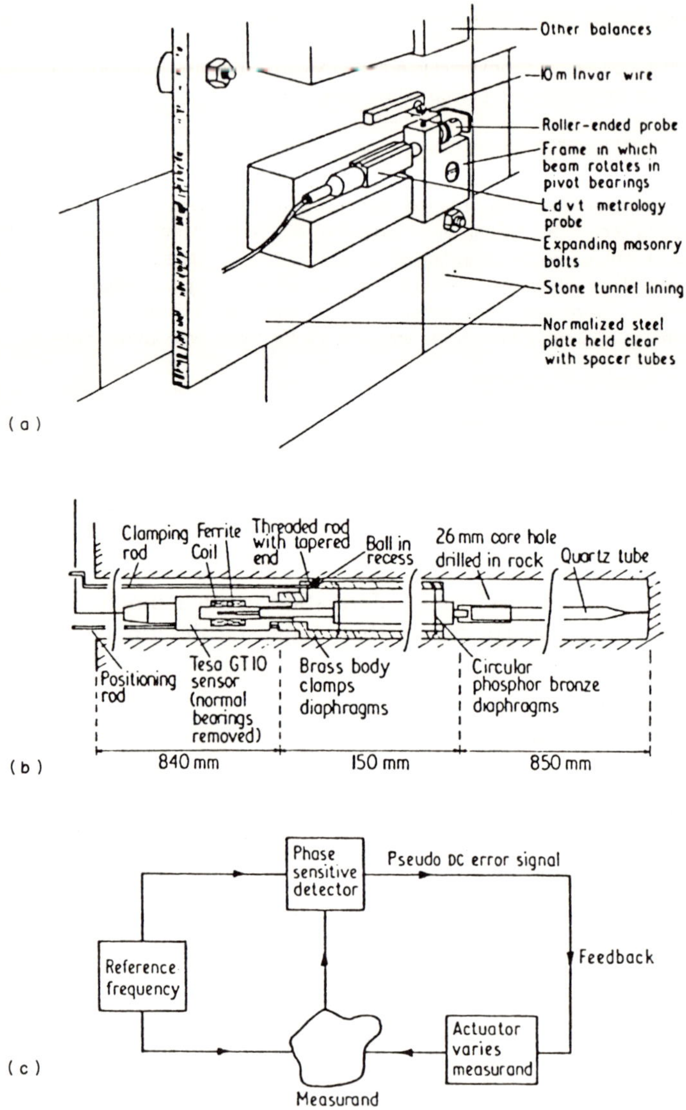

Figure 5.2 Iconography used in instrument engineering: (a) isometric schematic; (b) cross-section schematic; (c) block diagram with word description (but may be mathematical)

Mathematical models

Linguistic description soon acquires icons to make it more exacting. Icons show how organic building blocks are related and how entities, such as signals, flow. Physical models provide further organic relationships.

As the understanding of a topic develops it will become possible to apply mathematical relationships to the changing entities allowing the modelling

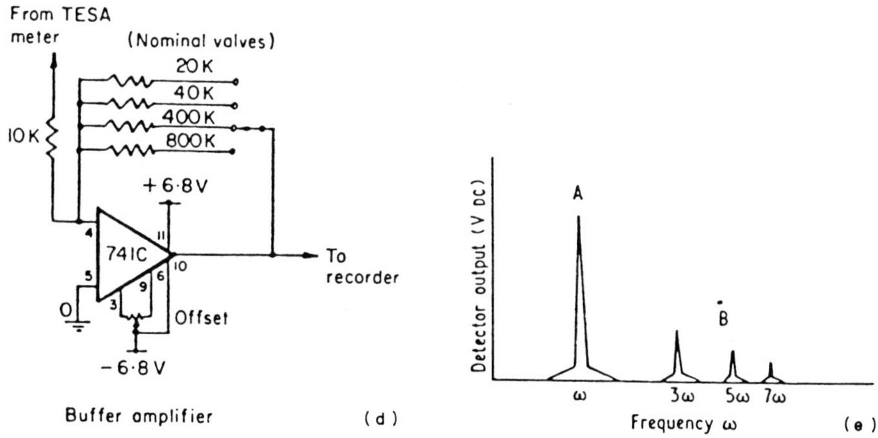

Figure 5.2 (*continued*) (d) electronic circuit diagram; (e) representational graph

process to better simulate the final system by providing numerical, quantitative understanding.

Working in the mathematical domain it is possible to run many different simulations of a system on common machines (computers are used). These models are relatively easy to set up. They can be run faster, or slower, than real time and they can sometimes predict results not previously known or expected.

Understanding of a topic reaches a high level of maturity when it can be *adequately* modelled with mathematical theory. In the development of an instrument system, models assist optimization of a design to extract the best from it. This chapter is mainly concerned with developing the mathematical kind of models for measurement systems.

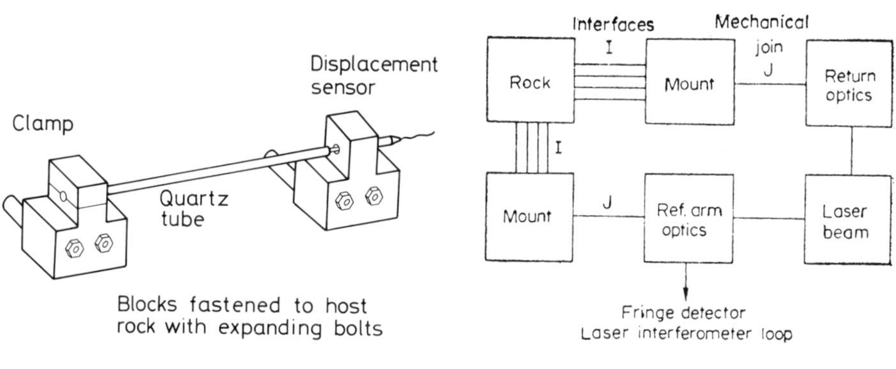

Figure 5.3 Physical model made to investigate long-term stability of the mounting arrangement for rock strain meters. (From 1973 *Geophys. J. R. Astron. Soc.*, **33**, 185–93)

Example 5.1 shows how each of the various models is employed in understanding a simple, commonly used, length sensor based upon the voltage potentiometer.

Example 5.1 Modelling the slidewire potentiometer length sensor

A straight, uniform, length of metal wire, having large electrical resistance per unit length, has a voltage applied across its ends. An electrically connecting contact is arranged to move along the wire. At any position of the slider its distance from a reference end can be sensed by measuring the voltage arising between a reference end and the contact.

The above *linguistic* description explains how the system is formed and how it operates. It is a cumbersome form of model in that it needs more words to better describe its shape, size and possible arrangement. A suitably trained person could, however, now make one and get it to operate. The principle has been modelled, as has the general organic arrangement, but the model lacks definitive information.

Icons are obviously needed. A drawing of the detailed manufactured device will eventually be needed but it is inappropriate to make one at this stage for we are not yet sure of the component sizes and materials needed. We also have no idea of the quantitative performance. Too many instruments are made before they are adequately investigated as models - with the result that they often do not work as hoped and need modification to obtain the required performance.

A more appropriate icon for this stage is a *schematic diagram*, as shown in Figure 5.4(a).

This simple diagram largely replaces the linguistic model. It now allows development of a mathematical model of its behaviour based on application of Ohm's law of d.c. electrical circuits (another mathematical model). The voltage appearing at the output terminals is given by the modelling equation given in Figure 5.4(b) - as are other system factors.

To illustrate the behaviour of the device, *without building it first*, a graph icon can be drawn as shown in Figure 5.4(c). It can be seen that the output V_{out} is proportional to distance d and that its magnitude depends upon a constant factor V_{in}/l set by system geometry and source magnitude.

These factors then allow the wire size to be chosen using power-dissipation laws (more mathematical models) and definition of the sensitivity needed.

Once each of these parameters, plus the actual final style, has been set it is then appropriate to draw an engineering drawing (another icon) allowing manufacture of a satisfactory final system such as that shown in Figure 5.4(d).

The impression might have been given that modelling allows complete understanding of a system. Remembering that modelling covers only *aspects* of systems and that topics contain an *infinite amount of latent information* leads to the realization that modelling can only proceed into a topic according to the resources that can be allocated. Modelling is a never-ending process that is halted when adequate information has been obtained.

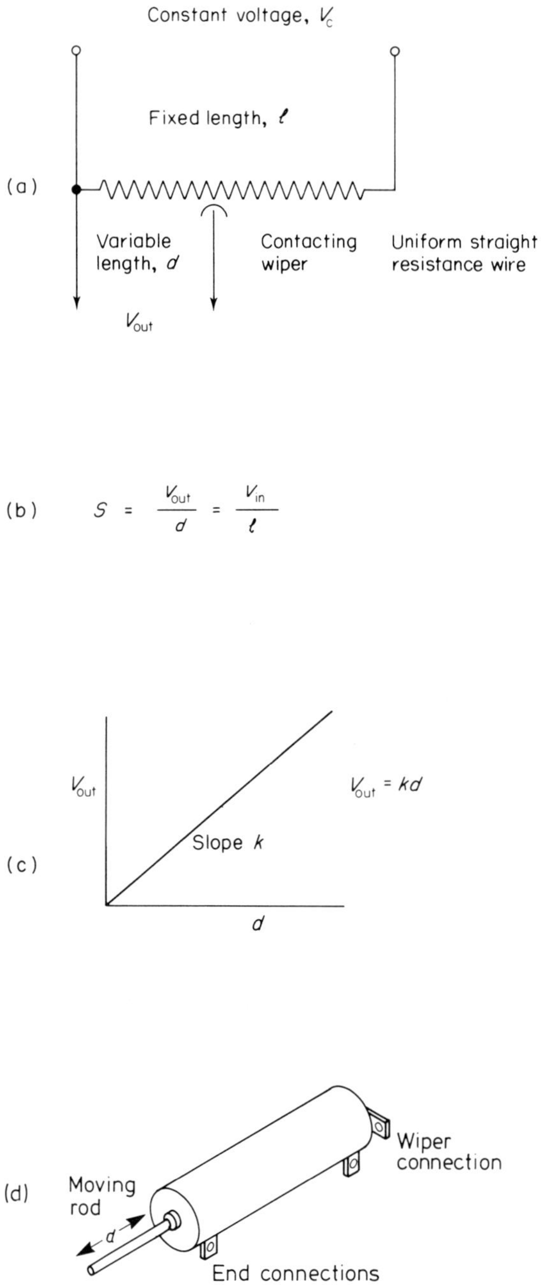

Figure 5.4 Models involved in understanding a slide wire length sensor: (a) schematic diagram; (b) mathematical model relating input and output signal; (c) graph of input/output relationship; (d) drawing of proprietary device

Models also need real-world data (see Section 1.3.1) to allow them to be run with reasonable bounds. A balance is needed between modelling in the abstract and real physical forms. No matter how much modelling has been undertaken there is always a finite chance that some important aspect has not been covered – the sobering fact scientists and engineers must always face when designing systems. Highly reliable and effective engineering systems, such as aircraft, television sets and wrist watches, are only such through the considerable expenditure of resource and extensive experience built up over the years.

In sharp contrast many measurement systems have not had such levels of development because they have not been developed in the same resource-rich circumstances. Modelling, especially the use of mathematical models, is an effective way to gain more system effectiveness with limited resources.

5.2 DEVELOPMENT OF A MEASUREMENT SYSTEM BLOCK DIAGRAM

5.2.1 Classification of system elements

From casual observation measurement systems will appear to be formed from combination of innumerable different building blocks combined in equally many different ways, thereby creating the impression that no underlying systematic development is possible. Most books on instrumentation provide *catalogue* style information in which are given explanations of how to measure this or that variable using the currently available technology. Commencing development of a model of an instrument is not assisted greatly by that *experiential* methodology for it is not immediately obvious where to begin model building.

Researchers of measurement system design methodology have presented several, somewhat similar, approaches to developing the architecture and block diagram of an instrument system. That given here is based largely on that published by Middelhoek and Noorlag (1981, 1982).

To begin a model building study it is necessary to break down the task, called *reticulation*, into subtask areas. As has been explained in previous chapters a measuring instrument is an *information* machine formed as a serial chain of building blocks.

These blocks carry out the signal-handling operations of sensing, storage, processing, transmission and display by manipulation of the modulations existing on an energy carrier that conveys the measurement signal information. A key classification needed is, therefore, one based on *energy forms* of the stages of the signal path. This is highly applicable to mathematical model building because the physical relationships of energy systems are well known, allowing instrument components to be modelled using existing mathematical relationships.

Figure 5.5 Functional block diagram of measurement signal path through an instrument system. (From Middlehoek and Noorlag, 1981)

The complete measuring instrument system can be reticulated into a *set of serially connected blocks*, each chain conveying one measurand's information as portrayed by the energy/information-sensing interface model given in Section 4.5.2. Carried on an energy carrier, the signal passes through a number of blocks in order to be processed as needed to form the measurement *mapping*.

Two prime kinds of energy operation can be identified; it can be *transduced* or it can be *modified* as shown diagrammatically in Figure 5.5.

An instrument *transducer* is a device that transforms an input signal of one energy form into an output of another energy form, doing so in such a way that a prescribed relationship is maintained between the input and the output signal.

An example is the mercury-in-glass thermometer in which heat energy is converted into mechanical energy, forcing the mercury to move along the glass as it expands or contracts.

The majority of, but certainly not all, transducers used today in testing, automation and control will have an electrical energy form of output.

In contrast the other kind of block is the instrument *modifier* which does not convert the energy form but does alter, or *modify*, the same form. An electronic filter, for instance, processes an electrical energy signal (more commonly called an electronic signal) by changing the frequency content distribution of the signal. It still provides an electrical energy output. Other examples of modifiers are the electrical transformer, mechanical lever and gearbox, hydraulic ram and heat exchanger.

Classification on an energy basis requires grouping of the various existing energy forms. Nine groups can be identified as an adequate (but not exhaustive) or exclusive) starting basis, these being:

- electromagnetic radiant energy (radio waves, through to UV);
- gravitational energy (attraction between masses);
- mechanical energy (forces, displacements);
- thermal energy (kinetic energy of molecules);
- electrical energy (electric and magnetic fields, currents);
- molecular energy (bond energy of atoms in molecules);
- atomic energy (binding energy between nuclei and electrons);
- nuclear energy (binding energy within nuclei);
- mass energy (mass in motion has equivalent energy).

From these, six groups can be identified as being generally relevant to instrument systems, namely:

- radiant energy;
- mechanical energy;
- thermal energy;
- electrical energy;
- magnetic energy;
- chemical energy.

(A simpler set - electrical, mechanical, thermal and chemical - is used in Chapter 6 because of space limitations. The experienced student of physics will appreciate that we do not know enough about the basic nature of energy to be able to give a clean classification. For example, thermal energy is the result of the mechanical motion of internal particles and thermal radiation is covered by mathematical models that are basically the electromagnetic radation laws. For convenience we are forced to make use of somewhat arbitrary energy groupings that derive more from how the branches of physics have developed than from truly arguable energy considerations.)

With this background we are now in a position to extend the system architecture given in Figure 5.5 to include the energy classification. The result is the more general system shown in Figure 5.6.

5.2.2 Matrix representation of measurement systems

The signal path of a system having more stages would move among a greater number of transducer and modifier columns than is shown in Figure 5.6. In

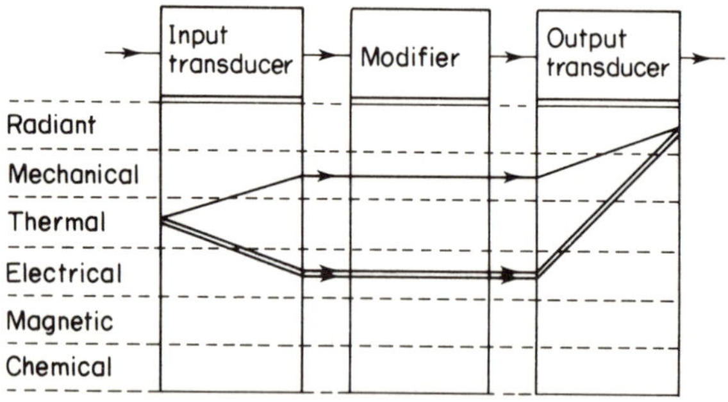

Figure 5.6 Generalized but simplified energy regime of an instrument. (The single line is the flow path for a mercury-in-glass thermometer: the double line is for a modern electronic thermometer with digital display.) (From Middlehoek and Noorlag, 1981)

principle it is possible to construct a diagram, of the kind of Figure 5.6, for complex systems that exist or to use the methodology to synthesize possible solutions to needs. Extra transducer or modifier columns are added, being placed in the appropriate serial position so as to obtain the required input-to-output conversion and serially accessible points as required. (This concept has the potential of providing the basis of future automated instrument design tools because it enables a generalized model of an instrument to be placed into a computer – but that is work still to be done in instrument design development.)

This iconic model of the (generalized) instrument system shows why numerous quite different solutions can be devised for the given measurement requirements.

To illustrate this synthesis procedure the architecture of a reasonably complex instrument is now developed in Example 5.2.

Example 5.2 Development of a relative humidity sensor

The measurement requirement is for an instrument to measure the relative humidity of air in a library in which valuable historical documents are stored. An instrument system architecture is needed that provides an electronic recording of control signal output and also a mechanically driven pointer to provide direct reading if the power to the sensor fails.

It is necessary to first transduce the chemical energy of relative humidity into a mechanical energy form to drive the pointer. A well-known transducer for this is a suitably stretched membrane in which moisture content change in its surrounding air causes sizeable, longitudinal, length change and force. This length change can then drive a pointer and also be converted into an electronic signal. A pointer system gives more visual acuity than a small displacement so a modifier is needed that changes the small mechanical displacement (about 0.2 mm at full scale) into a rotary equivalent. A lever driving a rack and pinion is chosen.

The small displacement of the end of the membrane can conveniently be transduced into an electrical signal by using the length change to bend a spring beam (called a *load cell*) on which are cemented electrical resistance strain gauges. Thus we need a spring element to form strain from the membrane's end motion – a modifier – which is followed by the strain-gauge transducer stage.

Change of electrical resistance of the strain gauge must then need modification into a voltage change, for which the Wheatstone bridge is usually employed – a modifier stage. The output voltage from the bridge needs to be adjusted to give a calibrated value; this necessitates another modifier amplifier in the serial chain. The electrical output path is then as needed.

The resultant system is shown, in matrix form, in Figure 5.7. A proprietary instrument using this architecture is that shown in Figure 6.1.

It is easy to see that many alternatives exist for devising such an instrument; indeed many variations of this example are marketed.

Availability of the system diagram then allows the required block diagram of the full architecture to be drawn (see Figure 5.8). Additional supporting blocks, such as the power supply, are also needed in a full system diagram, but are not usually shown.

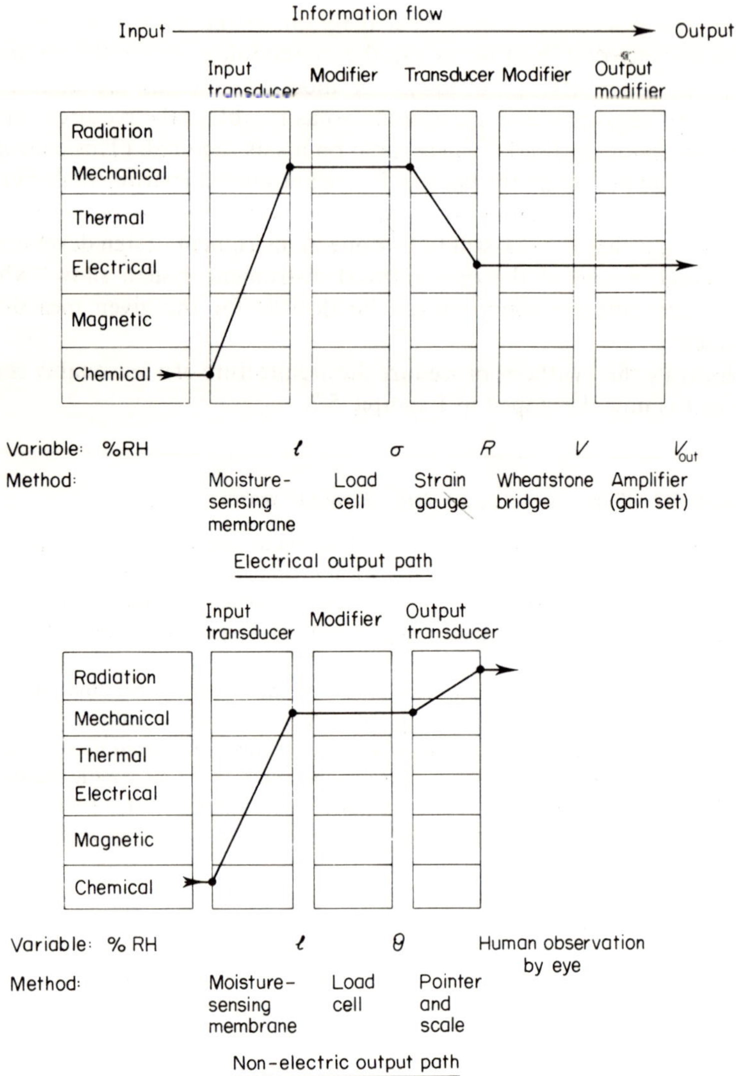

Figure 5.7 Matrix form of stages needed in relative humidity sensor.

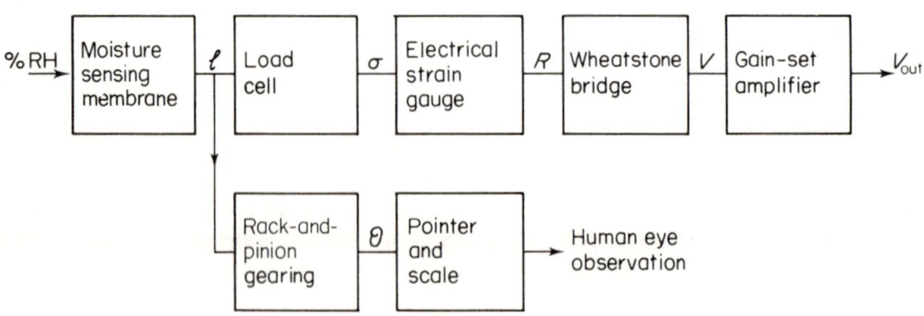

Figure 5.8 Block diagram of relative-humidity meter developed using energy-conversion matrix method

Once a suitable system architecture has been developed further work can then concentrate on modelling each of the blocks as detailed later in Section 5.3.2.

5.2.3 Self-generating and modulated stages

Study of the various blocks established in routine practice shows that they are of two main types of structure.

The first kinds are those that generate an output as the direct result of the input energy. These are best called *self-generating* types (but also called *passive*). Examples of these are the *photovoltaic cell* (which gives an electrical output when exposed to input optical radiation) and the *thermocouple* (which generates a voltage proportional to input temperature).

The other kind of sensor structure is that needing an *excitation* or *auxiliary* supply of energy for its operation. These are best called *modulating* types, for their operation makes use of a constantly provided energy flow that is modulated by the input measurand. (The term *active* is also used to describe these but that word has other meanings and, therefore, can be misleading.) An example of the modulating type is the photo-diode in which a *bias* electrical current is provided by a secondary energy source.

5.2.4 Symbology and notation for system description

The symbology, shown in Figure 5.9, has been suggested for the two self-generating and modulated types of system building block. The architecture for the relative humidity meter, developed in Example 5.2, is extended in this way in Figure 5.9(c).

The possible combinations of the input, output and modulating energies in applications can be portrayed with the three-dimensional energy/information space as given in the chart of Figure 5.10.

Self-generating transducers and modifiers, those blocks operating without an auxiliary energy source, lie in the bottom $x-y$ plane. Consider, for example, the thermocouple in which thermal information flows from thermal energy on the x-axis to information, transduced to use electrical energy, on the y-axis.

Modulating types exist in the $[x, y, z]$ space on the diagram according to which modulating energy source is used. A photoconductance sensor would have its input information at $x =$ electrical, its signal output at $y =$ electrical and its modulating energy at $z =$ radiation.

The energy/information space diagram assists recognition of transducer possibilities. It is clear that numerous different block structures are feasible candidates when developing a solution to a need. Whilst all known block structures can be fitted to the chart, there are still many positions to be filled with practical devices.

An instrument system building block, therefore, has three possible components - input signal S_1, self-generating or excitation energy E_1, output signal S_2 and output energy E_2. The difference between the self-generating and modulating types is a matter of how E_1 is supplied to the system (see Figure 5.9).

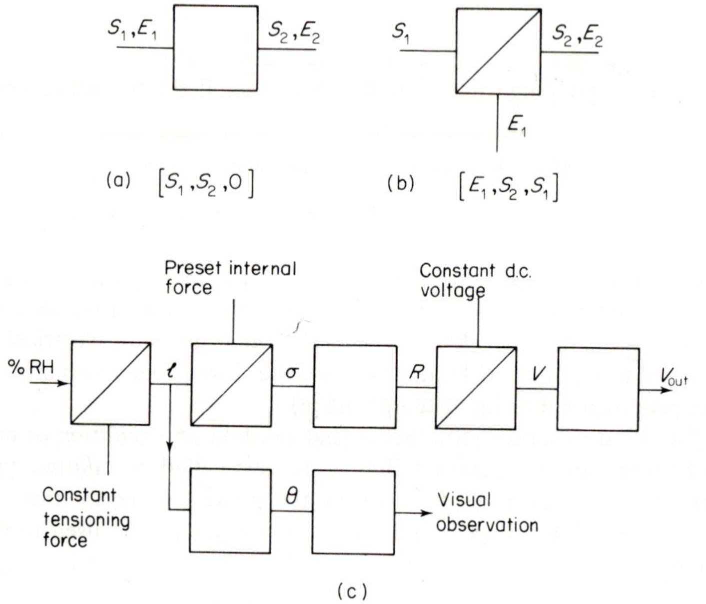

Figure 5.9 Symbology and notation for building blocks of an instrument system: (a) self-generating stage; (b) modulating stage; (c) extended relative-humidity instrument block diagram. (From Middlehoek and Noorlag, 1981)

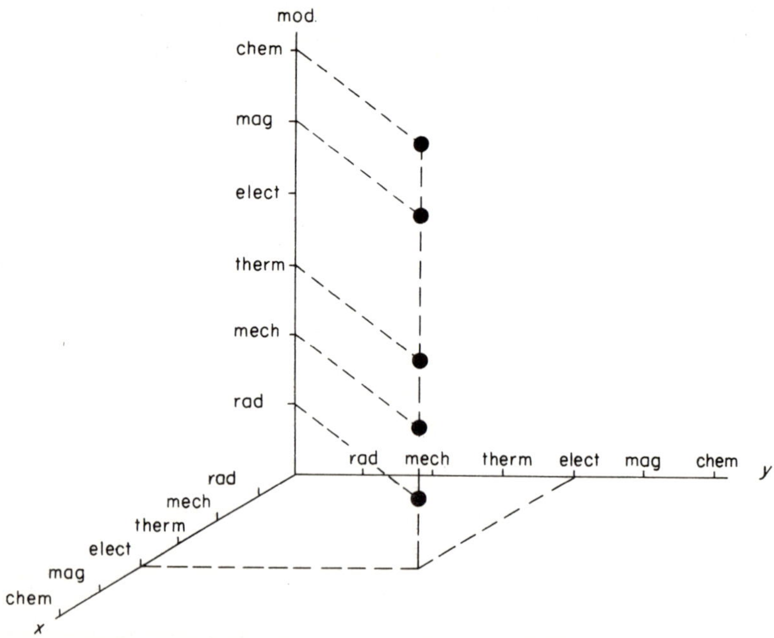

Figure 5.10 The three-dimensional transducer, energy/information, space. (From Middlehoek and Noorlag, 1981. Reproduced by permission.)

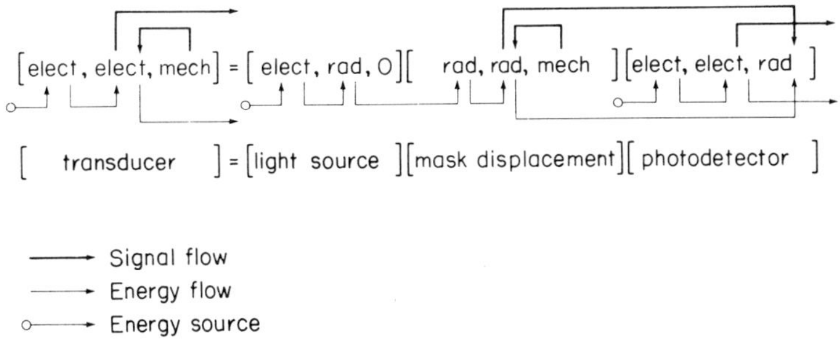

Figure 5.11 Energy/information space representation of an optical displacement transducer. (From Middlehoek and Noorlag, 1981. Reproduced by permission.)

Any instrument stage can, therefore, be represented by a notation $[x,y,z]$ in which x is the excitation energy form, y the input signal energy form and z the modulating input form. These are given in Figure 5.9.

Using this notation and methodology it is possible to portray the architecture of a chain of blocks with a diagram showing signal (information) and energy flows and energy sources. An example, shown in Figure 5.11, illustrates how the model is interpreted.

5.3 GENERATING FUNCTIONAL MATHEMATICAL MODELS

5.3.1 Introductory remarks

The mathematical model of a complete instrument system is developed in four distinct stages:

1. develop an iconic model of the measurement system with descriptors added;
2. develop a suitable architecture that is finally expressed as block diagram;
3. develop appropriate mathematical models for each block;
4. combine the individual mathematical expressions into a unified set giving a mathematical model of the full system.

The first two, more heuristic than mathematical, stages have been dealt with in Sections 5.1 and 5.2. Attention is now directed to the last two steps which require a procedure for developing a mathematical model of the system that enables *simulation.*

5.3.2 Mathematical modelling of individual blocks

Blocks selected using the matrix representation approach, given in Section 5.2.2, would be expected to have been selected partially because their overall physical behaviour is already reasonably well known in mathematical terms. From this knowledge a suitable mathematical relationship between the input and output can be assigned to start the mathematical model-building process.

Unfortunately, the *first-level* model will describe more the *ideal* than the *real* behaviour.

For example, Ohm's law

$$V = IR \tag{5.1}$$

(where V is the voltage across a resistor R when an electric current I flows through the resistor) is a first-level model of a modifier commonly used to measure large currents flowing in electrical power cables. A small-value resistor R is placed in series with the cable (or the resistance of a short length of the cable itself is used). The small, easily measured, voltage V appearing across this is related to the current I flowing by Ohm's law.

In practice the value R, of the resistor, may be temperature and voltage dependent, may vary with its mounting strain and with electromagnetic radiation levels – plus numerous other potentially important effects including variation with time for all such properties.

Realistic instrument block modelling requires development of the first level model into a more complex *second-level* model. In the resistor example the Ohm's law mathematical model could need extension to allow for many secondary, often very subtle effects. The more exacting the measurement mapping desired of a stage, the more detailed must be the second-level model. Deciding what secondary effects might need consideration is an ability that needs developing in order to achieve good instrument systems. This procedure is now outlined in Example 5.3.

Example 5.3 Development of a pipeline blockage indicator

A measurement system is to be designed to measure the solids concentration of pneumatically conveyed cement powder in a material handling system. A schematic layout is given in Figure 5.12(a). Cement powder is conveyed between two hoppers and it is necessary to monitor the concentration of solids in the pipeline in order to know when blockages are about to occur.

A system block diagram, Figure 5.12(b), has been chosen in which a capacitance-concept, concentration transduction stage is followed by an electronic modifier block to produce the electronic signals required in the central control room.

Solids concentration transducer

The principle of this transducer is straightforward. The electrical capacitance across two electrodes, between which the solids/gas mixture is flowing, is monitored. Any changes in capacitance are dependent on the geometry of the sensor, the dielectric properties of the flowing mixture, the flow regime inside the pipe and the solids concentration.

However, if the first three factors are constant the measured capacitance will be a direct function of the solids concentration. In that case, if a blockage begins to build to large proportions, the amount of solids passing through the transducer assembly will fall, reducing the capacitance.

Transducer design considerations

Capacitance transducers have the advantage of being simple to construct; they need be no more than two electrically isolated electrodes fixed to the wall of a conveyor pipe made of suitable material.

The main design parameters are the shape of the sensing electrodes and whether the electrodes should be insulated from, or in contact with, the flowing material. Simple mathematical models can be used to aid these decisions.

Transducer shape

It is desirable that the electric field across the sensing region inside the pipe is homogeneous. If the electric field varies throughout the measuring volume the measured change in capacitance due to a change in solids concentration will depend on the spatial distribution of the solids reducing the accuracy of the sensor.

A reasonably homogeneous field can be achieved for mixtures having low relative permittivity by using surface plate electrodes see Figure 5.13(a). The next aspect to be considered is the effect of the material used to separate the electrodes from the flowing material.

Insulating material

In order to determine the influence of the insulating material on the characteristics of the capacitance transducer, a simple mathematical model can be used (Figure 5.13(b). At this stage electrode shape is not important.

Since the fluid and the pipe are electrically non-conducting, the capacitance measured across the transducer C is given by

$$C = \frac{C_p C_f}{2C_f + C_p} \tag{5.2}$$

where C_p is the capacitance of the pipe wall, and C_f is the capacitance of the solids/gas mixture.

The two important parameters of the transducer transfer characteristic are its *sensitivity* and *linearity*. The simple model above may be used to *assess* the effect of the pipe wall material on both of these.

The *sensitivity* of the measured capacitance C to changes in the solids concentration, and hence C_f, may be found by differentiating Equation (5.2) to obtain

$$\frac{\partial C}{\partial C_f} = \frac{C_p^2}{(2C_f + C_p)^2} \tag{5.3}$$

This expression shows that the maximum sensitivity can be achieved when $C_p \gg C_f$, therefore when Equation (5.3) $\rightarrow 1$ the linearity of C, for changes in C_f, may be obtained by considering the two extremes:

When $C_f \rightarrow 0$, at low levels of solids concentration, then

$$\frac{\partial C}{\partial C_f} \rightarrow 1$$

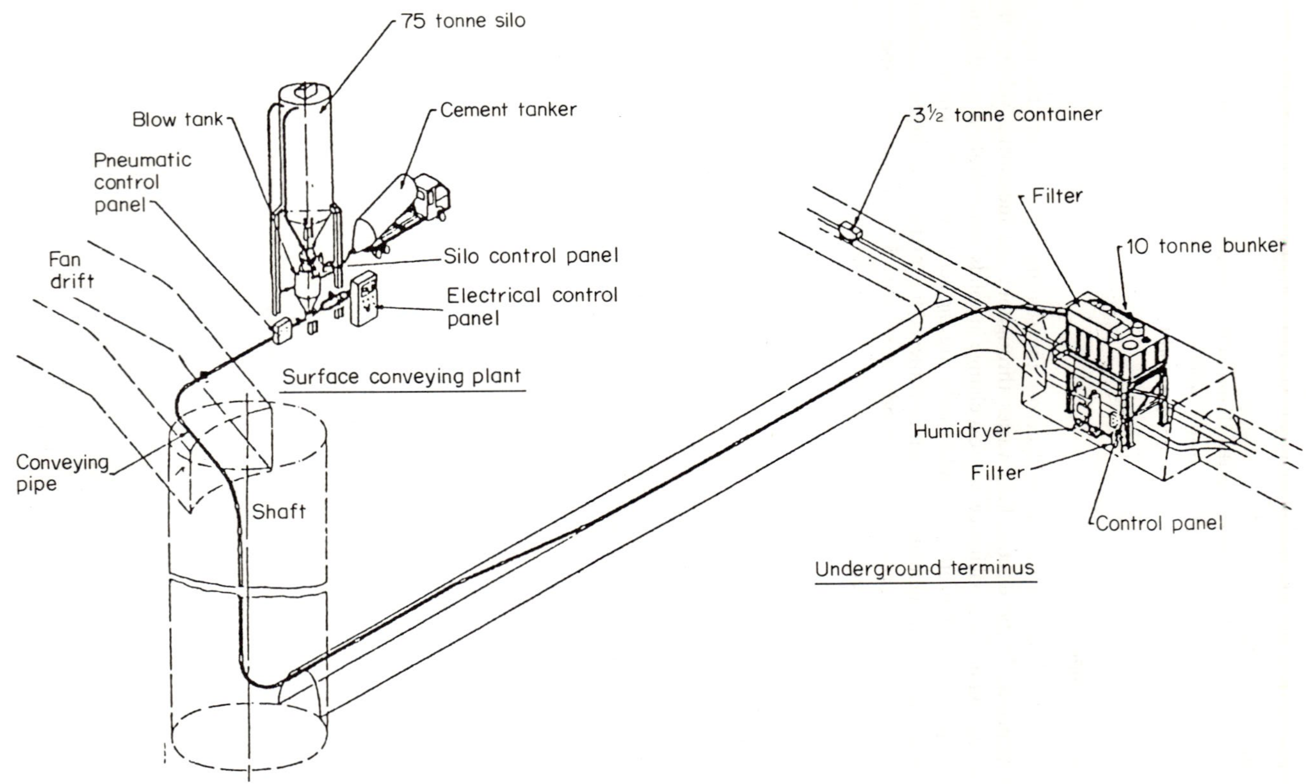

Figure 5.12 Schematic diagrams of pipeline blockage indicator: (a) schematic diagram of pneumatic conveyor system

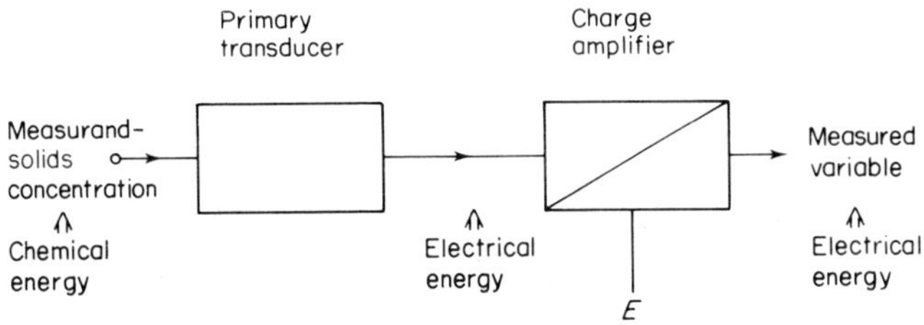

Figure 5.12 (b) block diagram of pipeline blockage indicator

When $C_f = C_{f\max}$

$$\frac{\partial C}{\partial C_f} = \frac{1}{(1 + 2C_{f\max}/C_p)^2} \tag{5.4}$$

From Equation (5.4) it is obvious that the best linearity will also be obtained if C_p is as large as possible (Brant *et al.*, 1985).

It can be seen that even simple mathematical models can tell us much about how to optimize the transducer principle and construct the best final sensor.

System transfer function

The relationship between the output of the measurement system (voltage) and the input (solids concentration) is of prime importance. These two quantities are related by the *system transfer function* which is of the general form

$$\frac{v}{S}(s) = KG(s) \tag{5.5}$$

where K is the *static sensitivity* or *steady-state gain* of the system, while $G(s)$ describes the dynamic performance of the measuring system. These can be modelled separately.

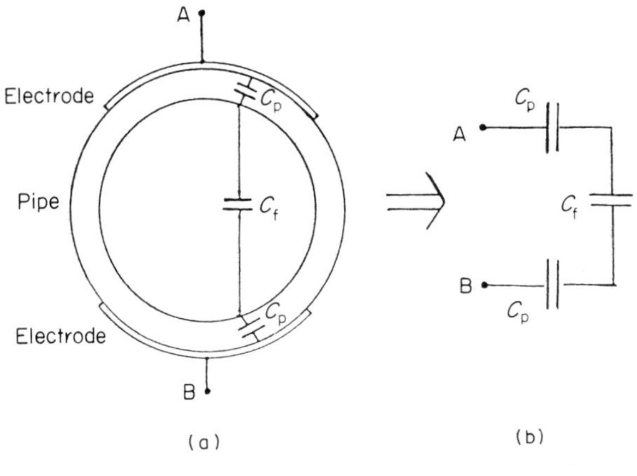

Figure 5.13 Models of blockage transducer: (a) physical electrode arrangement; (b) mathematical model

The static characteristic of the transducer

Many different approaches can be used to calculate the transfer characteristic of such a transducer configuration. Simple *lumped-parameter* electric models, such as that developed by Shu *et al.* (1982), can be applied with success.

For this model it is assumed that there is an homogeneous electric field between the two electrodes. To calculate the total capacitance between the electrodes the cross-section of the transducer is divided into (imaginary) horizontal slices, and the sum of the capacitance of these slices calculated (Figure 5.13). The capacitance of each slice is

$$\frac{1}{\delta C} = \frac{2}{\delta C_p} + \frac{1}{\delta C_f} \tag{5.6}$$

where

$$\delta C_p = \varepsilon_p L\left(\frac{a\delta\theta}{(a-b)\sin\theta}\right) \qquad \text{for } \theta_0 \leqslant \theta \leqslant \pi - \theta_0$$

$$\delta C_f = \varepsilon_* L\left(\frac{a\delta\theta}{2b\sin\theta}\right) \qquad \text{for } \theta_0 \leqslant \theta \leqslant \pi - \theta_0$$

substituting into Equation (5.4) and integrating from θ_0 to $(\pi - \theta_0)$ gives

$$C = \frac{\varepsilon_p \varepsilon_* aL}{(a-b)[\varepsilon_* + (a(a-b))\varepsilon_p]} \cdot \left\{-\ln\left(\tan\frac{\theta}{2}\right)\right\} \tag{5.7}$$

where

a = pipe outside diameter;
b = pipe inside diameter;
θ = angle subtended by electrode;
ε_p = relative permittivity of pipe wall;
ε_* = effective permittivity of solids/gas mixture.

In order to be able to calculate the relationship between capacitance measured across the transducer's two electrodes, and the solids concentration in the pipeline the effective permittivity ε_* of the solids/gas mixture must be calculated. This may be achieved using a mathematical model first developed by Maxwell (1873):

$$\frac{\varepsilon_* - \varepsilon_g}{\varepsilon_* + 2\varepsilon_g} = S\,\frac{\varepsilon_s - \varepsilon_g}{\varepsilon_s + 2\varepsilon_g} \tag{5.8}$$

where
ε_g = relative permittivity of gas;
ε_s = relative permittivity of solids;
S = solids concentration (0 to 1).

This illustrates another principle in the use of mathematical models – before developing a model study the literature to see what has already been done that could be adapted to the particular need.

Maxwell's equation, above, can be written in the explicit form as

$$\varepsilon_* = \frac{1 + 2SK}{1 - SK}\,\varepsilon_g \tag{5.9}$$

where

$$K = \frac{\varepsilon_s - \varepsilon_g}{\varepsilon_s + 2\varepsilon_g}$$

Thus Equations (5.7) and (5.9) can be used to estimate the static sensitivity and linearity for a particular transducer. Like many simple models, those developed here are limited in their validity. Maxwell's equation, given here, is valid only for particles which are small compared to the distance between them. In addition, Maxwell's equation and Equation (5.7) assume a homogeneous sensing field. However, they can still be used to give a qualitative guide to the expected range of measured capacitance and thus to setting up the following modifier stage.

If greater accuracy is required the potential distribution between the electrodes must be obtained by solution of the well-known Laplace equation. However, this means a move from the calculator to the computer as great computing power is needed for such mathematics. The simple approach, explained here, is often sufficient.

The dynamic characteristics of the transducer

The mathematical models developed so far can be used to estimate the *sensitivity* and the *linearity* of the transducer. These parameters are part of the transducer's static characteristic, and as such assume the measurement (solids concentration) to be constant or changing only very slowly.

In many cases the ability of a measurement system to respond to sudden changes in the measurand is very important, and to study this the dynamic characteristic of the measurement system is required.

This can be developed for the solids concentration transducer by starting with the simple iconic model of the solids flow to transducer interface shown in Figure 5.14(a).

This model assumes a pulsed change in the concentration of the conveyed material passing between the transducer's electrodes at a velocity v. If the field between the electrodes is homogeneous and does not extend beyond the ends of the transducer, then the corresponding change in the capacitance of the transducer is a pulse of width a/v and amplitude k. The value k is dependent upon the static sensitivity of the transducer and can be calculated using Equations (5.7) and (5.9).

If the original pulsed change in the concentration of the conveyed solids is short compared to a/v, then it can be regarded as a Dirac pulse. The dynamic characteristic of the transducer can, therefore, be shown to be (Hammer and Green, 1983)

$$\frac{\Delta C_T}{\Delta S}(s) = \frac{kv(1 - e^{-(a/v)s})}{a} \tag{5.10}$$

The amplitude response of this function is

$$\left|\frac{\Delta C_T}{\Delta S}(j\omega)\right| = ka\left|\frac{\sin(a\omega/2v)}{(a\omega/2v)}\right| \tag{5.11}$$

where ω is the frequency of the variation in solids concentration. If this amplitude frequency response is plotted (Figure 5.14(b)) then, as the resulting response shows, the transducer's electrodes act as a low-pass filter with a cut-off frequency ω_0(3dB) of

$$\omega_c = 2v/a \tag{5.12}$$

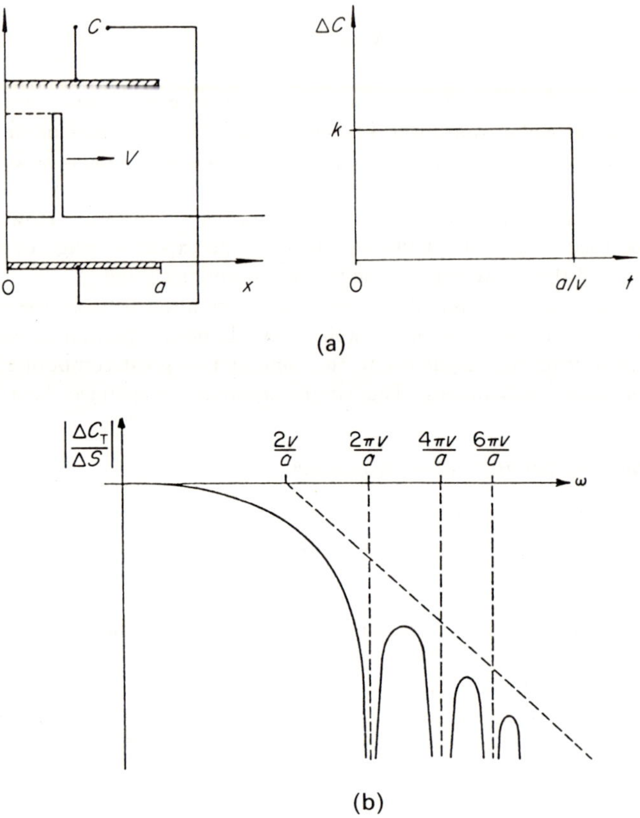

Figure 5.14 The transducer's dynamic characteristic (Example 5.3): (a) starting point of dynamic model; (b) amplitude—frequency response of the transducer

This relatively simple model shows that the measured capacitance of the transducer would not respond to sudden changes in the conveyed solids concentration. Only the average value of the solids passing through the transducer would be indicated.

Modelling the electronic modifier

Capacitance-to-voltage conversion can be achieved using a capacitance feedback amplifier, known more commonly as a *charge amplifier*. The basic charge amplifier has a circuit arrangement as shown in Figure 5.15(a).

Assuming for the moment that the amplifier has ideal characteristics (infinite input impedance, zero output impedance and infinite gain) then the voltage transfer function of the network may be written as

$$V_0/V_1 = -C_T/C_F \tag{5.13}$$

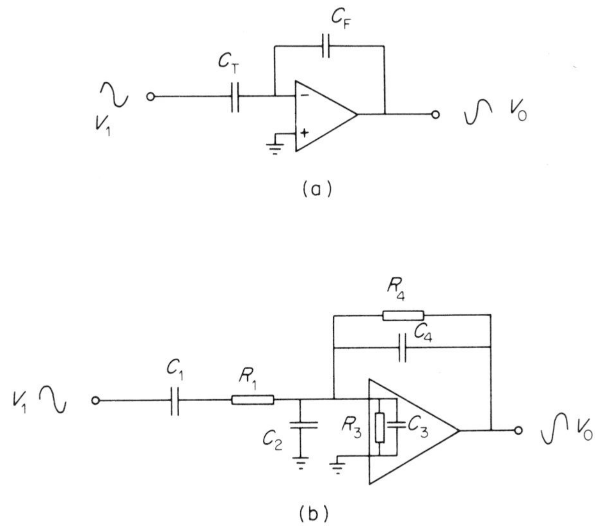

Figure 5.15 Charge amplifier: (a) schematic of circuit; (b) design model of circuit

or

$$\Delta V_0 = -V_1/C_F \cdot \Delta C_T \tag{5.14}$$

Thus if the feedback capacitance C_F and the peak amplitude of the input signal are constant V_0 will be directly proportional to changes in the capacitance of the transducer C_T. It should be noted, from Equation (5.14), that the sensitivity of the measurement $-V_1/C_F$ is independent of the capacitance of the transducer itself.

A further advantage of this configuration is that since the input terminal of the amplifier is held at virtual earth, then changes in stray capacitance to earth are rejected. This allows the use of long shielded cables between the sensor and amplifier (if required) without significantly affecting the accuracy of the measurement.

The fact that the charge amplifier has these desirable properties and is also simple to construct made it a preferable choice, in this case, to the more conventional methods of capacitance measurement, such as bridges or tuned circuits.

Charge amplifier design: general model building

The circuit shown in Figure 5.15(a) is oversimplified. For practical purposes a more realistic model is needed to allow for practical, second-level, less obvious factors that arise due to practical limitations. The extended model is shown in Figure 5.15(b).

This model takes into account the open-loop characteristics of the amplifier as well as the properties of the length of cable used between the transducer and the amplifier. In practical cases a feedback resistor R_4 is also necessary. This resistor is needed to discharge C_1 and C_4, since without it the amplifier output would gradually rise to become saturated.

The transfer function (from use of basic electrical laws and theorems to produce a set of mesh equations that are then reduced) of this network is given by

$$\frac{V_0(s)}{V_1(s)}=\frac{\dfrac{-R_4C_1s}{(1+R_1C_1s)(1+R_4C_4s)}}{\dfrac{R_4C_1s}{(A+AR_1C_1s)(1+R_4C_4s)}+\dfrac{R_4C_2s}{A(1+R_4C_4s)}+\dfrac{(R_4+R_4R_3C_3s)}{R_3A(1+R_4C_4s)}+\dfrac{1}{A}+1} \tag{5.15}$$

For this design the static sensitivity of the charge amplifier network at a constant frequency ω is required. This can be obtained by replacing the Laplace operator by $j\omega$ in Equation (5.15) as follows:

$$\frac{V_0(j\omega)}{V_1(j\omega)}$$

$$=\frac{\dfrac{-j\omega R_4C_1}{(1+j\omega R_1C_1)(1+j\omega R_4C_4)}}{\dfrac{j\omega R_4C_1}{(A+j\omega R_1C_1A)(1+j\omega R_4C_4)}+\dfrac{j\omega R_4C_2}{A(1+j\omega R_4C_4)}+\dfrac{(R_4+j\omega R_4R_3C_3)}{R_3A(1+j\omega R_4C_4)}+\dfrac{1}{A}+1} \tag{5.16}$$

This sinusoidal transfer function (a model of the amplifier) can be used to assess the effect of parameter changes on the gain of the network.

Charge amplifier design: parameter selection

A main reason for developing the realistic model was that it was hoped to allow selection of parameter values such that changes in closed-loop gain of the amplifier network would be solely due to changes in capacitance of the sensor. However, as Equation (5.16) shows, it is not possible for the gain of the network to be completely independent of the changes in the other parameters – the model, however, allows minimization of these effects.

Considering the parameters of Equation (5.16), R_3, C_3 and A are determined by the type of amplifier used. Consider, for example, a device based around an available operational amplifier (type TL061 made by Texas Instruments). This is a BIFET device with a very high input impedance (approximately 10^{12} Ω), a desirable requirement in this case.

Equation (5.7) can be used to estimate the capacitance of the transducer when either empty or full of solids (that is, a solids concentration between 0 and 1). Suppose, for a particular capacitance transducer, the empty capacitance is calculated to be 5 pF, then C_4 may be selected to be 5 pF in order to give a closed-loop gain of approximately unity when the sensor is empty. With C_4 fixed, the lower cut-off frequency (−3 dB) of the network is determined by the choice of R_4 since

$$\omega_c = 1/R_4C_4 \tag{5.17}$$

Selection of R_4 is a trade-off between the desirability of a low operating frequency (where it is easier to design low-drift oscillators) and the practical limitations set by the use of high-value resistors. R_4 is chosen to be 47 MΩ, thus fixing the lower cut-off frequency of the network at 677 Hz.

The choice of operating frequency ω is also a trade-off. The higher the operating frequency the greater will be the effect of changes in cable capacitance on the gain of the network. However, if the operating frequency is too low, then the gain of the network will be undesirably sensitive to drift in frequency.

Finally, the length of the coaxial cable used to connect the primary sensor to the charge amplifier will determine the significance of R_1 and C_2. If, say, the modifier were connected to the transducer with a 0.5 m coaxial cable, then R_1 would be negligible and C_2 approximately 50 pF.

Substituting all of the above parameters into Equation (5.16) the expected variation in gain of the network with varying sensor capacitance can be determined. As Figure 5.16 shows, an increase in voltage gain of around 12% can be expected as the contents of the sensor changes from all gas to all solids.

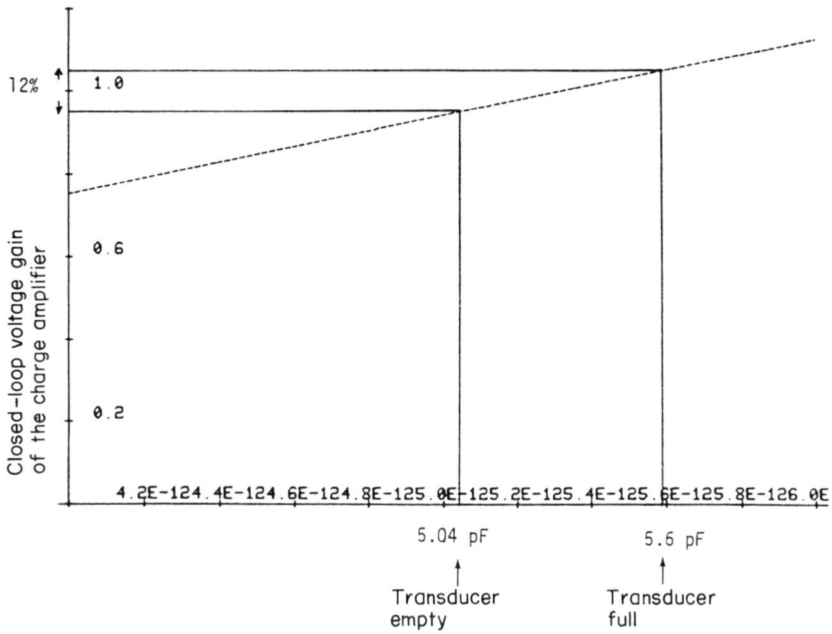

Figure 5.16 Calculated change in gain of charge amplifier as solids concentration varies within the sensor

5.3.3 Experimental determination of model parameters

No matter how detailed they are, mathematical models are rarely accurate enough to predict the absolute performance of a measuring system and usually contain a number of constants which must be determined by experimental means. The way in which these constants are established, for the actual transducer, will depend on the actual measurement system being modelled, but the following general approach can be adopted.

Each transducer or modifier of the measurement system block diagram can be represented by either a zero, first- or second-order equation. Higher-order systems can usually be approximated to a second order. From energy considerations it is usually possible to decide which order is appropriate.

Having classified each of the transfer functions into one of the above categories, a standard series of tests can be used to determine, experimentally on the actual unit, the unknown parameters of each block. These parameters are the stage gain, plus the time constant in first-order systems and the damping factor and natural frequency for the second-order case.

Constants needed for the static characteristics are found by static calibration. Time constants can be found by step frequency response methods, as has been discussed in Section 3.2.4. The various response types and their characterization are explained in Section 4.4. Such tests will also verify if the choice of order was correctly chosen.

5.4 SPECIFIC TRANSDUCER COMPUTER-AIDED DESIGN TOOLS

5.4.1 Using computers to assist instrument modelling

Essential ingredients for conducting a mathematical-modelling exercise are that

- a *formalized methodology exists* for conducting the modelling, and
- the *behaviour of blocks* can be expressed in suitable *mathematical terms.*

Consider Example 5.3. To develop a model of the pipeline blockage indicator it was first necessary to realize a likely system structure for both the capacitance transducer and the following electronic amplifier modifier; a flow path, for the energy and information, was set up. That done, the next step was to apply suitable equations to describe the behaviour of each block in terms of the wanted performance parameters of input signals, output signals and component signal levels.

Although not yet an extensively conducted exercise in instrument design there has been some work carried out using computers to assist this process. The generic name of this approach is *computer-aided design* (CAD) or *computer-aided engineering* (CAE). Its advantages have already been cited in Section 5.1.1. The essential benefits are reduced design time, improved performance and ability to be used by people with less training in this specialized area than the experts who produced the software. The more generalized aspect of using computers in the overall design of instruments is taken up in Section 7.4.

It is a logical development to use computers in this way because they can carry out procedural (and, more recently, non-procedural) tasks (the formalized procedure) performing any mathematical solutions (mathematical solution ability) needed after data is supplied to *bounded* generalized mathematical models to each specific case.

Many CAD software tools are available that can be used to assist the modelling aspect of instrument design. Two dominant alternatives are now described – the use of descriptive equations and the use of finite-element analysis.

5.4.2 The MEDIEM CAD package

In the 1930s the mathematical modelling of microphones and other acoustic equipment used in broadcasting led to the realization that the linear behaviour of acoustic systems (enclosures of air as springs, compressibility of air forming energy storage, energy lost by air friction) could be described with linear differential equations of the type covered in general in Section 4.4.3 and specifically in Section 4.4.6 (second-order systems). This work developed a concept of there being equivalence of a system allowing conversion into a common electrical energy domain and provided a means to model electrical, mechanical and acoustic systems in a coherent manner. This body of theory became known as the 'analogies'.

Initially the methodology (similar to that of Example 5.3) was applied by manual means – that is, set up the system architecture, seek out the suitable equations and conversion coefficients, provide the specific system parameters and run the model (often as an electrical analogue computer as this work preceded widespread digital computing) to see how the variables change.

In the early 1970s one group, that of the Measurement and Instrument Systems Centre, City University, London, began work that applied this methodology to eventually form a powerful CAD transducer and modifier stage design tool called MEDIEM (Multi Energy Domain Interactive Element Modelling).

The underlying basis of this work was a comprehensive development and extension of the 'analogies' which is described elsewhere (Finkelstein and Watt, 1983). This work formalized a rigorous classification of the various transducer types in terms of their 'through' or 'per' (an electric current or a mass flow pass through a block) and their 'across' or 'trans' (a voltage or force is applied across a block) nature and in terms of their energy-processing role (storage, conversion, source, interconnection and junctions). Having identified all possible generic building blocks and elements a coherent mathematical formulation was developed in terms of linear differential equations. The software capability was eventually extended to cover non-linear modelling.

A suitable means was needed to allow blocks to be joined with connections to produce a system architecture that could be modelled. This was done with *structure graphs* mathematical symbology.

In practice a 'functional' model, in which the input/output relationships are expressed, is first developed from a knowledge of the application. This is then converted into the 'structure graph' form which can be recognized by the computer. The necessary state equations are then set up automatically by the software. A data base must also be filled with materials and other physical data. The model is then run by exciting it with the required input functions using graphical plots to observe the behaviour of the system at chosen nodes. Example 5.4 gives the steps of this process, in relation to development of a thermowell for a thermometer, a simple task for the software.

Example 5.4 Designing a thermowell type thermometer using MEDIEM

In the application of temperature sensors it is often necessary to mount the sensitive element in a protective *thermowell*, as shown in Figure 5.17(a). Whilst it can provide

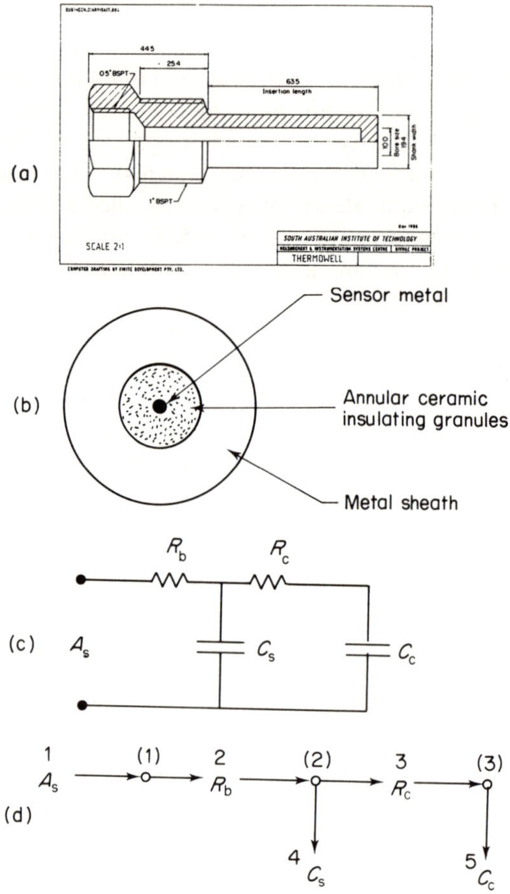

(e)

1, as , to , 1
2, r , fr , 1
2, r , to , 2
3, c , fr , 2
4, r , fr , 2
4, r , to , 3
5, c , fr , 3

Figure 5.17 Stages in modelling a thermowell: (a) the given physical shape and size; (b) cross-sectional functional model decided according to heat flow considerations; (c) electrical equivalent circuit (lumped parameter) of thermowell; (d) structure graph representing the lumped-parameter model; (e) table constructed from structure graph

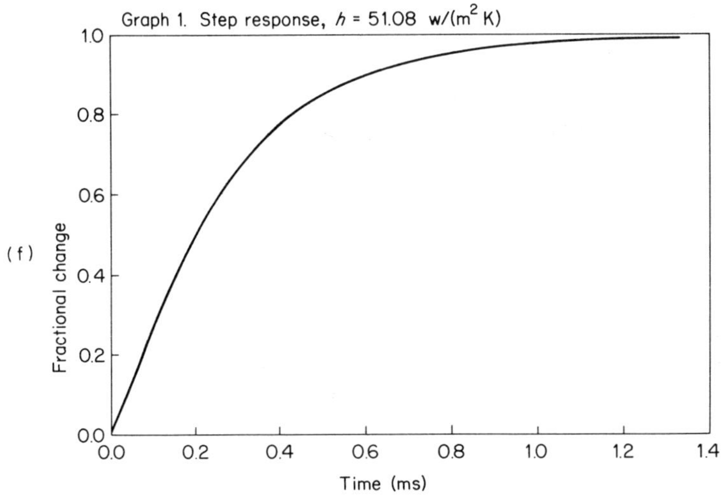

Figure 5.17 *(continued)* (f) step thermal response plot of thermowell

protection the thermal mass provides heat storage and the material possesses heat resistivity, implying that the response is at least first order and liable to be much slower to respond than the bare sensor. (Example 3.7, Section 3.24 and Example 4.6, Section 4.4.5 deal with different aspects of the practice and theory involved.)

It is therefore prudent, if not essential, to model the system so that its thermal response is known *before* installation. The modelling task is, then, to model a given shape and size thermowell of known material and fluid into which it is placed.

A cross-sectional model is first developed (by intuition and study of the original artifact – or the expected one if not yet built). This must adequately represent the heat flow situation and be such that the various materials and parts are clearly identified. Figure 5.17(b) shows the physical functional model. It is assumed that this two-dimensional model is adequate and that heat flow along the thermowell is not dominant.

The electrical equivalent circuit is then realized (Figure 5.17(c)) in a lumped parameter form. Source A_s, a voltage, represents the fluid step temperature input; electrical resistance R_b represents the boundary layer thermal resistance; C_s is the sheath capacitance; electrical resistance R_c represents the thermal resistance of the core; and C_c represents the capacitance of the core.

From this the *structure graph* (Figure 5.17(d)) is developed using rules (given in the software handbook.) These identify the element type, connections and the through and across variables. A table (Figure 5.17(e)) representing this graph, is then constructed from the graph and typed into the software.

The system then calls for material physical data for the material of the thermowell, sensor, heat-conducting material and fluid surrounding the thermowell. That completes the first file-handling, model-creation mode.

Once the state equations have been formed it then can be run, after user instruction about the appropriate input function – a step of temperature rise in this case, to plot out the response shown in Figure 5.17(f).

If the response is not as needed, the physical dimensions and material parameters can be modified accordingly.

The suitably trained person can set up such an exercise in around one hour, running each model in only a few minutes more. In this way two hours of work would produce results of about ten design variations. The traditional method of design, draw, construct and carry out practical tests, would need at least several days' work, plus special test facilities, to achieve the same results!

Whilst this example is of a relatively simple case the same degree of time and cost saving has been well proven for use in designing more complex instrument systems – such as bellows, pressure gauges, load cells and orifice-plate flow meters.

Use of descriptive differential equations is best suited to solution of lumped parameter systems. They can be extended to cover (but with difficulty) distributed systems problems by forming the latter as a set of distributed blocks.

An alternative is to use a quite different approach as follows in the next section.

5.4.3 Finite-element CAD packages

The *finite-element* method of analysis is a powerful, computer-based, analytical technique which was first used in the 1950s to solve structural engineering problems in the aircraft industry. The technique requires numerous calculations and was, therefore, limited in its use in those early times. However, with the development of inexpensive, powerful and fast computers it has now gained widespread application in such areas as thermodynamics, fluid mechanics, structures – indeed in any application involving need to know about the distribution of 'fluxes' or 'potentials' – for example, the thermal regime within an object heated at one end, or the stress pattern within a metal bar when loaded with a set of external forces. It has widespread application in transducer design because they also concern energy relationships within distributed systems.

The basic concept of the finite element method (FEM) is to take a defined region to be analysed (which is technically difficult in a continuous (distributed) form since the 'flux' entity exists as the result of an infinite number of points in the active space) and break the space down into a finite number of *elemental* regions (hence the name, finite element) each of which is mathematically more manageable when calculated individually.

After solution of these small areas their overall results are combined into progressively larger areas, re-adjusting the values of each region in a highly iterative manner until the whole system solution is in balance. Getting such computations to iterate toward a final solution is part of the problem – some do not and actually have unusable diverging solutions attempts.

The procedure is complex in itself but today many FEM packages exist that are very straightforward to operate provided the user knows what is being sought from the package. Use in measurement system design is now illustrated by a 'walk-through' of the use of FEM to model the electric field in a capacitance electrode system (see Example 5.5).

Example 5.5 Finite-element method used to model an electrode system

In the pipeline blockage indicator of Example 5.3 the simplest electrode configuration that could be used would be as shown in Figure 5.18(a).

The field potential between the two electrodes can be expressed in mathematical terms by using the Laplace equation (a complex level of mathematics that does not need to be understood to run advanced FEM tools).

The analysis is made simpler if the actual three-dimensional transducer can be reduced to an adequate two-dimensional system; here if the plates are assumed longer than their separation this is possible.

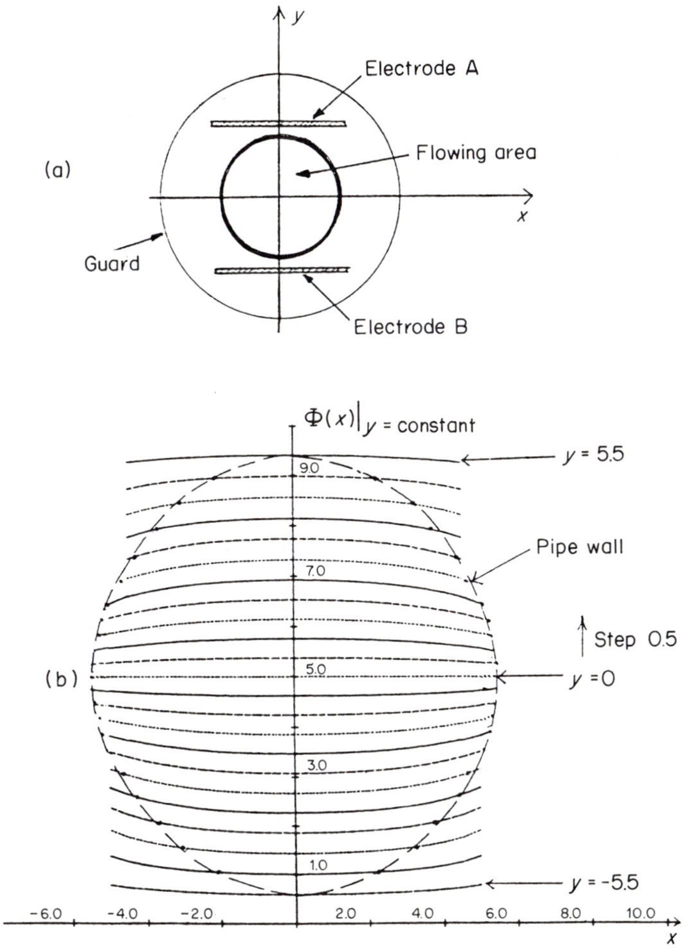

Figure 5.18 Electrode field modelling: (a) the parallel plate electrode system (the guard is held at ground potential, electrode A at + V volts, and electrode B at − V volts); (b) potential field distribution obtained with FEM

To obtain a specific solution to the generalized Laplace Equation it is necessary to also define

- the *geometry of the boundary* of the 'flux' system (in this case the electrode shape, including the *guard* ring – actually a cylinder – that is added to make the internal field less dependent on the influence of external fields and to give a more linear internal field distribution);
- the *electric properties* of the various materials in the system (here these are the permittivity of the material in the flowing space, the electrodes and the container wall inside the guard ring);
- the driving potentials (here they are the various voltages applied to the electrodes and guard ring);
- the density of the *mesh* of points that will suffice (obviously more points will take longer to solve, the extra time rising much faster than the increase gained in discrimination).

Once these parameters have been entered into the computer the field lines are calculated ready to display or plot, as shown in Figure 5.18(b), where equipotential field lines are plotted.

FEM is well known for its need for powerful computers to run the numerous iterations needed. Advanced packages will build and refine the mesh as the user views the display. A task such as this example would take about 30 minutes to perform, of which 20 minutes will be computer processing time.

FEM replaces many former methods of solving the Laplace modelling applications – these include the water tank and resistive paper plotting methods.

In Figure 5.18(b) it will be seen that the plot is quantitative (normalized voltages $y = 5.5$ and -5.5). This allows calculation, from charge density considerations, of the electrode capacitance *for the particular set of parameters* used for each computer run.

By running FEM operations many times, each with a parameter changed, it is possible to build up a set of plots that enable the designer to optimize shapes and sensitivities.

Unlike the 'analogies' modelling using descriptive equations (Example 5.4, Section 5.4.2) FEM does not allow operations such as differentiation to be used to establish maxima and the like. To obtain *design sensitivity* information a set of FEM runs must be made and the set studied visually.

To run an FEM study from a completely undefined 'flux' distribution starting point can be very time consuming. A technique that sometimes can be used is to estimate the approximate distribution first, using the FEM to refine it. This is called *probable flux path* modelling.

5.5 THE MATHEMATICAL MODEL AS A FUNCTIONAL PART OF A WORKING MEASUREMENT SYSTEM

So far the modelling of measurement systems has been justified by emphasizing that such procedures can assist development of better instruments. Models of instruments, however, are also of real value as a functional part of a measurement system. Example 5.6 explains the concept.

Example 5.6 Using a mathematical model as functional part of an oil/gas/water component measuring water

Consider the crude oil produced by an oil well reservoir. It often contains both gas and water components. It is important for the operator of a production facility to know which type of fluid each well is producing. Therefore an interesting but difficult

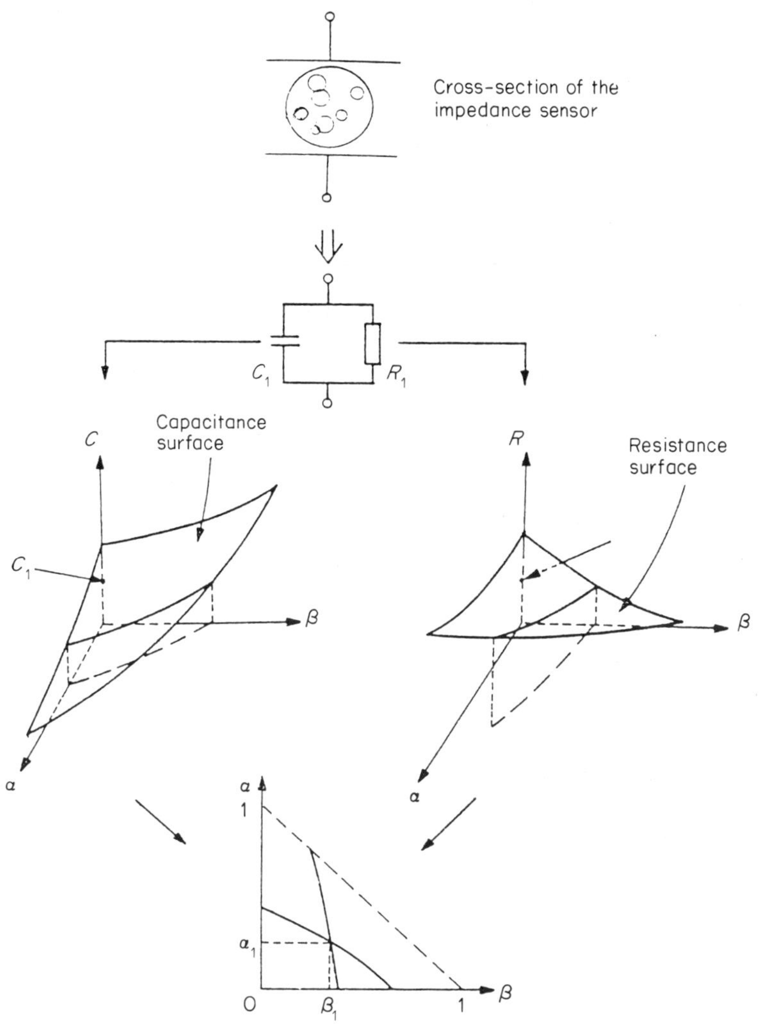

Figure 5.19 Using a model as part of a measurement system: (a) the relationship between impedance and component ratio

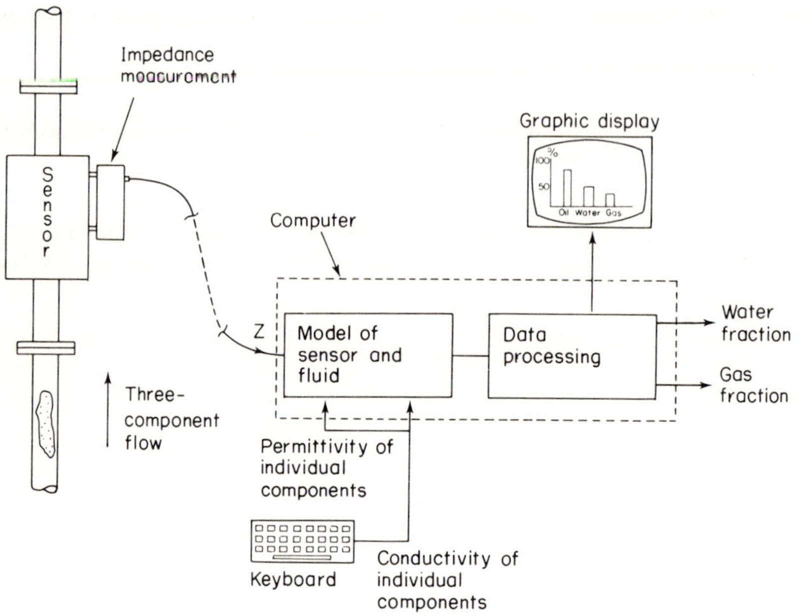

Figure 5.19 (*continued*) (b) block diagram of three-component measurement system

measurement problem arises, namely how can the component fractions of the oil/gas/water mixture be reliably monitored?

One solution is to use a method which makes use of the changing electrical impedance of the flowing mixture. The principle of this method is shown in Figure 5.19(a). Electrical impedance is measured across two electrodes, between which the oil/gas/water mixture is flowing. The impedance depends on the geometry and construction of the transducer, the dielectric properties of the individual fluids, the flow regime, and the fraction of gas (α)water (β) in the flow. However, if the first three factors are constant then the measured impedance will be a direct function of the component ratio.

The transformation between measured impedance and component ratio can be achieved in two ways. The resistance and capacitance surfaces shown in Figure 5.19(a) may be obtained from a multi-point on-line calibration and stored as look-up tables. This is a rather cumbersome solution and a more elegant (and practical) alternative is to model the relationship between measured impedance and component ratio. Inversion between the two can then be achieved using a small computer, as shown in Figure 5.19(b). A simple three-step calibration can be used to determine the electrical constants of the fluids being measured in a particular application.

This is an open-loop measurement system and so the accuracy of measurements will depend on the reliability of the model. Regular checks are required to establish if the process is still the same as the model being used.

Other examples of the use of models within the operation of instruments are the simple square-root calculations used in flow meters based on differential pressure measurements; look-up tables and characterized equations used to correct liquid content measurements of the storage tanks and the error-correction model now becoming common within *smart* instruments.

Such models can also be incorporated in hardware form if that is more convenient. The *standard eye* response is set up using a variety of optical filters set to produce a defined spectral response to light passing through the filter.

The preceding chapters have progressively developed a rigorous, theoretical and mathematical framework for studying the performance of measuring systems. It is, however, essential that instrument designers and users be aware of many practical issues: sensible application of parameters and knowledge of how instruments are designed and installed are the subjects of the following three chapters.

TUTORIAL EXAMPLES

5.1. Write short statements that describe the operation of each of the blocks forming the serial information chain of an instrument that is easily to hand. Suggest where pictures would have been a better form of communication.

5.2. For a nominated instrument system that is easily to hand, construct an iconic model of the system architecture using the matrix representation method. Label all stages and then develop the steady-state sensitivity expression after estimating the various input and output signal parameters.

5.3. The heart 'pulse rate' of a person can be measured by observing the changes in transmission of a beam of optical radiation sent through the lobe of the ear.

Develop a system architecture of an instrument to measure pulse rate in this way. Identify appropriate mathematical models for each block. Characterize the system with expected material properties and then run the model in a computer.

(This exercise would take several hours to complete. In some cases it might be better suited as a class discussion problem.)

5.4. Making use of books on transducers develop a suitable instrument to measure the viscosity of dough used in bread making. Develop the system architecture, decide what blocks are needed and appropriate mathematical models for each. Run the model, after addition of some practical coefficients, to discover its dynamic response and type.

5.5. Where MEDIEM, or a CAD package with similar linear equation modelling capability is available, use it to plot the response of the thermowell shown in Figure 5.17. The thermowell is to be placed into a vat making sugar syrup at a dilute state.

5.6. If finite-element software is available use the package to model the electric field distribution existing between the two parallel electrodes that are commonly used in fluid conductivity sensors. Use catalogues to obtain a specific design configuration. Once the field has been determined calculate the electrical parameters for the electrode.

REFERENCES

Abdullah F, 1985, Mathematical modelling aids transducer design. *Control & Instrumentation*, **17**, 73–75.

Brant M R *et al.*, 1985, Capacitance transducer measuring system for pneumatically conveyed cement powder. *Proc. Symp.* 'Reliable Flow of Particulate Solids', Christian Michelsen Institute, Bergen, Norway.

Finkelstein L and Watts R D, 1983, Fundamentals of transducers; description by mathematical models. In: Sydenham P H (Ed), 1983, *Handbook of Measurement Science*, Vol 2, Wiley, Chichester.

Hammer E A and Green R G, 1983, The spatial filtering effect of capacitance transducer electrodes. *J. Phys. E.: Sci. Instrum.* **16**, 438–43.

Maxwell J C, 1873, *A Treatise on Electricity and Magnetism*, Clarendon Press, Oxford.

Middlehoek S and Noorlag D J W, 1981, Three-dimensional representation of input and output transducer. *Sensors and Actuators*, **2**, 29–41.

Middlehoek S and Noorlag D J W, 1982, Signal conversion in solid-state transducers. *Sensors and Actuators*, **2**, 211–28.

Shu Ming T, Weinberger C B and Young H L, 1982, A simple capacitance sensor for void fraction measurement in two-phase flow. *Ind. Eng. Chem. Fundam*, **21**, 175–81.

PART B

PRACTICAL APPRECIATION

We notice that the geometricians are quite unable to apply their proofs in practice. When it comes to dividing a piece of land, or to any other operation on magnitudes and spaces, the surveyors can do it because of their experience, but those who are concerned with mathematics and the reasons for these things, while they know how it is done, cannot do it.

Accredited to Aristotle (384–322 BC) (as translated by T. W. Africa (1968), *Science and the State in Ancient Greece*, Wiley, Chichester.)

Ideas and theory alone cannot make effective instrumentation systems. The nature of measurement is such that designers must be very aware of relevant practice as each new measurement task is usually a 'custom' application.

The first two chapters of Part B develop the knowledge necessary for making the selection of appropriate hardware design parameters and for creating a design and then creating the actual instrument, by a systematic process.

The final chapter addresses the task of properly selecting and ensuring installed instruments are performing to expectations.

CHAPTER 6

Handling and processing information

6.1 THE ALTERNATIVE TECHNOLOGIES

6.1.1 The energy forms

It has been explained previously that information is conveyed, as signals, through the measuring instrument via the modulations existing on the energy (or to a much lesser degree, mass) state of the instrument system. As the signals traverse from inputs to outputs the information content is processed accordingly as discussed in previous chapters.

This chapter now delves deeper into the nature of the alternative energy forms that can be used in the implementation of an instrument assembly.

Many forms of energy can be used in measuring instruments, the choice depending on the particular application. For example, high-speed processing of a visual image can be achieved using a hybrid combination of optical and electronic information forms.

Example 6.1 Energy forms in a humidity-measuring instrument

In the humidity sensor, shown in Figure 1.3 and repeated here as Figure 6.1 with its organic parts labelled, the initial form of the energy (chemical) is converted into a mechanical equivalent with the hygroscopic membrane. The information about humidity remains in mechanical form up to display as a rotating pointer where the user sees it in the optical domain.

The information is also converted into electrical form by use of resistance strain gauges placed upon the bending beam deflected by the lengthening membrane.

When observed by a human the mechanical energy (pointer position) is read by the eye which is an *optical* transducer sending signals to the brain by *electrochemical* signalling and processing.

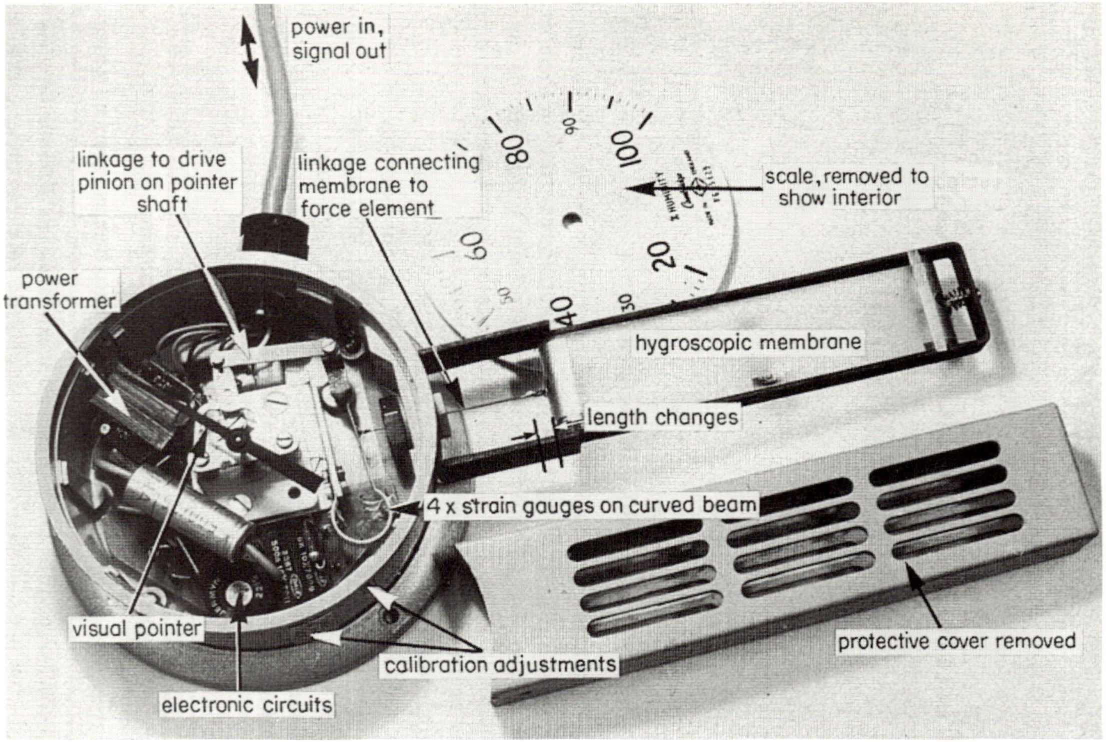

Figure 6.1 Organic structure of a humidity meter. (HBMS, figure 15.3) (Reproduced by permission of Kent Industrial Measurements Ltd, St Neots, Cambs, UK.)

When recorded by a chart recorder the electrical energy output signals are used to produce mechanical energy (pen position) with mass transfer (ink on paper) as the memory function.

For convenience (see Section 5.2.1) it is generally workable to classify the energy forms as electrical (including electronic), mechanical (including fluid), thermal (more correctly radiative and including optical) and chemical. Consideration will show that these are not uniquely separable; energy forms cannot truly be divided into distinct classes at the pragmatic application level of understanding.

The relative merits of each form, when applied to instrument use, are shown in Table 6.1. It is left to the reader to develop a lasting awareness to which is best to use in a given situation by learning the characteristics of each. Stein (1970) is relevant reading.

In addition to the heuristic knowledge given in Table 6.1 there exist widely applicable mathematical ways that can be used to describe the behaviour of each in a universal manner. Principal areas are network theory, the 'analogies' and Fourier techniques. Chapters 3–5 cover that aspect in more depth.

6.1.2 Abstract simulation of the various energy forms

All physically realizable measurement systems are concerned with energy; however, there is no single universal mathematical-modelling approach. The 'analogies', for example, cannot be applied to each form to the same extent of applicability. Other forms of modelling are better to use in particular cases.

Electrical, mechanical and fluid situations work well as a group especially if the systems behave in a linear fashion. Compressible fluids (being non-linear) give rise to the need to extend the 'analogies'. Optical systems can make use of some of that theory but tend to be better modelled using the Fourier methods commonly applied in electronic engineering.

Suffice it to say that, by use of a collection of mathematical methods, signal behaviour in each energy domain can be formalized by suitable approaches allowing them to be studied in the abstract state before actual hardware is manufactured. The methods also often allow a mixed system to be considered, in its mathematical representation, as a unified model.

6.1.3 Mixing energy forms

The mix of energy forms to be used in a particular instrument application will be decided by consideration of the particular system parameters present for each case. Although the electronic form is most commonly the best choice there will usually be need to use others if only because the original signal in the real-world state is most likely to be of another form. Instrumentation design and operation needs much wider knowledge than electronic engineering provides.

It is sound practice to combine energy forms in the serial information pathway through an instrument (see Sections 5.2.1 and 5.2.2). That often produces the best overall information system design.

Table 6.1 Comparison of use of energy regimes in information-machine applications. The emphasis is on ability to process and transmit information. Each group provides sensing and actuation

Parameter	Electrical (inc. electronic)	Mechanical	Thermal (inc. optical)	Fluid	Chemical
Mathematical modelling usefulness and applicability (analogies apply)	Excellent if linear	Excellent if linear	Low accuracy and only partially able to use 'analogies'	Excellent, but often non-linear	Poor, not related to other groups
Level of sophistication possible per unit resource expended	High	Poor	High in optical regions	Poor	Barely yet applicable
Theoretical aids relevant to instrument development	Well developed	Well developed	Well developed	Medium	Not yet developed
Ease of change in development model	Fast, very low cost	Slow, costly	Medium speed, costly	Slow, costly	Medium speed, medium cost
Relative cost for given information handling	Cheapest of all applicable tasks	Dearest	Cheap for high bit/pixel density	Cheap but only applies low bit density	Not yet relevant
Size for given information handling density	Smallest in use, microminiature	Small miniature	Small to bulky	Bulky	Potentially smallest
Skill and training level needed to develop average instrument	Medium hand skills and systems understanding	Good skills, low understanding of systems	High understanding (high skills, not readily available)	Specialized knowledge but low skill	Not yet known

Time to design, build product at low volume run	Days to months	Months to years	Months to years	Short, if use kits	Not yet relevant
Prototyping aids	Plug-in boards, CAD, PC boards, good component availability	Construction kits, ready-made componentry	Optical benches, components available	Rapid join devices available	Not yet relevant
Maintenance capability of final instrument	Very good if designed for such	Often not easy to repair rapidly	System easily maintained but highly susceptible to environmental influences	Medium	Medium
Obviousness of principles to relatively untrained persons	Not at all apparent	Largely self-evident	Often apparent but misleading	Not apparent	Not yet understood
Rate of technological change in instrument use	Traumatic, 1–3 year relevance and continuing to speed up in all areas	Virtually stagnant but maintaining relevance, lost momentum in 1950s	Steady change in 'modern' optics, classical, maintains relevance with small change	A burst in 1950–60s, stagnating	Awaited future dominance if biological systems provide example
Ability to handle and process information	Vast	Very poor	Vast	Very poor	Not used?
Time of development of discipline in terms of information application	150 years	6000 years +	400 years	30 years	Yet to come

(*continued*)

Table 6.1 *(continued)*

PARAMETER	ELECTRICAL (INC. ELECTRONIC)	MECHANICAL	THERMAL (INC. OPTICAL)	FLUID	CHEMICAL
Closeness of first model to production version using prototyping aids	Very close, if not identical	Distant	Close	Close	Not yet relevant
Speed of response (depends on device and its design)	0–10^8 Hz serial mainly, parallel in development	0–10^4 Hz serial/parallel	'Speed of light' limit, parallel mainly	0–10^3 Hz	0–10^2 Hz is likely, parallel use
Areas of best utility in information machines	Electronic data processing	Physical support systems, trivial processing	Image processing, chemical instruments	Fluidics, actuators, pneumatic sensors	Barely applicable except for chemical analyses
Typical elements	Transistors, capacitors, conductors, resistors, switches, ICs, LSI, VLSI microprocessors	Gears, frames, supports, linkages, shafts, belts	Lenses, prisms, mirrors, gratings, stops, polarizers	Fluidic, pneumatic devices	Yet to be revealed

Each energy form must be carefully interfaced as the information traverses the system so as not to degrade the information content of the signal from that desired at that point.

The practice of sensing systems is well represented in many 'catalogue' style books such as Mansfield (1973), Noltingk (1988), Norton (1969) and Sydenham (1984).

More depth is now developed about each of the above energy regimes.

6.2 ELECTRICAL REGIME

6.2.1 Direct current systems

Although the junior energy discipline, electrical energy knowledge is highly developed. Electrical network behaviour is of relevance because it underpins electronic theory and application.

Electronic (the information handling use of electrical energy) technique has become the dominant information-processing methodology due to its ability to support manufacture of highly reliable, small, and exceedingly complex, sophisticated systems with a reasonable expenditure of effort.

The energy variables here are the *voltages* V that force electric *currents* I to flow. These have counterparts in the other energy forms; for example, force causes motion in mechanical systems. The relevant electrical theory is that pertaining to how currents flow through *networks* as the result of the voltages existing to cause the current flow.

At their simplest level there are the direct current (d.c.) laws - Ohm's law, Thévenin's and Norton's theorems, Kirchhoff's laws - that allow simulation and hence study of circuit networks in their abstract form. These laws apply when the system state is operating in a linear, superposition, fashion. These laws also apply to other energy forms, in fact, many were discovered by study in other disciplines. Today they tend to be associated with electrical engineering.

They describe system conditions when the energy distribution within the network has settled to a static state. They explain how currents and voltages at various places in the network are related as the result of complex paths made up of components (resistors having *resistance* R) that oppose the flow of current.

These laws also have been extended to explain where the energy dissipates in the various components. The resistor can only dissipate energy by converting it to the thermal form; it cannot store electrical energy. It is this factor that restricts the usefulness of the d.c. based information machine. Many sensors and processors make use of d.c. circuitry to good effect.

Example 6.2 Measuring position of a shaft with a potentiometer

A potentiometer is an electrical device in which a mechanical input changes the position of an electrical contact along a fixed resistance that is placed along the traverse according to some law - linear and logarithmic are commonly used. If attached to the end of a turning shaft, then it can be used to give shaft position - see Figure 4.11. These are used for partial circle angle sensing when low cost is desired. (The linear form has been discussed in Examples 4.5 and 5.1.)

Assume a reference voltage of 10 V d.c. is applied over the ends of the resistance of 1000 Ω that is spread uniformly around 300° of rotation. Simple d.c. laws enable the sensor performance to be predicted and the system designed for a long life.

Ohm's law, applied to a voltage divider, can be used to show that the voltage change per degree of rotation is 10 000 mV/300° = 33.3 mV/degree. This is the system sensitivity - provided the signal is read with a device that does not electrically load the potentiometer at a significant level.

Laws about energy generated allow the power dissipation that will take place in the potentiometer to be calculated as

$$\text{watts} = 10^2/1000 = 100\ \text{mW}$$

A suitably large potentiometer would have to be used to avoid its overheating and destruction.

6.2.2 Alternating current systems

In many types of instrumentation it is vitally important that the dynamic behaviour of the signals are understood, for these fluctuating states are the important signals that convey information.

By operating the electric network energy in a constantly modulated manner, that is, as an alternating current (a.c.) system it is possible to greatly extend the usefulness of electrical energy to information machines. Example 6.3 illustrates how it is used in a position measuring device.

Example 6.3 Alternating current sensor – the linear variable differential transformer (lvdt)

A commonly used displacement sensor is the linear differential transformer (lvdt) - see Figure 4.30 in Example 4.15.

In a normal transformer an iron circuit is added to closely couple the windings, in which case the transformer converts voltages closely as the inverse of the turns ratio. In this application, however, it is the variable coupling formed by moving the iron core that makes it a useful sensor. An a.c. signal must be used to bring the transformer action into operation.

The performance of the system can be analysed by use of a.c. theory with regard to predicting the output given knowledge of the input excitation voltage and frequency and coil inductances (which is not easy to calculate to the required precision for instrument use).

Carrier modulation and demodulation theory enables the output of the detection system to be predicted, as done in Example 3.15.

When a.c. currents flow in coils (called inductors) or are fed to the terminals of insulated conducting plates (called capacitors) they can be used to add, or deplete, electrical energy stored in those two components. This behaviour requires further laws of explanation, for the d.c. laws do not always apply when electrical energy interchange is taking place.

Primary a.c. laws are those that relate the currents and voltages of changing energy states in networks comprising the resistor, inductor and capacitor. The behaviour of the a.c. system is more complex than that of the d.c. system but again there exists well-organized knowledge of how to model what is taking place.

These fundamental laws provide the basis for many more that allow the behaviour of components, formed into special networks, to be explained. A level of explanation higher than this is to be found in Sydenham (1983) where reference to yet deeper explanation is given.

Basic a.c. laws relate to how networks having a.c. states behave as the three basic components - resistors, inductors and capacitors - are combined. These are known as the *passive components.*

Passive component a.c. systems provide more ways, beyond d.c. methods, of enabling the information content of a signal to be varied in a systematic manner. This primarily arises because the *apparent resistance* or *impedance* (the capacitive X_c or inductive reactance X_L) to current flow in the inductor and capacitor alters as the frequency of the alternating current varies. The phase of the current variations, with respect to the voltage causing them, also varies.

Figure 6.2 shows how the reactance of an inductor X_L and a capacitor X_c change as the exciting frequency f varies.

If an inductor and a capacitor are connected together a useful performance feature arises. When the reactances of each are equal in magnitude their phases will be opposite; the effect is to produce a circuit that has overall resistance to current flow of zero, or infinity, depending on whether connection is in series or parallel and how they are connected to the source. The frequency at which this

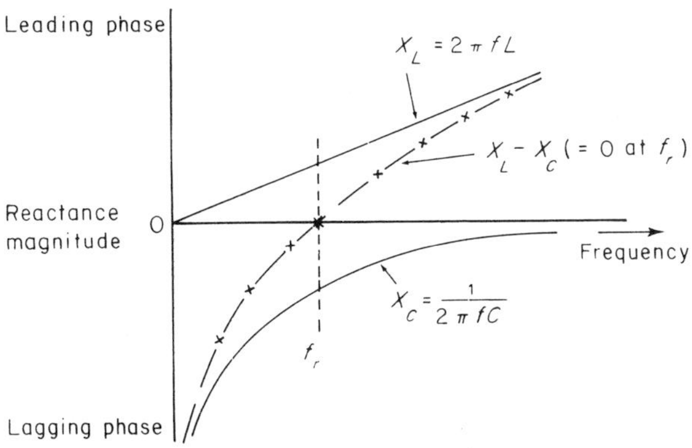

Figure 6.2 Reactance variations of the inductor and capacitor as frequency varies. (HBMS, figure 22.4)

occurs is called the *resonant frequency* f_r. These are second-order systems (see Section 4.4.6).

Information is conveyed as fluctuations. These complex time-varying signals can be broken down (in theory and in practice) into sets of simultaneous signals of sine-wave, time-varying, form. Each will have different amplitude and phase relationships. This is the essence of Fourier theory.

The ability to consider signals in terms of fundamental sine waves is highly developed in electronic engineering. The methodology has applicability to other energy forms, notably mechanical and optical to varying usefulness.

6.2.3 Active network elements

Whilst much can be achieved in information handling with electrical energy sources using networks comprising only passive components the information-processing possibilities of electronic systems burgeon significantly when *active devices* are added.

An *active element* is one in which electrical energy is supplied from a source to enable the energy level of the signal to be increased.

The most commonly met active element is the *amplifier*. A popular misconception is that the amplifier magically increases the energy level of an incoming information signal. The magnitude of the input signal is apparently increased by using that signal to form a different replica from a source of energy, called the *power supply*. The operation of amplifiers of any type – the concept applies on all energy regimes – is shown in Figure 6.3.

The most common active component in electronic systems is the *semiconductor transistor*. The relationship between the input controlling base current I_b and the output voltage V_{CE} and output current I_C are shown in Figure 6.4. Study of this set of characteristics shows how the system can be used to amplify input signals or convert them from currents to output voltages.

As well as providing the means to amplify small signals active devices are used as the basis of a very wide variety of information-processing functions which are thereby realized in a close to ideal manner. For example, use of an amplifer in a

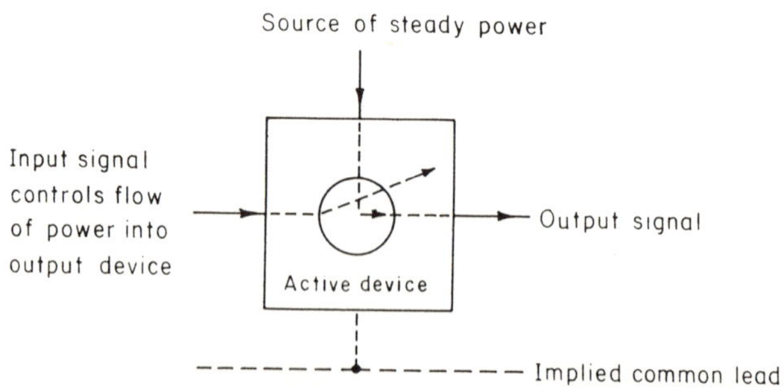

Figure 6.3 Generalization of how an amplifier increases the size of a signal. (HBMS, figure 22.5)

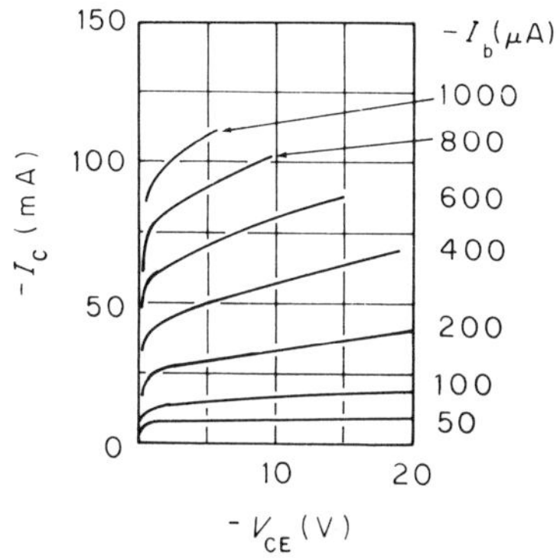

Figure 6.4 Characteristic curves for a typical transistor. (HBMS, figure 22.6)

special passive network (called the *operational amplifier*) enables improved amplification fidelity or, as another example, near perfect rectification.

The frequency content of signals can be altered using filters; sine-wave and complex wave-form signals can be generated; power supplies can be made with stable voltage or current output as the demand on them varies.

6.2.4 Other componentry

To make practical electrical/electronic circuits it is necessary to use many other devices such as switches, wires, terminals, mounting boards and indicators. These are well developed and easily procured at low costs. Their use to form networks requires only a modest amount of hand skill and support plant. Individual elements are easily changed. Many are made variable in value by turning a screw, this allowing circuit performance to be tuned after assembly. Figure 6.5 shows a collection.

6.2.5 Building-block approach

If each new information-processing system had to be developed from basic components upward their cost would be enormous.

Making use of the above-mentioned relatively few basic passive, active and supplementary components the discipline of electronic engineering has progressively developed commercially available building blocks that cater for the often called-for higher-level functions.

These *chips*, as they are called, are most commonly supplied in *integrated circuit* (IC) form, packaged in plastic. They are mounted on printed circuit boards. At the higher level of integration there exist chips built with over 200 000 circuit elements. The system designer only needs to know how they are used, not how every component operates inside the chip.

The electronics supply industry constantly increases the range and complexity of available circuits. Functions divide into two main groups according to the

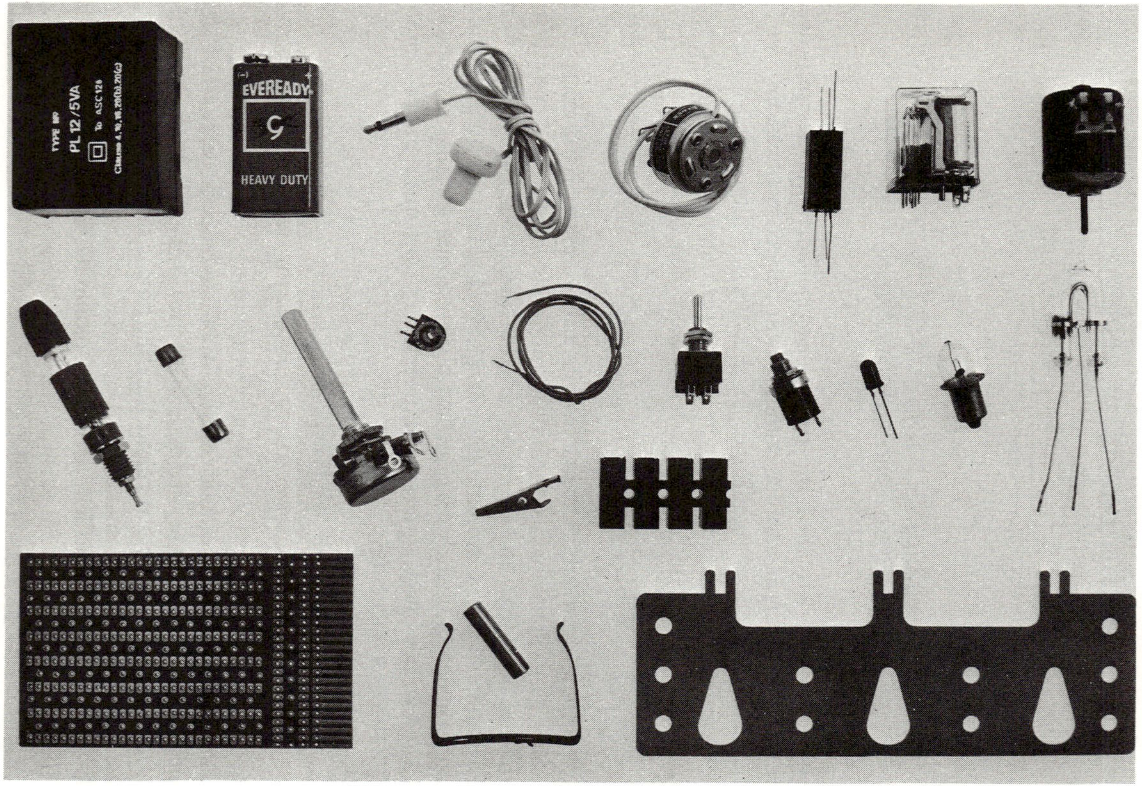

Figure 6.5 Collection of typical supplementary parts used to form electronic information systems. (HBMS, figure 22.7)

nature of the signal they handle–the *linear* and the *digital* families. Both look similar in their packaging. Figure 6.6 shows facets of the technology involved in linear and digital electronics.

The term *linear* (also known as *analogue*) derives from the fact that these circuits are concerned with signals that are continuous over a range. For example, the output signal from a common linear gain stage would swing over a 24 V excursion with every voltage increment possibly representing different information, say the different temperature of a water bath.

Linear circuits are mainly based on the use of high-performance operational amplifiers (op-amps) that have very high gain (order of 10 000 or more), this being a condition for their use with the feedback needed to form a whole variety of linear signal operations

Using the basic op-amp connected with a few passive components, it is possible to provide hundreds of different analogue signal processing and supply functions. A few examples are gain stages to amplify signals, converters to change signal forms from voltages to equivalent currents, power-supply regulators, signal wave-form generators, and filters that alter the frequency content of signal.

Digital signals, in contrast, carry information not as the variation in magnitude but as a coded use of *on–off states*. A typical digital signal changes, virtually instantly (in nanoseconds) between 0 V and 5 V, the serial sequence of these transitions indicating part of a digitally coded signal. Morse code was one of the first digital codes to be used in electrical engineering.

The digital family, although now most extensive in the level of developed options that are available, are all composed from a small number of generic functions. These are the family of *gates* that perform Boolean algebra logic operations (AND, OR, NAND, NOR, EXCLUSIVE OR, NOT), the *multivibrator* family (monostable, bistable and astable) and the *converter* family that change linear signals into a digital form in a limited manner (comparator and Schmidt trigger).

Digital blocks are combined to form computational machines wherein literally millions of such devices are employed.

6.2.6 Electronic systems development

The highly systematic and ordered nature of electronic systems has made possible progressively easier means to assemble systems. This, however, has allowed them to become more sophisticated, meaning that their top-flight design still needs considerable expertise.

A circuit assembly is designed as follows. First a system requirement is generated in terms of what signals are involved and what must happen to them. This is then progressively transcribed into a system schematic diagram formed of available building blocks that perform the necessary information-processing task for that point in the system.

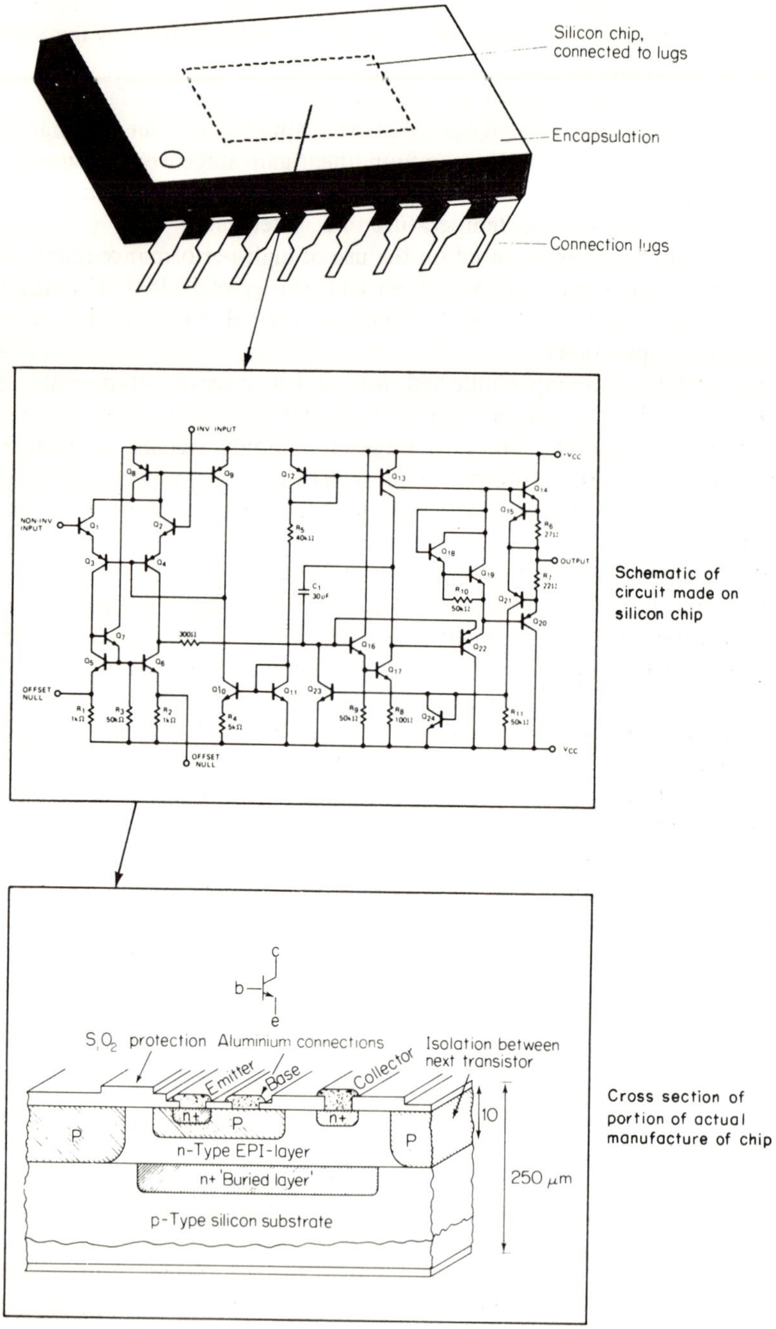

Figure 6.6 Technology of linear and digital ICs. (HBMS, figure 22.8) (Middle part: reproduced by permission of NS Electronics Pty Ltd.)

Each block is then designed in detail for the given application which will need the values of components to be calculated for the particular task. Text books, manufacturer's application notes, past experience and, with growing importance, CAD software, can then be used to aid this process.

As each block is detailed and component values sized it is then built into the appropriate form of technology.

It could be an assembly of IC packages mounted on a *printed circuit board*; a *hybrid circuit* made from printed and baked paste wiring and components, along with unpackaged ICs all mounted on a typically ceramic substrate; or it could be as a specially commissioned silicon chip, application-specific integrated circuit (ASIC), designed with CAD software and made at dedicated fabrication plants called *foundries*. The latter technology groups into those ICs that make use of already largely completed assemblies (semi-custom) and those that allow every part of the design and fabrication to be varied (full-custom).

As a block is made it is tested for its functionality and then added to the whole system, testing performance of the whole as each part is added. Interfacing, that is, getting each block to work correctly into and out from another, is a major, but systematically soluble, concern in system design. Modern electronic chips make this reasonably straightforward by including interfacing circuitry.

The ease at which each block can be changed or added facilitates rapid design but also can allow a certain slackness in prior consideration of suitability. Good planning and prior design is a better path than a 'build-and-try' approach.

When the whole system appears to be working adequately an 'engineered' version is usually then needed to streamline assembly, parts placement, maintenance and other cost and use factors.

The low cost of complex electronic consumer products has led to the incorrect belief that *all* electronic systems will be of low cost, perform well and possess high reliability. Unfortunately this only applies when the volume of production allows extensive development and testing. Small production number systems cannot be made at the same low cost per unit and will be considerably more expensive to develop. This is particularly true for instrument systems as they rarely are made in volumes exceeding thousands. Even so, quite adequate electronic systems can be assembled by persons having relatively little formal training in this field.

6.2.7 A simple electronic instrument example

As an example of electronic information processing practice consider the pressure gauge (Georgaras, 1986). This makes use of available resistance strain gauges that are interrogated with a d.c.-excited Wheatstone bridge.

Example 6.4 Electronic aspect of a pressure gauge

Figure 6.7(a) shows the general construction of a pressure gauge. The block diagram for the information flow from the pressure entering the input port to the output electronic port is given as Figure 6.7(b).

(a)

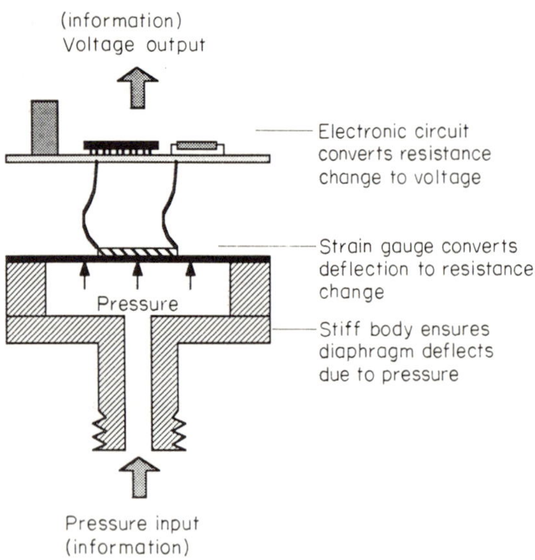

(b)

Figure 6.7 Pressure gauge as example of electronic assembly: (a) pressure gauge without cover; (b) information flow

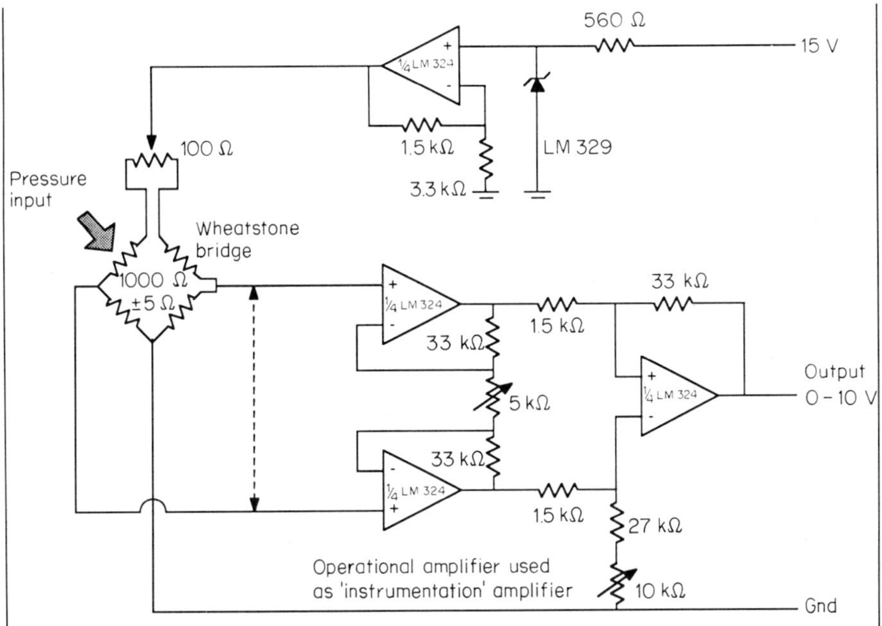

Figure 6.8 Circuitry of pressure gauge shown in Figure 6.7

The information flow begins in the mechanical-fluid regime and is transduced into equivalent signals in the electrical regime via the strain gauges mounted on the sensing diaphram that deflects under pressure change.

The electrical Wheatstone bridge is used to detect small differences in strain appearing as voltage changes. The voltages are amplified with *op-amps* (three are used together to obtain superior amplification) and then are conditioned into a form suited to the signal transmission standards of the process industry. The Wheatstone bridge and op-amp circuitry are shown in Figure 6.8.

A printed circuit board (like that shown in Figure 7.8) is used to carry the proprietary components and to allow firm reliable connections to be made between them. Linear circuits, of which this is one, often require adjustment of values to obtain correct operation. The trimming potentiometers, needed in this case, are clearly visible.

Space does not permit more to be explained here; Chapter 22 of Sydenham (1983) builds on this, leading into the literature. Tran Tien Lang (1987) is a relevant companion volume to this course. The rapidly changing nature of electronic technology requires constant updating of design knowledge to ensure the latest methods are used.

6.3 FINE MECHANICAL REGIME

6.3.1 Basis of mechanical design

Compared with the electronic regime the mechanical one is far less powerful from the processing aspect. As a general rule it is not practical to use mechanical systems to process other than small amounts of information.

However, it is important that the characteristics of this regime be understood, for all systems have to make use of it. For example the electronic transistor must be implemented with materials which behave in the mechanical, as well as electrical, regime.

Instrument design requires its mechanical regime to be considered as part of an information-processing machine. Here information is conveyed through the system as signals represented by changes of force or motion.

Even the apparently benign framework must be carefully designed, or its changes of shape, caused by thermal effects or forces from the mechanical signals, will give rise to erroneous mechanical changes that look like wanted effects.

A dominant objective when designing mechanical information machines is to create a system of adequately controlled movements.

An extended object in space has three degrees of *translatory* and three degrees of *rotational* freedom. As signals are modulations of motion (sometime forces are used, being the *dual*) the aim is to devise a structure of frames and mechanism that move in a defined manner.

Examples are the one degree of translation of a microscope travelling on a measuring slide, the three degrees of translation for a three-dimensional coordinate-measuring machine and the one degree of rotation of roundness measuring machine.

Two different philosophies for mechanical design exist – *kinematic* and *elastic.* Whilst being quite different in approach they can be combined in a design.

In the *kinematic design* philosophy the aim is to locate all parts relative to each other, whilst allowing degrees of freedom as needed, by connecting points together without significant elastic deformation. Figure 6.9 shows three examples of kinematic couplings. In Figure 6.9(a) three rotations are possible for the ball, whereas in figure 6.9(b) and (c) no freedom to move is provided.

A simple theorem that allows calculation of the number of points of contact was enunciated by Strong in 1938. He defined

> kinematic design is correct when a body in contact with another has at least $6 - n$ points of contact where n is the number of degrees of freedom existing.

If the system has more than $6 - n$ points of contact it has *mechanical redundancy* and the assembly must invariably then also be invoking elastic design principles which need further consideration.

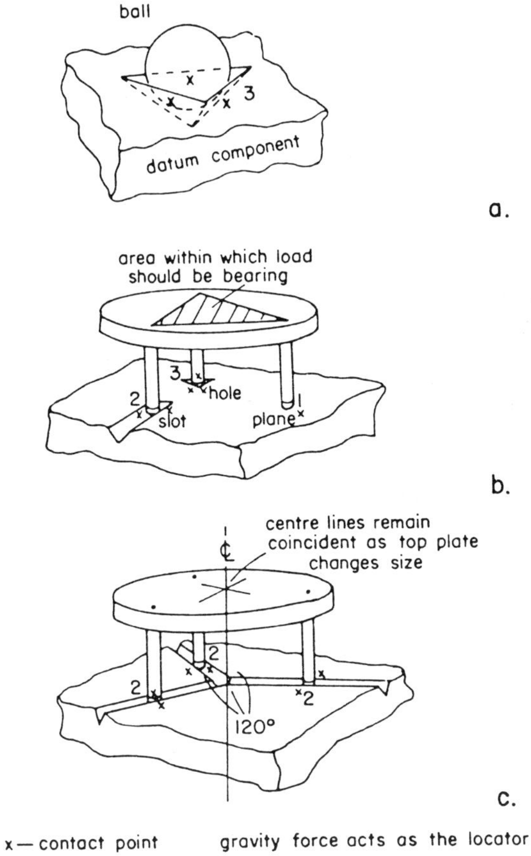

Figure 6.9 Kinematic couplings: (a) all three translations are prevented; (b) hole–slot–plane in which all freedoms are prevented; (c) slot–slot–slot variation of case (b). (From Sydenham, 1986. Reproduced by permission of Instrument Society of America.)

Example 6.5 Kinematic support of dividing engine turntable

In one type of *radial-line grating generator* a disc-shaped dividing turntable is rotated by a small d.c. electric motor. The motor turns the 400 mm diameter, 10 mm thick steel disc via a driving wheel that is bearing directly on the outer upper region. This *friction gearing* enables the gear ratio to be infinitely varied.

On the driving-motor shaft is mounted an optical encoder disc that provides 1000 pulses per driving-wheel revolution (wheel diameter of 20 mm). Each pulse, consequently, corresponds to nominally 1/17 000 of a full circle; it is used to flash a line on photographic film fastened to the disc. The number of lines exposed around the full circle is decided by the radius setting of the driving wheel and the division ratio of the shaft pulses. For the dividing action to be precise and accurate in its line placement the system mechanical gearing ratio must be as uniform as possible. Being friction based any variable torques in the disc support arrangement will vary the instantaneous place at which each line is exposed.

A kinematic mount was used as follows. The disc was supported at the centre, underside, with a small point bearing (controlling the three translations). The drive wheel was set to push down on the upper surface to obtain 'grip'. Two more supports were needed on the upper surface, at 120° apart from the drive wheel, to control the final two rotations and maintain the disc held in place on the underside point bearing.

Provided these two upper supports have low friction in the direction of motion (wheels are used instead of points) the system has the required single freedom – rotation of the disc. The system is non-redundant and therefore is a true kinematic support arrangement.

The system could divide the circle to a repeatability of 1/22 000 of a circle repeatability and all parts, except for small point bearings, were made with low-precision machine tools.

In the *elastic* design alternative the components are joined together as a contiguous unit obtaining the required degrees of freedom by virtue of elastic deformation of the system that is allowed to arise in the appropriate place. This alternative is restricted to motions, relative to the structure's size, that are small. This is primarily because the usable proportional *elastic limit* (point beyond which permanent deformation takes place) in materials is rarely more than 1 part in 1000 of the deformation.

Example 6.6 Elastic elements used in a humidity meter

An example of the use of elastic mechanical components is to be found in the humidity meter shown earlier in Figure 6.1.

Conversion of the linear motion of the changing length of the membrane to an electrical form is achieved by placing strain gauges on the curved spring that also tensions and controls the movement of the membrane. Elastic guides are often simultaneously used as measuring elements called *load cells* whilst performing a constraining role.

The membrane changes length only by about 1 part in 100, so an elastic tensioning and measuring method is suited. It is also compact and allows use of well-known sensing of the elastic element.

Other membrane, or fibre, humidity meters exist that use capacitance and inductive sensors to sense the small length changes.

Where the elastic method can be used it provides a low-cost, high-performance solution.

When creating mechanical systems care is needed in deciding the necessary system elements as changes are not as easily made as they are in the electronic information machine. Interfacing of one stage to the next usually requires careful design of each part for it is not practical to later add extra elements virtually at will.

Throughout the development of the mechanical engineering researchers have sought, without worthwhile results, a building-block approach to mechanical systems. Mechanical design of instruments still must use many parts that are customized to each application. Exceptions, such as the Swiss clockwork watch industry, are rare. This means that the design of mechanical parts of an instrument can be slow compared with the associated electronic parts. Furthermore, design changes tend to be conservative because of the cost and time constraints inherent in the mechanical aspect of instruments.

6.3.2 Componentry of fine mechanics and its design

It is assumed here that the reader will be familiar with basic mechanical design including how to assess loads and stresses in components.

Parts that are an assembly in an information machine must keep their shape to maintain fidelity of information. Their design tends to be decided more by the allowable *deformations* than by *strength* factors. It is essential that parts are not stressed into the plastic region as they will then permanently change shape.

Constructional members are the most basic element of a mechanical system. Where they are fashioned to have large length-to-thickness ratios they need special consideration under a compressive load. Examples are columns, beams and plates. When used in tension the slender members can be loaded higher than when in compression; if made too slender they will buckle with compressive load well before failure arises because of excessive load.

Linkages are members joined together to provide the required degrees of freedom. An example is the pantograph, which provides duplication of a motion at altered size.

Analysis of linkages can become very complex. Consider working out the individual motions of four links joined in line with a one axis bearing at each joint. The movement of a robot arm involves linkage analysis.

Two types of *joints* exist for coupling members together without allowing any freedoms across the joint — those that are made to be *separated* and those that are *permanent*. Many kinds exist and each must be designed carefully to provide an adequately strong and stiff joint between members and structures. They must also often provide accurate positioning as the joint is made.

Joints that allow some degrees of freedom, called *bearings* or *pivots*, are used. The ideal bearing is one that does not possess friction. Examples are the pivoting action of the coil assembly of an indicating meter and the bearings supporting shafts carrying gears in a gearbox.

Whilst bearings must allow movement they must often still provide an adequately *stiff* joint that controls system geometry. Lost motion in this way can mean information being corrupted. For example, play (such as slop or excessive compliance) between the nut and lead screw of a measuring machine will provide incorrect dimensional measurements of parts placed in it for inspection.

Numerous types of bearing exist. Mechanical bearings are either of the

- *friction* type, where the two surfaces slide on each other, or
- *rolling* form where hardened balls roll between contacting surfaces.

Ball bearings have *coefficient of friction* values (0.001 is common) much less than friction bearings, but do not move as smoothly for very slow rotation rates.

Magnetic, oil and air film bearings are also used where very low friction, high speed or smooth motion are involved. They are much more costly than simple mechanical forms and may require support systems in operation.

Instruments often need conversion of rotary motion to linear, and vice versa, and amplification or attenuation of movements. Gears, cams, linkages and other mechanisms can be used. Again their design must ensure that they provide accurate motions with varying load and with time as they wear.

Amplification ratios here must be generally preset in the design as they are costly to change once installed. Variable gearboxes and gear-wheel changes are an inferior method for tuning a design as they are expensive and lack precision.

Transferring information in mechanical form over long distances is seldom practical as the members have to be of considerable dimensions to control their flexibility. Railway, trackside, signalling wires cables are examples.

The dynamics of the elements must also be considered as their mass gives rise to increasing deflections and resistance to movement as accelerations increase. Only by careful attention to reduction of the mass and intertia of parts and to optimizing their stiffness can mechanical systems be made to respond at frequencies exceeding around 1000 Hz.

Design in this regime tends to first realize, on paper, the required functions of the respective parts of the system. Parts are then designed from structures, members, linkages and other components to perform the task, ensuring that each unit interfaces to the next as they are designed. As the whole assembly will be three dimensional and probably have to fit into a pre-given space, considerable effort is needed in conceptualizing the whole before final details of the parts can be set.

6.3.3 Materials a limitation

The performance of the mechanical aspect of an instrument is highly dependent on the physical properties of materials used – perhaps more so here than for general mechanical design because the low-volume instrument market has not provided resources to develop the special, more ideal, materials it would like to have available.

The purpose of a measuring instrument is to perform a processing function on the information embodied in a stage. A selected mix of materials is used to obtain a design that enhances its primary information-processing role whilst depressing unwanted effects.

The most prevalent influencing parameter is temperature change for this alters many properties, notably dimensional size and elasticity.

When fine detection and faithful information handling is needed, the adequate reduction of unwanted effects usually requires compromise between such effects because of the lack of suitable materials.

Example 6.7 Designing a spring element for measurement use

Consider the design of a high-precision helical spring used as the load-cell element of a weighing scale. The spring element would, at first sight, appear to need a performance that is independent of its temperature of operation – but this is not so!

Temperature changes would change the size of the support structure. As weight is measured here as a change in spring length, compared with the support structure's size, a movement in the zero position would arise if the temperature of the weighing scale changed altering the scale's calibration. So, a zero *temperature coefficient of length* of spring is not the choice after all (unless the structure has a zero temperature coefficient – which is very expensive to make!).

What is needed is a spring material with a thermal coefficient of length that matches that of the structure. As a common structural material is steel, that appears to be the material to use for making the spring. However, this is not necessarily so either!

Materials with various thermal coefficients of length exist (Figure 6.10), but their elasticity constant (Young's modulus) varies with temperature. The magnitude of this effect is given by the *thermo-elastic coefficient* for a material. It generally is negative, that is a spring becomes less stiff with increasing temperature. Unfortunately ordinary constructional steel is quite poor in this respect and thus is not a good precision measuring spring material. So what material do we choose for the spring we need?

To date no material with a thermal coefficient of expansion the same as steel is known that has a low thermoelastic coefficient, so there is no solution to this design problem by way of selecting a suitable material to leave no residual error.

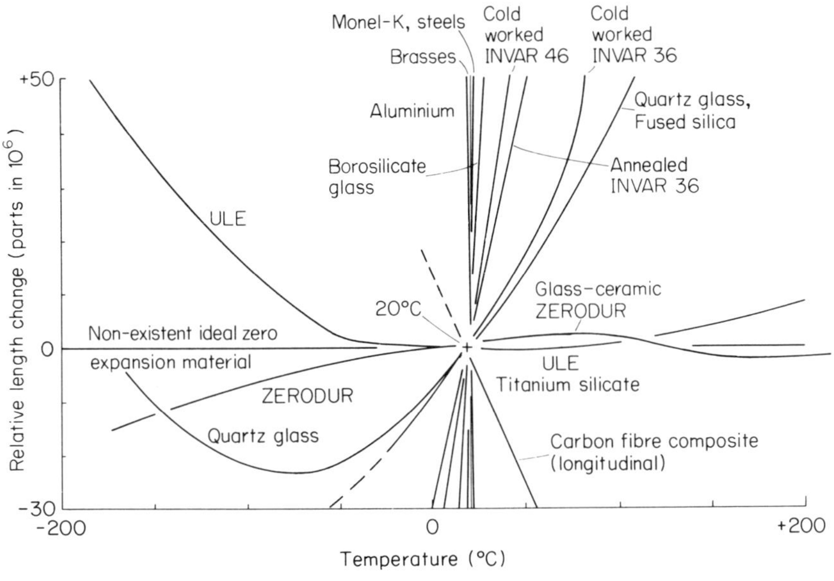

Figure 6.10 Thermal expansion of materials used in instruments. (From Sydenham, 1986. Reproduced by permission of Instrument Society of America.)

Special, relatively uncommon, materials do exist that have a reasonable combination of these two thermal coefficients. These, provided their properties are understood, allow a compromise design to be developed that can then be compensated by other means.

Consider that temperature effects have been overcome using the special material that best suits this need. That is still not the end of the design task for there exists another important limitation to be accommodated. Spring materials must have the ability to return to precisely the same place after each weighing excursion has been undertaken as loads are applied and removed.

All materials exhibit some degree of *mechanical hysteresis* (the property of not following the same characteristic with cyclic input). Special alloys exhibiting almost no hysteresis have been developed for making springs – but they do not have the required value temperature coefficients, and so on, as the designer chases one effect after another in the search for the perfect, error-free, instrument.

Fine mechanical design is further complicated by the fact that the declared coefficients are seldom actually fixed values. Coefficients are normally dependent upon such factors as

- history of mechanical stressing of the material (work hardening will extend the elastic range of a spring but not the elastic modulus);
- long exposure to changing temperature which will relax out built-in stresses, giving a more stable material;
- annealing, which will significantly alter hysteresis and magnetic properties;
- the respective influencing parameters: the temperature coefficient of length depends on the temperature of operation, as shown in Figure 6.10, as does the temperature coefficient of elasticity.

If the application has large magnetic fields present, calls for high corrosion resistance, very low drift with time, low cost, easy availability of stock, rapidity of manufacture, ease of cutting and many other needs, then the apparently simple mechanical element can become a major design and manufacturing task. It is, however, attention to these second-order details that makes or breaks a design. A useful check list of available material physical effects is to be found in Sydenham (1986).

In the spring-design problem discussed in Example 6.7 the zero shift effect can be greatly reduced by the use of a suitably sized support of a second material that is arranged to move the fixed end of the member as temperature changes. Bimetal springs are often used as compensating devices in this way. One way to accomplish this is covered in Example 6.8.

Example 6.8 Thermal compensation with re-entrant mounting arrangement

Figure 6.11 shows a laser interferometer return reflector mounted to reduce the movement of the datum – the rear optical surface – as it changes size with temperature.

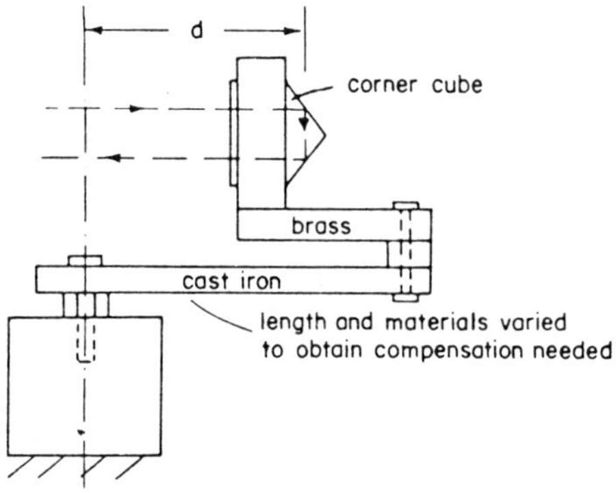

Figure 6.11 Thermal compensation by re-entrant geometry. (From Sydenham, 1986. Reproduced by permission of Instrument Society of America.)

The reflector corner cube must be mounted a distance *d* from the mounting point to allow room for other optical elements. Its base (the brass bar) changes length with temperature change, giving rise to significant error because this distance is an arm of an interferometer where a fringe displacement corresponds to around 350 nm.

A bar of zero thermal expansion (Super Invar or quartz derivatives) could be used but was not practical due to availability, cost and machining difficulties.

By choosing brass and cast iron the *re-entrant* form of mount was rendered almost thermally insensitive whilst still being of a short length.

Another means to reduce mechanical system error is to make use of *active compensation*, in which a sensor detects the influence parameter magnitude, and a suitable system alteration is effected to cancel it out. This compensation can be added at other places in the instrument information flow path. For example, electronic signals representing temperature changes in the size of an element can be used to alter the gain of a following amplifier.

Despite the importance of materials in design and operation there is little available review literature to assist designers. Chapters 2 and 7 of Sydenham (1986) help identify relevant published materials that gives in-depth reports on materials, their properties and error reduction.

6.3.4 A strain-meter example of fine mechanics

To put many aspects of fine mechanical design of instruments into perspective the parameters of a seemingly simple device are considered. Attention to material properties and how to compensate for error took the performance level of an inexpensive strain meter to a point where a 10 m length instrument could measure nanometer (10^{-9} m – molecular size) displacements arising in rock over periods of years.

Example 6.9 An inexpensive, highly sensitive, earth strain meter

Geophysical research uses strain meters to observe strains induced in the solid (in the rock - not sea) earth by the varying gravitational attraction of, mainly, the Moon. Figure 6.12 shows the schematic form of one such instrument.

The design requirement is to provide constant tension in a horizontally suspended wire, monitoring the change in length dl as the rock on which the strain meter is mounted alters length. The maximum strain magnitude is only 1 part in 10^7 and the discrimination must be to 1 part in 10^{10}. A typical strain signal variation magnitude is 1 μm. The wire employed is typically around 10 m in length and 1 mm in diameter. Discrimination and stability, over years, needs to be within around 10 nm.

As the wire has elasticity (from its elasticity modulus plus the effective compliance of the *catenary shape*) changes in tension alter its length, creating apparent strain error. This is controlled by maintaining adequately constant tension.

Temperature errors in the instrument are avoided by mounting the instrument in a tunnel deep in the earth (temperature varies there only 1 mK over a day and 10 mK over a year.)

Gravity is selected as the tensioning force as it is adequately constant when used as a second-order action in tensioning the wire. Dynamic performance does not need to follow movements of faster than several second order.

The balance design is arranged such that it supplies the tensioning force and does not increase the detection sensitivity. This is done by sensing the dl quantity immediately in line with the end of the wire. Mechanical gain, of around 10, could have been obtained by sensing movement of the end of an extended balance mass but then the position of the pivot would be a parameter of the sensitivity expression.

The wire, due to the combination of Young's modulus and 'catenary droop' effects, acts as a relatively weak spring. Friction in the pivot will cause error in indicated strain as the wire changes length because of friction changes in the balance pivot. Design of the pivot (and sensor mounting) is, therefore, critical as length changes of only 1 part in 10^{10} are to be measured.

Friction forms of pivot for which parts are available are unable to carry the 40 N forces needed to obtain the necessary catenary.

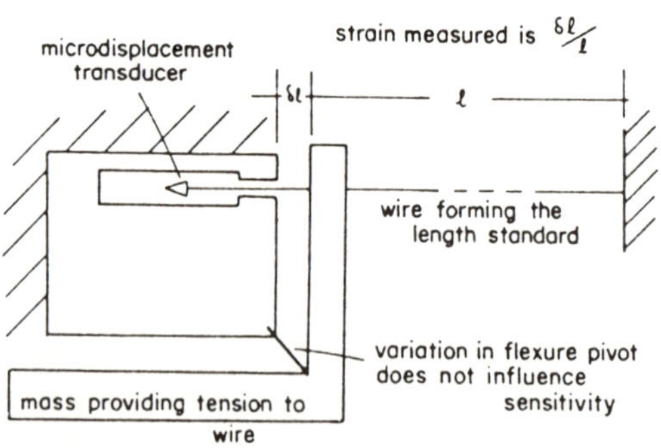

Figure 6.12 Schematic form of a mechanical nano strain meter. (From Sydenham, 1986. Reproduced by permission of Instrument Society of America.)

Ball pivots (special small forms exist) come close to being suitable from the loading standpoint but their very low-speed (d*l* changes at less than 1 μm/h) performance exhibits excessive *breakaway frictional* effects.

The case is obviously suited to use of a simple single flexure pivot arrangement because the rotation is very small, the forces in the balance pivot always act in one direction keeping the pivot in tension (thereby avoiding column buckling effects), it is easy to manufacture and it exhibits virtually no frictional effects.

The direction of the pivot must be placed along the line of resolved forces. Two strips are needed to ensure only one degree of freedom of the balance moving part. The tensioning system of forces controls other potential freedoms.

The overall cross section of the pivot must be able to carry the tensioning force without working in the plastic zone. The width of the flexure is dictated by the bending moment produced by the strips as the balance rotates. This must be small enough not to cause length changes in the wire as the balance rotates.

The physical size of the strips must be strong enough to allow assembly and installation. A clamping plate is necessary, between the bottom mass and the body, during manufacture and installation.

The flexure materials must have low mechanical hysteresis, be purchasable in adequately small purchase-order values, and have good spring properties to obtain a reasonably high yield point.

The two flexures are made from 0.05 mm thick phosphor bronze with a working area width of 2 mm. They need to be made in a jig to form the shape which has widened ends to allow them to be clamped.

The balance is assembled as a whole and the flexures added last. Once installed the clamping plate is carefully removed.

Operation is such that stops are not needed in use but the whole strain-meter instrument has be enclosed to exclude flying insects, spiders, people and breezes from disturbing the wire.

It is essential to use a non-contacting sensor which has the necessary discrimination ability (about 1 nm). In this case inductive lvdt sensors (such as that described in Example 4.15) are appropriate using the flexure system to support the moving armature of the sensor without contact.

Selection of a mechanical method (laser interferometers are also used) for this application is based on the fact that it can be designed in detail and its performance accurately predicted before manufacture. It is of very low cost compared with alternatives – at least 1/500th of the cost of a laser system. Its operation is intuitively obvious, making maintenance and installation easy.

Further detail on the mechanical regime of instruments is given in Jones (1988), Chapter 21 of Sydenham (1983), Sydenham (1986) and Trylinski (1971).

6.4 RADIATION REGIME

6.4.1 The forms of radiation used in instruments

Radiative energy is used in many forms of measuring instrument. It is applied in acoustic (mechanical pressure waves), and electromagnetic (EM field radiation) forms. Its usefulness stems much from its ready ability to provide an energy carrier linking the observed subject with the measuring instrument, and from its ability to be modulated to form carriers of information.

As the mechanical forces involved in establishing this link are virtually zero, measurement can usually be obtained without accountable loading of the observed energy system.

When radiation is sent to interrogate the observed subject the method is said to be *active*. In some cases the observed subject emits radiation that can be a measure of its properties. These are called *passive* methods.

When the observed system can be modified to assist the measurement - the addition of a reflector, for example - the method is known as *cooperative*.

Examples are mechanical motion being monitored by the movements of an optical reflector in an interferometer arm (active, cooperative method), temperature of a body being observed as the passively generated wavelength of radiation emitted (passive, non-cooperative method), and water flow rate being measured by the travel times of sound waves in the fluid (active, cooperative method).

Radiation is also generally easy to modulate by electronic means using one of many transduction effects.

For purposes of explanation the acoustic spectrum tends to be divided into *sound* (human ear response range of frequencies) and *ultra-sound* (above the upper response of the ear, at around 20 kHz, to megahertz.)

At the lower end of the electromagnetic (EM) spectrum are the very low frequency (VLF), radio frequency (RF), very high (VHF) and microwave frequencies all usually associated with radio technique.

Above these (*wavelength* is more commonly used as the unit) is the optical spectrum (beginning with the longer wavelengths of infrared through the quite narrow visible spectrum to the shorter ultra-violet radiation). Above this exist X-ray and other ionizing radiations.

Measurement systems can be constructed using all wavelengths and although their design may appear to be quite different because of their respective technology, a similar underlying basis exists for understanding the behaviour of each.

6.4.2 Basis of design

Design of radiative systems seeks to set up a linking energy beam between observed object and the observing instrument. This is then modulated in active systems, or used as it received, in passive systems, to convey measurand information. Beam properties such as variation in amplitude, phase or polarization can be used.

The system design involves understanding how to set up a suitable radiation link, choice of an input signal transduction method to add this to the radiation where needed, transmission of the radiation and its transmission zone behaviour, its receipt and detection and ultimately its processing into the required output form.

Radiation can exist in two forms. *Coherent* radiation is where all energy in the beam has the frequency of individual 'packets' in arising in phase. Within *incoherent* radiation the packets are not in phase. The former is essential for methods involving the principle of interference.

Radiation is theoretically characterized in terms of its three-dimensional properties in space as well as the one-dimensional time property. The full mathematical treatment can, therefore, be very complex. Often simplification can be obtained by making use of system symmetry.

Electronic engineering has developed methods for using the region below the optical bands whilst optical engineering has covered the infra-red to ultra-violet range. The overlap between each is considerable but as yet no truly unified formal treatment has emerged for radiative systems.

Ionizing systems form a class of their own, for coherence is rarely valid and the particular nature of ionizing radiation requires counting technology in its measurement uses.

The ultrasonic method makes extensive use of coherent systems theory; many such systems have counterparts in the optical region.

Measurand names of variables positioned within the two radiation spectra, along with their standards of units and calibration, are also not unified set. To confuse matters further appreciation of visible systems first developed using the human eye as the sensor. This has led to the psychometric approach based on subjective testing of people's vision. The *radiometric* system, used more in scientific and engineering work, can be related to a useful but not exact, degree. The problem is that the signal-processing part of the eye cannot be separated from the sensing eye, nor can the output of the eye be tapped into to measure energy levels.

Radio antennae, detector and transmission theory has covered much of coherent radiation modelling from the time-frequency viewpoint with some extension to spatial properties.

Optical physics and engineering, largely through the *optical transfer function* (OTF) approach, has provided understanding of the spatial frequency aspects of coherent radiation. The modulation behaviour of an optical system can be described with a spatial frequency response plot as shown in Figure 6.13. The *maser* and better-known *laser* devices were the prime stimulus for development of OTF theory, especially when the laser provided the long awaited, long coherence length, highly monochromatic, narrow beam, radiation source.

The classical, so-called *Gaussian*, theory of optics is based on the concept of tracing the paths of pencil rays permeating through various radiation elements and paths. It is able to explain much of coherent (at both image and wavelength level) systems behaviour but only on a point-by-point basis. To understand how whole coherent images combine at the wavelength level needs the use of three-dimensional Fourier theory applied to the spatial domain.

Image-based measurement systems concerning wavelength coherence require highly trained persons in their development. Similarly, detailed design of radiation system elements using Gaussian methods is also a job for the expert. Special software is used to design lens systems. Sophisticated systems, such as a quality zoom lens, cannot be calculated by long-hand methods as they would require more than a person's lifetime to process.

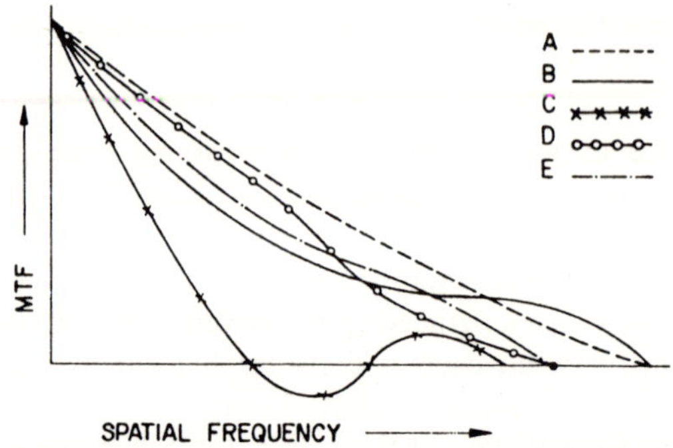

Figure 6.13 Response of an optical system can be described by its modulation transfer function: (a) best possible diffraction limited performance of a particular example; (b) central obstruction added; (c) defocused state; (d), (e) different designs with same requirement. (HBMS, figure 23.6)

With a few exceptions design of radiative systems must be from an analysis viewpoint. An apparently suitable system is first proposed. Its performance is then analysed using theoretical and experimental means. Iteration is then applied to tune the design.

Before outlining the nature of a radiative system design some appreciation of the componentry is needed.

6.4.3 Componentry

Whilst the size, materials used and necessary element treatments vary with wavelength of operation it is possible to appreciate componentry in a generalized manner from their functional standpoint. Practical considerations make low frequency and X-ray based methods both physically large. The smallest systems arise in the optical region.

Components can have radiation source, detection, transmission, processing and storge roles, these often being obtained simultaneously within a single element.

A mechanical framework is required to hold parts in adequate spatial relationship, this becoming increasingly exacting as the wavelength shortens. For example, in optical systems, where interference effects arise (as they do even in the design of a simple lens) precision is often to 20 nm tolerances.

Materials used vary widely depending upon wavelength of the radiation. At low EM frequencies large structures (antennae) and electronic componentry are involved.

Sonic radiation is typified by its transmitters and sensors which are mostly conducting-layer-covered ceramic materials.

The microwave zone is characterized by the extensive use of metal wave guides and dish-shaped antennae that act in the same way as large mirrors gather and focus radiation.

Infra-red, visible and ultra-violet devices use optical parts such as lenses and mirrors, the choice of materials where transmission of radiation is involved being critical.

X-ray instrument systems are characterized by the large size of electrical insulation needed for the high voltages involved in their generation and in the extensive shields placed around instruments and operators. Radioactive sources are small but require large shields.

The performance of a radiation source is characterized in terms of its

- radiation output (that is, its energy output);
- input drive requirements;
- spectral response;
- spatial energy distribution;
- source size;
- coherency length;
- polarization;
- likeness to defined types of source such as the black body.

Physical parameters can also be important as some sources can be large and need additional support equipment such as cooling.

As many factors (such as capture geometry, spectral response, wavelength and bandwidth of operation and energy capture efficiency) decide the sensitivity of a particular detector, the optical instrument industry has developed the concept of *detectivity* – a normalized way to rank detectors regardless of their actual details.

As a guide, the physical size of the detector element will be of the order of the wavelength of operation with the additional size factor covering mounting and cooling requirements.

To obtain improved sensitivity detectors are often cooled to well below room temperature. This is done to reduce thermally induced noise signals, as explained in Section 4.6.4.

Transmission of radiation can be as *open radiation.* This applies when the beam is already well collimated and does not need guidance to prevent the radiation from being dissipated into an ever-widening spatial domain – or when widely dispersed radiation is required. Examples are ultrasonic radiation through fluid to measure its flow rate and laser radiation in an alignment system.

The alternative is to contain the radiation in some form of *guide.* This is practical for short wavelengths, for the guide dimensions must be of the order of the wavelength. Optical fibres can easily convey optical signals, wave guides are used in microwave systems, but VLF signals would need enclosures of kilometre size.

Generally links suffer signal attenuation of amplitude at differing rates with wavelength, phase shift of wavelength and polarization, broadening of polar distribution and pick-up of unwanted radiation. The *transmission path* must be treated as a dominant element in a radiative system design.

Processing components are primarily selected on their ability to perform a certain information-processing role. They then must be assessed in terms of their radiation transmission properties, as explained above, for they form part of the overall energy link.

Processing functions, with selected examples, include the following (see Figure 6.14):

- guiding relation (prisms, lenses, mirrors, fibres);

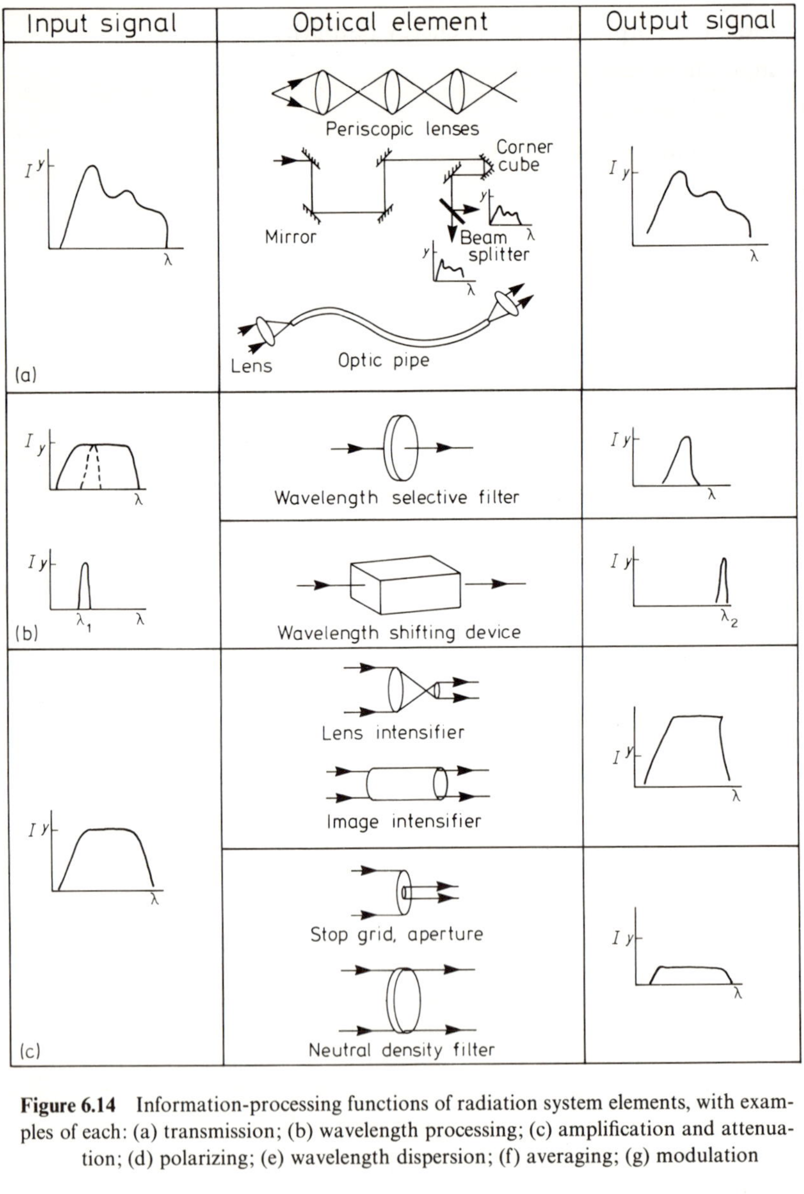

Figure 6.14 Information-processing functions of radiation system elements, with examples of each: (a) transmission; (b) wavelength processing; (c) amplification and attenuation; (d) polarizing; (e) wavelength dispersion; (f) averaging; (g) modulation

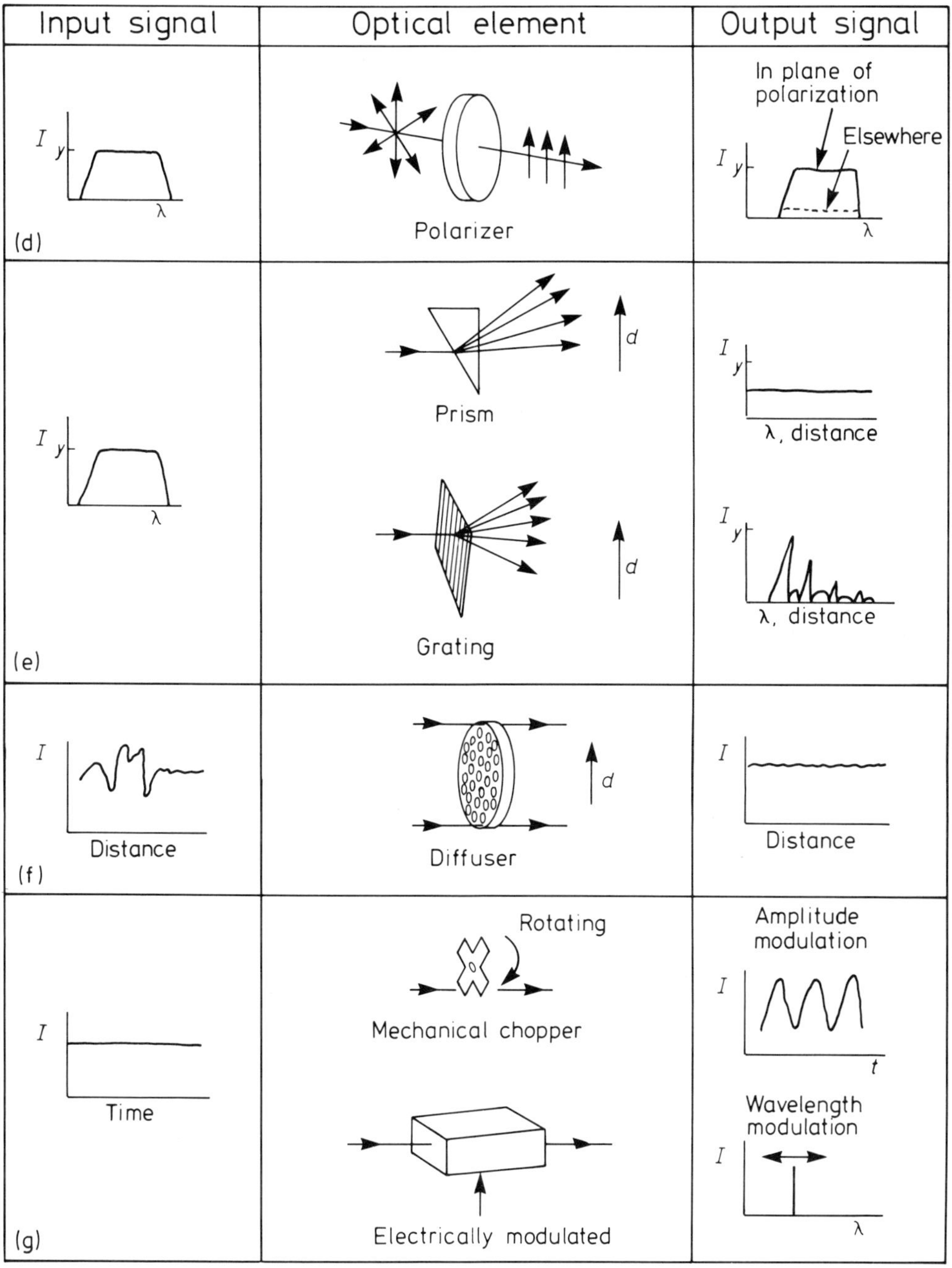
Input signal
Optical element
Output signal
(d)
I y
λ
Polarizer
In plane of polarization
Elsewhere
(e)
Prism
d
λ, distance
Grating
(f)
I
Distance
Diffuser
(g)
Time
Rotating
Mechanical chopper
Amplitude modulation
t
Electrically modulated
Wavelength modulation
λ

- wavelength processing (through selective spectral transmission attenuation, interference effects and active elements);
- amplifiers (laser device, travelling wave tubes);
- polarizers (Polaroid material, reflection from surfaces);
- wavelength/frequency changers (lasers and crystals);
- geometrical devices (stops, pupils, zone plates, gratings).

The availability of these functions varies greatly with the wavelength of operation because of limitations in available technology.

6.4.4 Practical application

A critical parameter in radiative measurement system development is usually the choice of wavelength (or frequency). It is usually limited by the allowable size of the measurement hardware and by the wavelength discrimination required.

With so many forms of radiation-based measurement systems it is impossible to give a fully representative example in use on measurement in all regions. The design factors of the following optical example, however, apply to many systems.

As an example of how radiation systems are theoretically considered – modelling is a vital method with radiation systems because they are expensive to build and not always amenable to experimental methods of development – the reader is now taken through the main points of the theoretical exercise given in Hudson (1969), beginning at p 419. Here are also added practical considerations of an application.

Example 6.10 Detecting the presence of persons by their emitted radiation

A security system is to be developed to detect human intruders by sensing their naturally emitted infrared radiation.

This being a *passive, non-cooperative* sensing situation, the first need is to estimate the particular radiation-energy level and wavelength distribution (this spectral plot is called the *signature*) emitted by a 'typical person'.

The received energy must be used efficiently to produce an output electronic signal. In this application the *total energy level* of a person is a suitable radiation signal to sense.

Radiation emitted from a human has a well-defined, but variable, spectral irradiance that can be expressed in mathematical terms. (Generalized formalization is the starting point of the Hudson derivation.) As most influencing parameters are wavelength-dependent the estimation of the energy levels in the system's *through path* is mathematically complex.

A reasonable approximation is to assume the person emits radiation as though being a point source radiating into space. For this the spectral radiant intensity at a sensor is inversely proportional to the square of the radial distance from the person.

Without assistance a sensor would only intercept a very small part of the total omnidirectional spherical envelope of radiation emitted by a person (about 80 W is radiated and less than a microwatt is received). It is, therefore, usual to add some form of lens or mirror system to enlarge the capture aperture. This gain block takes the system capture to about 40 000 larger. As will be seen, much of this is lost within the optical system but even so a surprisingly simple, well-designed, optical system can be so sensitive that it can detect, from Earth, a candle flame level of emission from the Moon!

The transmission path through the optics to the sensing surface will dissipate energy in surface reflections, transmission losses and further beam broadening in the system. Truly parallel collimation is not possible. Each untreated optical element surface will lose around 5% of the passing energy. Special, costly, treatment of surfaces and use of appropriate materials can reduce these losses significantly.

At the sensor a few milliwatts of radiation energy will fall on the detector, but the sensor will not make use of all energy reaching it. The sensitive area will not be exactly matched to the incoming beam dimensions; the surface will not absorb all of the radiation falling upon it. Furthermore the electrical output response will not be the same at all wavelengths. The conversion factor for the radiative-to-electrical transduction will be temperature and device dependent.

The sensor also possesses electrical performance wherein response falls off with increasing signal frequency. A moving person, being relatively slow (1 Hz bandwidth) compared with electronic signal speeds, will obtain the best sensor response level.

By the above design process it is possible to make a *throughput estimate* by mathematical integration of the product of all loss and gain factors. Such studies give reasonable estimates when the bandwidth transmitted is narrow but need much more mathematical study when the bandwidth of received radiation is large.

The integrated mathematical product of all effects leads to an estimate of the signal output expected from the detector. To make this meaningful as a useful measurement signal it is necessary to form a ratio comparing it to the noise-energy modulations that can be expected - the SNR.

In this case noise will mostly arise in the electrical part of the detector. Stray optical energy can largely be eliminated by good system design, for example using optical filters to remove shorter wavelengths than infra-red and by an appropriate field of view.

In practice it is possible to obtain useful estimates (but not exactitudes) of the S/N ratio and the formulation enables the designer to tune the design for best effect. Usually improvement can be obtained but at diminishing rate of return for effort expended.

The above procedure also enables the *theoretically best* system to be calculated, a useful benchmark that prevents excessive resources being applied. It also provides the framework for comparing different systems performing the same task. Hudson provides further exploration of the above leading to the so-called range equation that can be used to compare systems.

Understanding of radiation systems can be expanded by use of books on telecommunications, optical and infra-red engineering, and on ionizing radiation. Prasad *et al.*, in Sydenham (1983) and Levi (1968, 1980) cover optical technique and Hudson (1969) infra-red systems.

6.4.5 System development

Radiative systems development requires a considerable degree of prior planning and design consideration as the radiation is not always visible and it cannot easily be sensed due to its dispersed nature. A system arrangement is first synthesized. The performance should then be estimated prior to manufacture as components and mountings are expensive.

Some degree of modularization of parts exists in each region of the EM spectrum but the limited supply of materials restricts design choice. In the microwave regions standardized constructional 'plumbing' is available with which to test ideas. In the optical region well-developed, but also expensive, 'optical bench' systems can be purchased.

Good complementary mechanical design, manufacture and testing support, and often electronics, is needed to develop radiative information systems.

6.5 CHEMICAL REGIME

6.5.1 General approach

Chemical systems concern energy relationships and mass. Their difficult measurement needs are characterized by energy reactions taking place in small volumes, at rapid rates and for short time scales. Their overlap into living biological systems also makes their treatment different to the physical state. Their complex composition makes measurement descriptions complicated.

Metrology of chemical processes in concerned with classification of species, assessment of relative proportions and distribution, determination of reaction rates, with physical properties and a host of other variables.

As physical material is always involved, differentiation of chemistry from the discipline of physics is a much debated issue. From the measurement systems development viewpoint chemical measurements do not possess unique instrumentation fundamentals. The fundamentally oriented instrument systems person can contribute here provided that considerable applications domain knowledge is available.

Different forms of energy interception are used to learn about the state of chemical systems. Optical regime methods are used extensively as they ably permit the effects of energy reactions of short-lived processes taking place at a microscopic scale of size to be monitored without loading effects. Almost all radiation methods discussed above have found use in chemical measurements.

Measurement information can be interrogated in the form of the changing absorption, emission, wavelength shifts or polarization in spectral bands of radiation that are passively produced or that are used to excite the process. In this class optical regime design (from infra-red to ultra-violet) is widely applicable.

Another metrology class, that of *electro-chemistry*, is based on electric charge energy effects. Voltages arising across chemical volumes or current flowing through bulk quantities can yield information. Here electrical and electronic methodology , discussed in Section 6.2, applies.

6.5.2 Separating the constituents

An important requirement in chemical measurements is that the would-be measurer must first be reasonably sure which substances are present in an often 'mixed bag' of constituents.

Chemical measurements, like all others, must proceed on the basis that they seek to establish the quantity of expected substances using techniques appropriate to sense them. An assay of a mineral sample for gold could well overlook the presence of other valuable minerals unless the assay addressed their presence. The problem is that many constituents will be present and it will not always be obvious what they are. It is often necessary to pretreat the sample to remove unwanted constituents that can mask the metrology effects being sensed.

Constituents must be separated to assess their presence and magnitude. The relative positioning of these individual components - as the result of a separation process - in some form of spatial representation provides a fingerprint of the sample. An example is given in Figure 6.15.

In the *mass-spectrometry* class of instrument an electromagnetic field arrangement is used to deflect the constituent elements of a molecule, according to their mass/charge ratio, into different spatial places at which sensors are placed. The quantum nature of atomic mass makes these places predictable and calculable, allowing species identification.

In *chromatography* the sample, in liquid or gaseous state, is passed along narrow tubes called columns. Constituents travel at different velocities, thus arriving at a sensor at different times. Figure 6.16 shows the schematic diagram of a typical system.

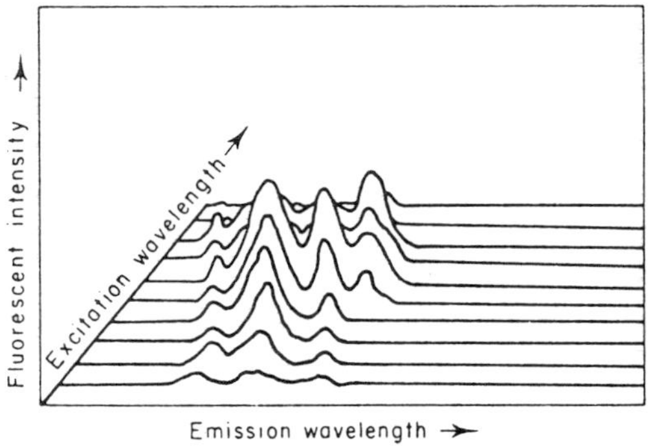

Figure 6.15 Isometric signature of the fluorescence spectra of a sample. Individual two-dimensional traces are used. (HBMS, figure 27.18)

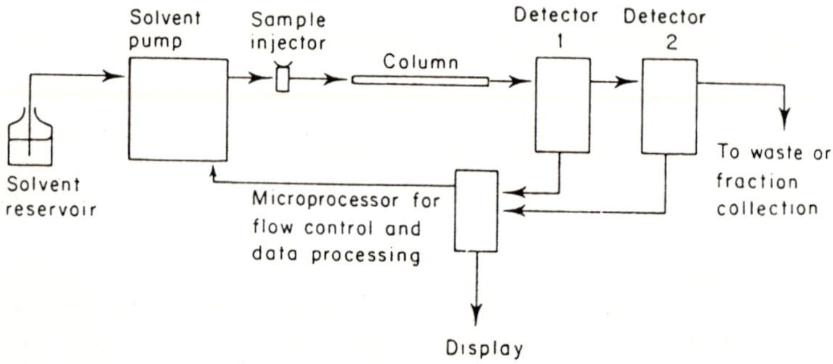

Figure 6.16 Schematic diagram of a liquid chromatograph. The solvent addition stage and second detector are enhancements of the basic concept. (HBMS, figure 27.26)

The dispersion effect can also be obtained in gel materials and absorbing substances, such as paper, wherein they migrate to different distances in the carrier.

Radiation-sensing methods produce a wavelength spectrum output signal. This spectrum is spatially dispersed in spectrometers to allow identification of the parts of the spectrum from its *fingerprint*.

6.5.3 Detection

Having separated chemical constituents into their own unique location the energy process associated with each is made to fall on a sensor for measurement of its appropriate parameters. The complex nature of many chemical measurements has made electronic output sensors mandatory to allow automatic rapid recording of the fingerprint.

Detectors mostly used in chemical instruments sense either

- properties of appropriate radiation
- electrical parameters of a chemical reaction
- colours of substances.

Their operation and design follow the practices of electronic, fine mechanical and radiation engineering.

Examples are total radiation energy sensors, where energy interactions of passing substances cause electrical resistance changes by thermal effects; the photomultiplier that detects photons providing electronic gain; the solid-state optical detectors that directly convert photon energy into electric signals; ionizing radiation counters such as basic gas ionization and scintillation enhanced devices; the ion-capturing Faraday cup; electrical conductivity monitors; voltage and current detectors and magnetic induction coils.

Given many system building blocks and different ways to combine them allows extensive variation in chemical measurement instrumentation. The following example gives an impression of just one form of chemical instrument.

Example 6.11 Measurement of pollutants in air

To control the quality of air in the environment it must first be measured. Many cities have networks of *air monitoring stations* that send data back to a central control room where appropriate action can be implemented if excessively high levels of any offending gases are detected.

In a monitoring station air is drawn into measuring instruments by a small pump. Knowing that the gas will be air, plus certain expected pollutants, appropriate sensors are chosen that can detect and measure pollutant gases such as SO_2, NO_2, NO, O_3, CO and H_2S.

If an unexpected gas should appear it will be ignored. It also might *poison* the sensors. Regular calibration is used to keep the system in a known working state.

Incoming air is split and fed to each sensor; one unit for each gas is needed as multicomponent analytical methods, such as the mass spectrometer, are inappropriate here.

Various sensing methods may be used. Consider, for instance, the *coulometric* cell. In use of this liquid-based detector the gas to be analysed is bubbled through an electrolyte. The chemical change resulting from the presence of a pollutant is compensated by an electric current in the cell that is proportional to concentration. Each cell has to be made specific to a certain gas through use of specific gas selection filters placed in the line feeding the sensors. Peltier cooling cells, turned on by detectors of falling liquid level in the cells, condense atmospheric water into the cells to maintain the cell electrolyte level.

Chemical sensors often need considerable additional plant to support operation. Coulometric cells must first have their electrical output zero set to provide one calibration point. This is done by passing the incoming gas through 'all pollutant capture' filters. They must then be fed with *standard pollutant gas* to set the other end of their scale. They are then ready to be used.

This sequence takes many minutes to complete; several valves have to be operated, lines adequately purged and new situations stabilized. These cells drift about 2% in 24 h, requiring daily recalibration.

Given enough back-up support, such systems can operate without attendance for several months. Eventually, however, standard pollutant gas supplies, poisoned filters, eroded electrodes and leaking valving require maintenance.

Chemical instruments involve considerable support engineering and maintenance. Lifetimes can be as short as minutes, and drift rates such that hourly calibration is needed.

The practice of chemical measurement is reviewed at chapter length by Watts in Sydenham (1983), where more references are to be found to assist deeper understanding. A detailed exposition of the design aspects of a complex modern chemical instrument, an advanced form of spectrophotometer, is given in Hewlett-Packard (1980).

TUTORIAL EXAMPLES

6.1 For a particular instrument of your choice, and using Table 6.1 write a short statement identifying the energy forms used and the key design parameters that made that combination a good choice.

6.2 Investigate, and report on, why beam chopping finds application in radiation sensing methods.

6.3 Using mathematical explanation to enhance your answer, explain why three low-performance op-amps can be used together to produce a high-performance instrument amplifier.

6.4 A theodolite is to be used to measure bearings of several benchmarks on a land mass to establish earth settlement.

Suggest, with justification and some engineering detail, a quick-to-use support by which the theodolite can be replaced in exactly the same position each day. It must allow for thermal effects.

6.5 For an existing optical instrument of your choice develop the outline of a model of its radiation throughput allowing for all loss components. Identify the special features needed for elements of the system. (A full mathematical description is not required.)

6.6 On-stream analysis of the energy content of coal flowing into a ship for export is needed at a port. Develop a suitable system at the conceptual level. Justify all system components and your choice of technology at each point.

REFERENCES

Georgaras C, 1986, Design and development of pressure gauge. Student Project Report, School of Electronic Engineering, South Australian Institute of Technology.

Jones R V, 1988, *Instruments and Experiences*, Wiley, Chichester.

Hewlett-Packard, 1980, *Hewlett-Packard Journal*, **31**(2) (whole issue containing 7 papers).

Hudson R D, 1969, *Infra-red Systems Engineering* Wiley, Chichester.

Levi L, 1968, *Applied Optics*, Vol 1, Wiley, Chichester.

Levi L, 1980, *Applied Optics*, Vol 2, Wiley, Chichester.

Mansfield P H, 1973, *Electrical Transducers for Industrial Measurement*, Butterworths, Sevenoaks.

Noltingk B E (Ed), 1988, *Instrumentation Reference Book*, Butterworths, London.

Norton H N, 1969, *Handbook of Measurements for Electronic Measurement Systems*, Prentice-Hall, Englewood Cliffs.

Stein P K, 1970, Sensors as information processors. Research/Development, June, 33–40.

Sydenham P H, 1983, *Handbook of Measurement Science*, Vol 2, Wiley, Chichester.

Sydenham P H, 1984, *Transducers in Measurement and Control*, (3rd edn), Adam Hilger, Bristol.

Sydenham P H, 1986, *Mechanical Design of Instruments*, ISA, Research Triangle Park, North Carolina, USA.

Tran Tien Lang, 1987, *Electronics of Measuring Systems*, ed J McGree, Wiley, Chichester.

Trylinski W, 1971, *Fine Mechanisms and Precision Instruments*, Pergamon, Oxford.

CHAPTER 7

Creating measurement systems

7.1 INTRODUCTION TO DESIGN

7.1.1 Definition of design

This chapter introduces the reader to the procedures by which an instrument system is generated. Appreciation of this is essential in the instrument field, for measuring systems are very much designed, built, installed and used by persons not specially trained in this activity. The accent in this chapter is for the person in the small-group situation.

Design is the process whereby a virtually infinite set of possibilities gradually are selected and converted into specific results that allow a task to be solved to set requirements. It relates to organized endeavours in general, but here consideration is restricted to measuring instruments. A pictorial representation of the design process is shown in Figure 7.1.

Design must start from a basis of existing knowledge. It makes use of this knowledge in two ways.

First there is the *experiential methodology*, the way that most instruments are designed (but not necessarily the best). This way modifies someone's past experience for a similar task. Much of the measurement science and instrument engineering literature is of this form.

The better alternative, *systematic design*, is to use a methodology based on a structured understanding of the measurement process. This new approach is emerging through assistance of computer-aided design in mechanical and electronic engineering. It promises extensive improvements in the way by which solutions to measurement needs are obtained.

There exist two interpretations of the idea of design. Both are needed but they have not come together where instruments are created to the degree that they should.

The first kind is mostly concerned with externally evident aspects such as ownership esteem, appearance, ergonomics and overall compliance to its required task – generally known as *industrial design.*

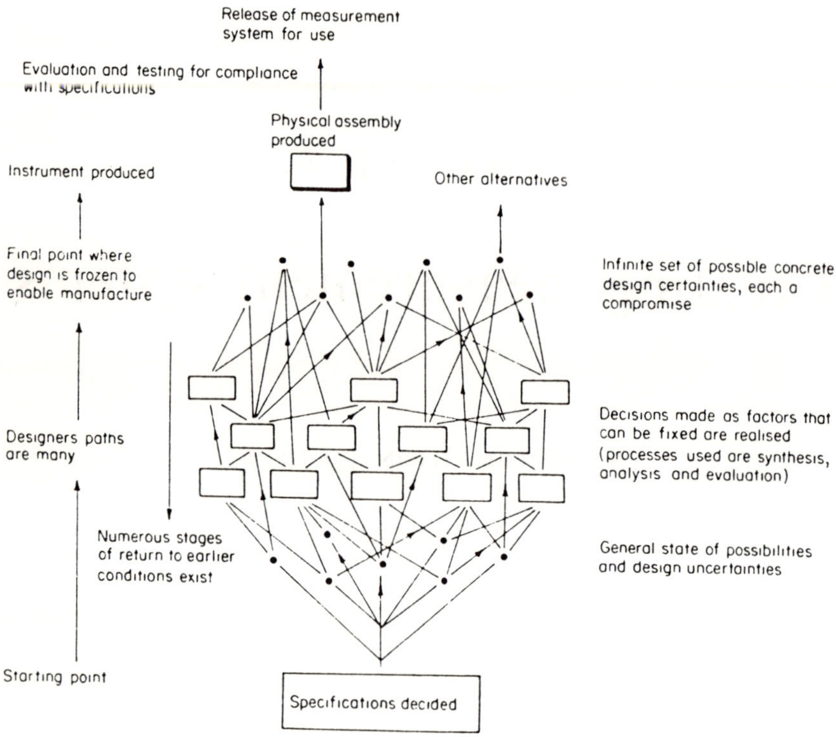

Figure 7.1 Design is about moving from a state of generality to one of specificity. (HBMS, figure 15.2)

The other kind is *engineering design*, which concentrates on the internal aspects such as mechanical, electronic and other deeply technical knowledge.

Neither is sufficient without the other. This chapter seeks to present an overview that is placed between the two. It provides understanding of the less promulgated *management of the design process*. The engineering and industrial design of an instrument are each well accounted for elsewhere and the reader will generally be trained in one kind or the other.

7.1.2 The design process

When the design task is complex, as that of instrumentation often is, there is need to manage many diverse aspects that are varying with time. Table 7.1 gives a checklist of design-process parameters. The design manager has to steer all of these toward the requirement within given performance, cost and time constraints.

Just how this is done best is not at all clear. The traditional, more widely held and adopted, experiential view has been that good designers 'just know' how to do it. Their intuition, based on experience and subconscious processes, tells

Table 7.1 Checklist of design-process parameters

Accommodation	Materials and components
Availability of sub-units	Mechanical design
Commissioning and acceptance	Modularization
Complexity	Optimization
Computer aids	People
Decision making	Performance
Delivery	Post-delivery performance and care
Designer's environment	Product
Documentation	Quality control; in design, in production
Environmental factors, by instrument	Project scheduling
Environmental factors, on instrument	Reliability
Ergonomics	Societal factors
Finance	System organization and partitioning
Future needs	Time scale
Interchangeability	Training of operators, designers and manufacturers
Interfacing	Validity of design
Inventiveness and innovation input	Variations/tolerances
Manufacturability	

them what will work best, which is the best principle to use, what to make and use and which appearance is likely to be more acceptable.

The more recently formalized view, emerging from engineers who have sought out the systemic nature of engineering artifact development, is that much of that experience *can* be identified and placed in a rigorous framework to assist less experienced persons to reach a reasonable design performance level in much less time.

Whatever the view held, it is clear that each kind of person will have need to use intuitive skills that defy objective study. The intuitive designer can often leap over formally needed steps of a structured path to reach good conclusions much more rapidly than by ordered procedures. The weakness of this approach is that there can be little real degree of certainty that they have produced a near *optimal* solution. The design is not easy to justify and it is not in a form that allows others to carry it on efficiently.

The use of modern computing methodologies, the availability of powerful computers on designers' desks and the ability to embed knowledge into computers now seriously challenges the intuitive approach – at least as it is used at present.

Design has two distinct facets – that of *realizing* a schema of a suitable artifact and that of progressing the item from idea to hardware. The latter, *project management* of instrument design, has been largely neglected in training and the literature.

The birth point of an instrument ranges from an intuitively generated invention through to a systematically, market-researched, conclusion of a need and solution space. In either case there must be some event that *pulls* or *pushes* the flow of resources to carry out an engineering design.

Innovation is the name given to the process that carries the idea through the design stage to production. This is a managerial function requiring flair, drive, risk taking and numerous other factors not normally taught to industrial or engineering designers. It is partly teachable; training in innovation and entrepreneurship is now well established and ordered.

Design may appear to be only relevant in the early stages of the product, or service, life cycle. It is, however, vital throughout that lifetime, for production success depends much on good initial design and, later, the ability of the designers to solve on-going changes and to produce updated designs.

Instrument products are characterized by their

- need for multidisciplinary activity;
- usually large electronic content;
- relative simplicity of design path;
- low per product item value;
- low volume manufacture;
- need to follow rapidly changing implementation technologies;
- short commercially viable lifetime;
- necessity for short design time.

Their designers will, all too often, have had little dedicated training in instrument engineering, industrial design or entrepreneurship. The purchasers of instruments will too often be insufficiently familiar with what they really need from the product.

The life cycle of an instrument product can take from one to many years to reach application. It passes through the steps shown in Figure 7.2.

Two examples, at each end of the manufacturing spectrum, are now given. The first, Example 7.1, is a case study of a 'typical' small-instrument product requirement. The other, Example 7.2, is of a far more challenging product that needed the development of numerous areas of engineering.

Example 7.1 Development of a water-level sensor for a washing machine

To effect automatic control of a top-loading clothes-washing machine it is necessary to sense water level so that agitation does not begin before there is enough water in the bowl.

The manufacturer required a sensor that produces *proportional* electronic output instead of the traditionally used diaphragm, that operates an on–off switch when moved by the air pressure of a rising column of water in a sensing, *stilling*, tube.

The first step was to establish an appreciation of the current system so that a suitable principle and system design could be generated. The manufacturer decided that the current diaphragm drive and sensing tube should be retained and that the existing switch body be modified as little as possible to keep costs down. Financial arrangements to carry out the development, including ownership of intellectual property, were worked out.

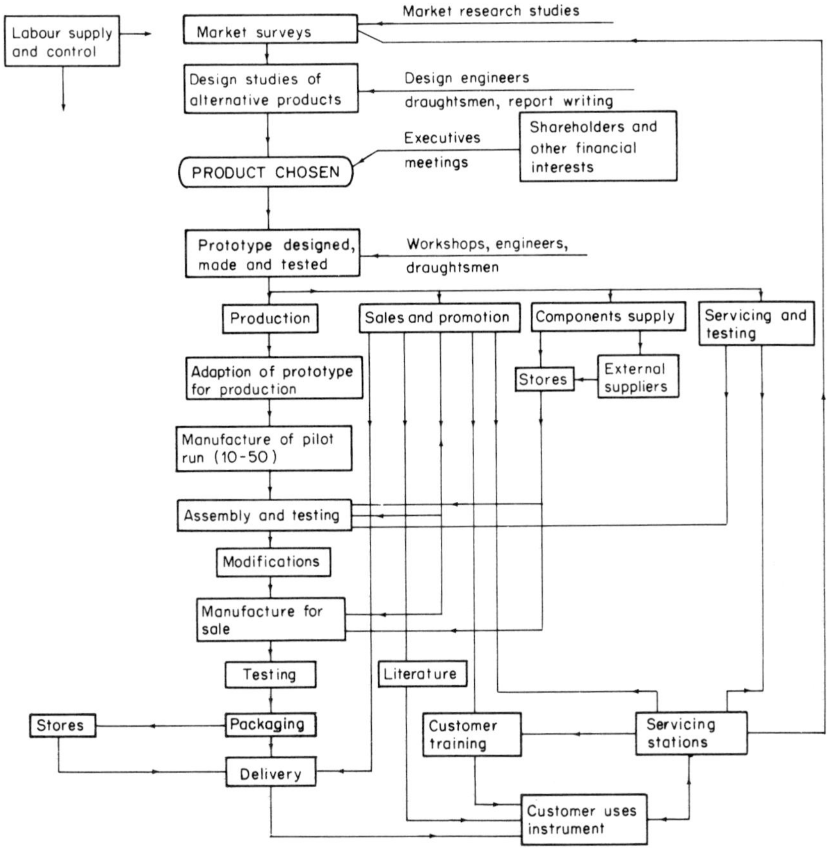

Figure 7.2 Life cycle of instrument from idea to product. (From Sydenham, 1986. Reproduced by permission of Instrument Society of America.)

A *think-tank* (see Section 8.3.1) identified that the most likely method would be to use a linear variable differential transformer (lvdt) driven by the movement of the diaphragm. This step took about two weeks to organize and report to the client; it included a quotation for the cost of further work.

The next phase was a larger contract to build and test the idea in a simulated situation. This involved modifying existing sensor bodies adding a suitable spring (found by testing – the maker's drawings did not give design information) and lvdt sensor, with circuitry, which had to be designed to suit the space available. Testing of a crudely constructed unit confirmed the general approach was suitable. A second report suggested the next step and its cost. This stage took 2 months to complete.

Next needed was an engineered evaluation model that would better suit the production line. This took 8 months to design, build test and deliver. Several identical units had to be made to provide for field trials.

A suitable manufacturer was then located who could build and supply these at 30 000 per month. This step required modification of the original specific evaluation prototypes to suit low-cost, mass manufacture. It took some 9 months to produce the first preproduction models, another 2 months to field test them and then 3 months more to set up the factory for routine production over subsequent years.

This project took about 24 months to move from a *first request* to the *in-service* unit, a typical time for this type of 'known technology' problem.

Today, common electronic consumer products, such as the ubiquitous tape-deck/radio/alarm in its many variations, can be developed to mass-market release in just 6 months! That, however, is not always the case, as Example 7.2 shows.

Example 7.2 Development of a complex consumer instrument product – the compact-disc player

Compact-disc (CD) players, now so commonly used to replay audio and video recordings, were devised as prototype models at least as early as 1965. By 1969 preproduction models and trial manufacturing plants had been established. They were released on the specialist professional market around 1975, arriving in the trade houses of the domestic consumer market around 1980.

Design of these involved many new areas of engineering. Much of their design had to break many new barriers and make use of yet unproven detectors and materials. The tracking reading head follows the micrometre dimension tracks with its six degrees of freedom controlled to within micrometres, doing this at over 1 kHz bandwidth and without making contact.

The disc material (called the *medium*), recording process and replication process posed major challenges due to the (then) revolutionary digital recording nature and very high storage density. The discs has to be reproducible at a reasonable cost. They had to be robust and have a long life. The system had to perform with dirty, scratched and deformed discs. The electronic circuits were also major developments as considerable processing had to be packaged into a small case. Standards had to be established for the disc formats and system signals.

Thousands of person-years of resource were mustered and administered to bring this product into the drawing room.

Of all the many factors of instrument design the *lead-time-to-product* has become the most significant. Commercial success is necessary for survival and information technology products can be designed and made with ever shortening times as streamlined methods are developed and the pace of new component development increases.

7.1.3 Design infrastructure

To create a design, human and other resources must be set up within a suitable support *infrastructure*. Whereas, once created, that system might be able to produce products in less than a year, setting it up to perform well can, itself, take a year or more. As a rough guide a small product, to be made in tens to hundreds at a time, needs about two to three years to go from conception to preproduction versions. It will then take one to two years more to the product launch and, it is hoped, supportive sales from then on.

During that period the costs will be at least equal to the salaries of some four to five people, with their on-costs and overhead for their place of work, consumable items, parts for developing and testing prototypes and a whole range of supporting services such as market research, accounting, communications and legal advice.

The popular image of the individual genius developing a giant corporation within five years, when starting with only personal resources in a garden shed, is not the norm.

A short account of the resources needed is now given.

Personnel

Core day-to-day staff needed will include enough scientists or engineers to develop the design detail and supervise prototype manufacture and testing. They will need the support of skilled technicians or engineering associates.

Financial matters will need constant management. In the small enterprise that task will be carried out by one of the trained staff using the services of an accountant and auditor on a part-time basis. A large organization will be able to support an accountant on the staff.

A considerable amount of record-keeping, filing and correspondence always exists. Office secretarial services will be needed, perhaps on a part-time basis as a small enterprise grows. Use of computer-based office operations reduces the need for a secretary but does not eliminate it.

There are also many low-skill tasks to be carried out, for which a technical assistant is suited.

Regardless of the size of the operation there will always be need for external consulting services to some extent. These would include such tasks as carrying out market research, developing a business plan, putting industrial design into the product, preparing publicity material and printing it, and so forth.

Finance

An instrument design enterprise can, being a relatively low consumer of capital, equipment and components in its initial stages, begin on a 'shoe string' but that path is not one engendering confidence for the staff and the all-important creditors. Planning should be used to get the operation going on an increasingly sound financial basis.

A clear *cash-flow* plan, showing how the cash will flow to meet costs and incoming invoices as each month (better still each two weeks) passes, is essential. If the group cannot meet their bills they will lose their credit opportunity with suppliers, speeding up their demise.

It is quite normal for new instrument developments not to see returns from sales for two to three years. Finance to cover this period must be obtained before it is needed. Figure 7.3 shows the resource flow typical of a small instrument enterprise.

Once the prototype is proven, taking the operation into the production stage usually requires some 5–10 fold increase in finance. Production tooling and plant, increases in staff, promotion, marketing, packaging and the cost of the stock that must be produced to meet demand on the sales launch make new and expanded financial demands.

Once the product is delivered income will become available but it will not include enough profit to allow expansion from that source alone. Much of any surplus will go into loan principal and interest repayments, and into plant and

Aspect	Seed 1	Early start-up 2	Start up 3	Growth 4	Cash flow 5	Mature 6
Management	Incomplete	Incomplete	Incomplete	Partial	Full	Full
Team	Inventor	Inventor	Inventor and other	Team	Team	Team
Product	Idea (design concept)	Incomplete prototype	Partially tested	Fully tested	Improved	New product
Business plan	None	Embryo	Incomplete	Needs refinement	Next stage	Next stage
Development costs	Some	Substantial	Substantial	Some	Nil	Nil
Production costs	Nil	Nil	Nil	Some	Substantial	Substantial
Sales income	Nil	Nil	Nil	Some	Substantial	Substantial
Net profit	Nil	Nil	Nil	Nil	Some	Substantial
Expected rate of return before tax % per annum (approx)	100+	80+	75+	60+	50+	40+

Figure 7.3 Techsearch Table, modified by P. H. Sydenham. (Reproduced by permission of Techsearch Inc.)

accommodation. More finance is needed to expand production if all has gone well. This step will usually make even bigger financial demands.

The consoling feature is that a well-planned and financially sound (in the long term) business venture will not find it hard to attract financial backing. Entrepreneurs consider the financial aspect to be one of the easier parameters to organize. It should be remembered that all design and manufacturing enterprises possess the same kind of factors and, therefore, face similar problems.

Accommodation

The staff, manufacturing plant and stock need suitable accommodation. This can range from converting a domestic dwelling to renting purpose-built incubator space in a technology or science park.

The choice will depend on many factors. Location with respect to necessary external supporting services makes technology parks, like that shown in Figure 7.4, attractive for instrument development, for they can usually offer a high level of casually and contracted academic consultants, test and other equipment each of which can reduce the need for up-front capital for the new enterprise. Another factor will be proximity to a suitable work force.

As the new enterprise grows according to its business plan there will be need for expansion of the accommodation. Developmental work usually needs special facilities. An electrical testing product might need access to a low-impedance mains connection. A farming tractor instrumentation system development might need fields for ploughing trials.

Instrument development needs some office space for the manager, designers, secretary; laboratory space for technical development and an assembly, testing and dispatch area. Stock is usually small in size but still requiring some storage space. Special plant, such as a wave-soldering machine, electronic board assembly and light metal work will need manufacturing space.

Instruments are often developed using subcontractors to produce sub-assemblies which are specified by, and assembled into the whole system by, the enterprise.

This *cottage industry* method has the advantages of allowing rapid product development for it avoids the time needed to acquire staff and plant (provided the region has this to offer). It also allows competitive pricing and gives stock just when needed, thereby reducing the financial burden somewhat.

Disadvantages of this method are that confidence needs to be developed about the suppliers and vice versa. Their failure to deliver systems and components of adequate quality on time becomes the enterprise's problem.

Another factor regarding accommodation will be proximity to markets. Instruments are intrinsically high-value products meaning that their freight cost by air is not overtly significant, making instrument developments viable in relatively remote locations. For example, a leading optical scale maker set up premises in the heart of the Black Forest in West Germany.

Figure 7.4 Example of incubator accommodation – Technology Park, Adelaide. (From Technology Park, Adelaide, South Australia.)

Equipment

Virtually all instrument developments will need at least some level of in-house electronic circuitry, light metal and plastics fitting, assembly, finishing and testing capability.

There is usually need for some special plant, examples being static and dynamic pressure test rigs for pressure transducer development and temperature-controlled baths for temperature sensor development.

Environmental testing is eventually needed, requiring special test enclosures. As these are expensive their use might be hired.

The modern enterprise will also need adequate computing power ranging from personal computers to main-frame capability. These are needed for electronic, mechanical and optical design and testing aspects and also for administration, finance, correspondence and report writing

Effective communications with suppliers dictates the need for telephone, Telex and Fax facilities. A photocopier and typing system are also needed along with filing cabinets.

A significant part of this cost can be reduced by taking space in rental incubator locations, such as that shown in Figure 7.4, as they provide a ready-made administrative support infrastructure that is paid for as used.

Product concept

An important factor for success is the product chosen. The entrepreneurial nature of product development in a free-enterprise climate always carries with it a degree of risk.

If a need is well researched and proved to exist, if its costing and sales potential are well developed, and if many more factors are attended to in a properly orchestrated business plan, the risk is greatly reduced.

This more formal development path is well documented and is used by many successful enterprises. But it is also possible to succeed by the *seat-of-the-pants* method. The choice will depend on the management of the operation.

Not always appreciated is the fact that the product idea is a relatively small factor in the overall business plan development. Ideas acquired from the *inventor* path often have difficulty separating the inventor from the project so that its real market potential can be exploited.

A case study is now presented, as Example 7.3, to illustrate this process.

Example 7.3 *Aquaculture Controls Pty Ltd comes to life*

An opportunity arose to supply instrumentation needed with aquaculture ponds (where fish are farmed in captivity) being set up by land farmers seeking new products. Development of adequate instrumentation and control of water condition required expertise they did not have. An instrument expert approached an entrepreneur and they decided to set up a private firm to supply systems as *turnkey* (designed and built ready to go) contracts.

They registered their small enterprise to obtain purchasing discounts and lines of credit from suppliers. A major loan was arranged, using a well-prepared business plan, to finance the first two years of operation – the time at which they predicted sales should start to produce cash-flow income.

Premises were rented in a business park as that gave them a 'good' address, provided ready contact with useful complementary firms, use of a central office service and room to expand as they grew.

They *retained* (that is, paid a fee for regular consultancy) an agriculturalist consultant, expert in fish breeding and development. A part-time office secretary and an accountant were appointed.

The first steps for patenting their ideas were carried out with help from a lawyer.

Defined tasks and consulting contracts were let on special aspects – brochure layout and artwork, pond design, overseas sales studies being needed.

After a few months, having carried out considerable research into the engineering and the market needs, they were ready to build a *prototype* (the first) system and set up their test pond. Technical staff were appointed to develop the instruments. Contracts were let to form a pond – this required permission which was obtained by convincing the park management that the pond would be an attractive asset if made as part of the landscape.

Whilst the development of the pond and prototype system was in progress – which needed more staff appointments – an application engineer was appointed and trained ready to go out to farms to seek out the first customers. These early clients would be given systems at much reduced costs as they would be involved in field trials. Sales literature, application guides and users' handbooks were developed.

Patents were taken into their consolidated stages.

One year after commencement the test pond was stocked with fish, the prototype system was in use in it and in six client ponds. The system product was apparently going to be a success so more funding was arranged to support manufacture of 30 units at a time.

After two years sales were still increasing. More space was required. More salespersons were needed along with technical staff and installation staff expansion. This was the time to 'go public' by floating a *public company* on the share market in order to raise the quite considerable capital needed to expand to the next level of operation.

The above (imaginary) account cannot cover the many heartaches and traumas that would arise but it does give an impression of how a small instrument firm begins life – and almost all big companies of today started that way. Hewlett started his business in a home garage, Packard joined him shortly after that when they moved into a local science park. Sola International, a spectacle lens maker now in many countries, started out as a three-person firm with an idea of how to use plastics developed during World War II as cheap and robust spectacle lenses.

7.2 MANAGING DESIGN

Much has been written about management of enterprises in general but little is targeted at instrument-producing enterprises. Peuscher, Chapter 28 in Sydenham (1983a), a paper by Sydenham (1983b), Sydenham (1985a) and Chapters 3

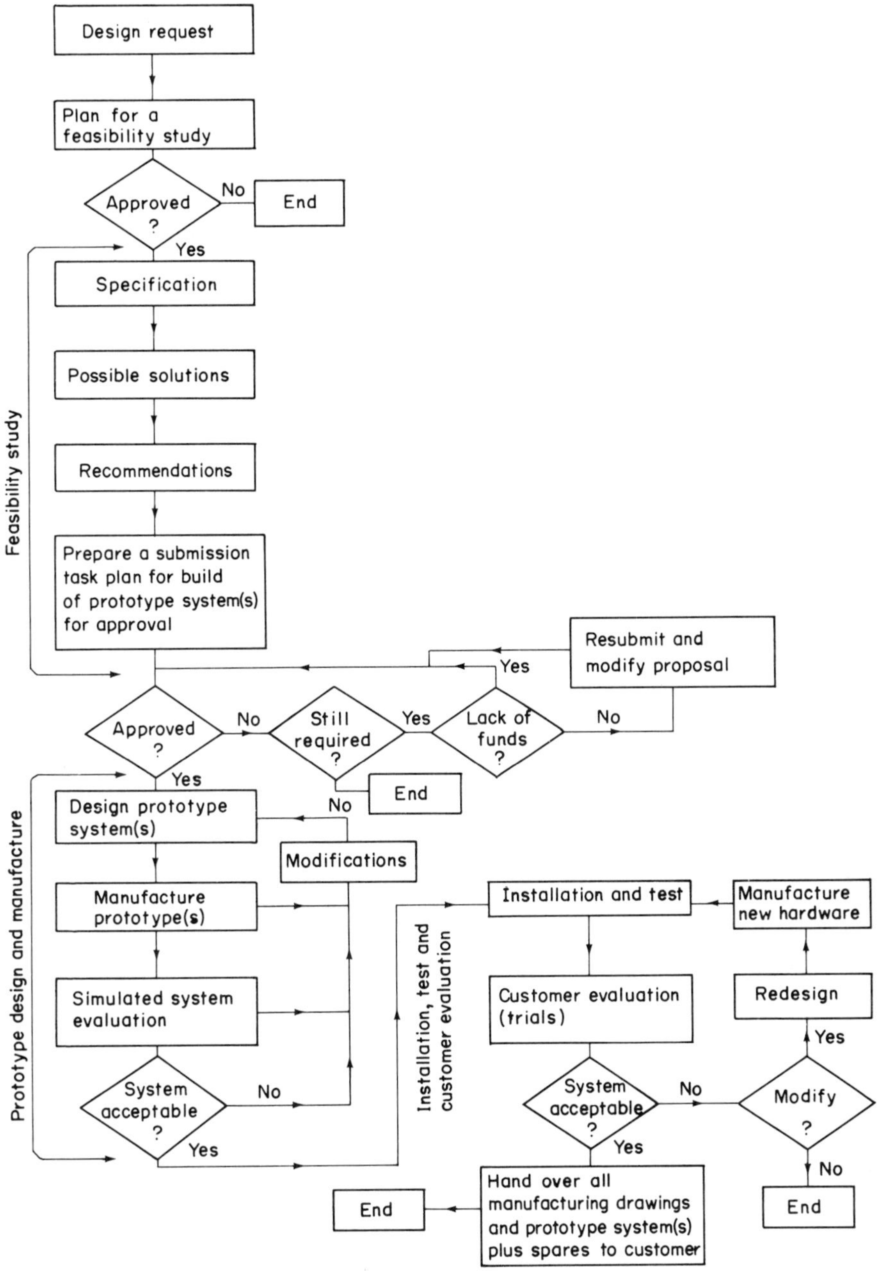

Figure 7.5 Flow chart showing process of design from idea to product ready to begin its production stage

and 8 of Sydenham (1986) each give more detail of aspects of the design process and its organization.

7.2.1 The design-process flow chart

Design, as a process, follows well-ordered procedures that can be expressed in flow-chart form. Figure 7.5 shows the main steps in getting from the idea to the manufactured artifact at a one-off level, or where it is ready to go into the product manufacturing stage.

Good engineering management is essential at all times in this process for it is expensive and time consuming to retrace steps. The flow chart only expresses the salient points of the process and it is highly iterative. The process will often need to go back to earlier stages because each step requires assumptions and decisions in order to make the move to the next.

The nature of engineering requires decisions to be made with varying degrees of uncertainty. As the process unfolds these are seen in a new light, often showing that some needed revision. An aim of design is make decisions that are doubtful as seldom as possible. Doubt is usually resolved by giving more attention to that aspect.

7.2.2 Reticulation of the design task

Effective management of the design process is based on breaking the whole into identifiable parts. This is called *system reticulation*, or *partitioning.*

Partitioning can be achieved by recognition of aspects of the problem. There exist numerous aspects and the design-process manager selects them to suit the particular task in hand.

A convenient way to express the results of reticulation is to use a *tree diagram.* Figure 7.6 is one classifying principles used in gravity meters.

Tree diagrams find use where classification forms of reticulation are needed. Examples are principles that can be used, as shown in Figure 7.6, the relationship of tasks such as design features to be remembered, and how ideas can be improved by breaking down each critical design feature into lower-level features.

They are also useful as the basis of *functional analysis*, wherein the functions of an instrument are broken down into increasingly smaller tasks allowing each to be catalogued for future reference, described and worked upon as entities.

Another reticulation method is the use of *block diagrams* that show the relationship between system building blocks and entities associated with them. A *flow chart* is an example. Entities commonly expressed this way are the flows of information, mass or energy at the most abstract level, movement of personnel and materials at the operational level, and the form and nature of signals at various points.

A third reticulation form is typified by *utility analysis* (discussed in some detail in Section 8.3.2). Here the problem, such as which competing instrument or purchase or which design principle to adopt, is studied to reveal the key

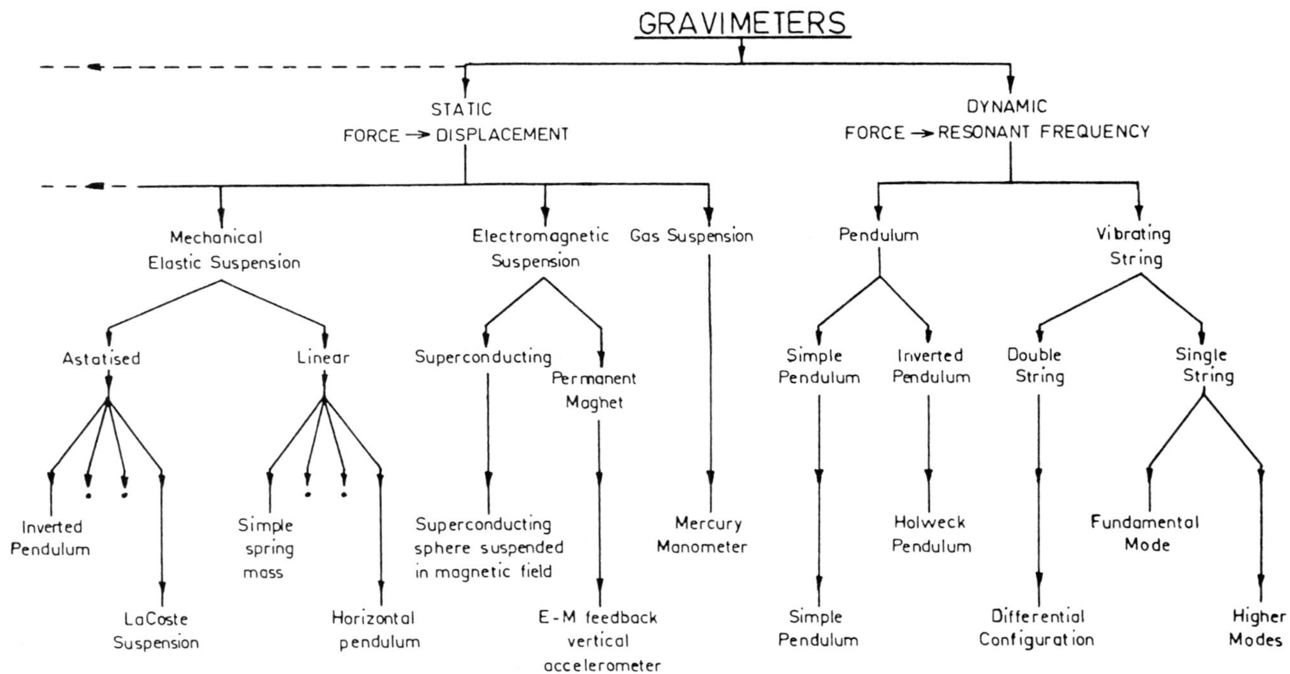

Figure 7.6 Tree diagram classifying principles used in gravity meters. (From Hugill, 1984.)

parameters. How the *utility*, or usefulness, of each varies with change in magnitude is expressed as a *utility curve* for that parameter.

The *candidate* options that appear to satisfy the need are then compared by taking the particular value for each utility variable and forming a collective combination (the simplest is summation of all values for each candidate) which indicates which is the best option to use.

Reticulation is discussed with respect to instrument systems in Bosman (1978), in Peuscher's chapter in Sydenham (1983a) and in Sydenham (1986). A good introduction to its power at generating ideas by *structured reasoning* is given in Marshall (1978).

7.3 SPECIFYING THE NEED

7.3.1 Knowledge-needed basis of engineering requirement

As both the instrument designer and the client needing the instrument do not always have enough knowledge about each other's knowledge domain, the development of a new instrument is a case of interfacing two poorly matched factions together.

Many meetings may be needed to clarify the need in order to prepare a suitable engineering design task.

The *measurement process algorithm* (MPA), discussed in Section 2.3.4, can provide a guiding path to getting a true representation of the task in terms of engineering requirements and suiting the next step of the design cycle. The MPA flow chart and method can, with modification, also be applied to guide meetings and work-task considerations.

It is far more efficient to develop a *system requirement* at this stage, before hardware and other resources are committed, than to have to make changes later in the design process.

Important features that must be resolved are what measurands are to be measured and how are they to be combined to give the user the knowledge sought from the measurement system? This entails study of the *metrological variables* of each measurand - see Example 2.5.

Incorrect knowledge will be obtained if the environment of the system includes *influencing variables*. It is often important to the final success of the system that influence measurands are also sensed to facilitate their correction.

There will be many more system design constraints. The important ones in the early stages are to estimate what cost and time ranges are involved. People who request special instruments to be built do not always recognize that custom, low volume, designs (which are often the need in this field) take considerable time and finance to develop.

Relevant further reading here is Finkelstein (1977), Bosman (1978) and Sydenham (1985b).

7.3.2 Specifications

At critical stages throughout the life cycle of the instrument there arises need to carefully define what result is needed and how it will be produced and verified. Such definitions must be committed as written and drawn statements called *specifications.*

Many different specifications are needed to cater for the varied aspects of the process and product. These are used to ensure the plan devised is followed and that each serial step will allow success at the next.

The first specification needed describes the *operational requirements.* It is essential that a project rapidly develops this statement as the result of early discussions between the client and the designer.

The operational requirement specification avoids stating how the instrument will be constructed to perform its task – that is a later result of design considerations. It describes what it should do and expresses how it will be used. It should not introduce any more constraints than are clearly relevant, see Example 7.4.

Example 7.4 Operational requirement for a garden tap water timer

The [numbers] are added to allow comment. A statement would be:

'A timer unit is needed that can control the flow of water from a domestic tap according to a length of time duration set by the user [1].

It must have adjustment from 5 minutes to 2 hours, be easy to set and alter, and indicate the period still to run, once set. It must be reasonably fail-safe in operation [2].

It must be able to be fitted to the tap by a normally capable householder [3]. Operation must not produce water needing disposal [4].
The form and shape must be attractive and easy to manufacture. A lifetime of satisfactory operation of two years is needed under bright sunny conditions [5]. It must operate in any global garden location where garden hoses are used [6]. It will be made from plastics where possible and be affordable by most households [7].'

Each such statement will embody some constraints; various points of the statement (those numbered) are worth noting:

[1] The way it is to be done is not dictated – it could be electronic, water driven, solar operated or mechanically spring driven – each has been made by various manufacturers.
[2] It is obvious that failure to shut off on time can cause a flood. It is probably impossible to make it absolutely reliable but this factor must be given high attention.
[3] It will sell best if able to be put into operation by anyone.
[4] Initial considerations have found that one marketed design uses a water bleed to drive a timer system. This water has to be dumped somewhere. This is to be avoided to make the unit more saleable.
[5] The design life must be specified for the use of appropriate plastic and additives can target this precisely, keeping price to a minimum.
[6] World sales are needed to support large-volume production.
[7] This general statement bounds the design possibilities and dictates plastics as this is the main business of the manufacturing enterprise.

This first written specification is usually prepared by the designer as he or she has greater incentive. To both parties it provides a first *contract of understanding*. Without it both parties, now devoting time to more adequately learn about the need, will often dispute what was agreed in earlier meetings.

Specifications must be approved by *both* parties. They then provide a firm foundation to begin design or purchase. The act of expressing conclusions in written form clears up many factors, sharpening the task definition.

The custom nature of instrument development – each new application brings to the task its own peculiarities – means that probably neither the designer nor the client has a perfectly clear understanding of the task. For this reason most specifications will need to be updated as the design path is traversed.

A balance is needed in specification writing, for too much detail can make the description overly cumbersome and restrictive, yet too little will leave the task under-defined. Experience helps get this right. The adage '$pecify $anely or $uffer' must be kept firmly in mind.

The next specification to develop is that describing how the task will be achieved in terms of the technology; this is the *functional* and *technical requirements* specification. It is written in engineering terms, for by this time in the process study will have begun to firm up just how the instrument will be structured at its top level. This is the stage at which the technology that is to be used is decided.

Some other specifications that are needed cover:

- implementing and checking the design;
- manufacturing and quality;
- factory testing;
- acceptance by the client;
- how it will be documented;
- packaging arrangements;
- installation;
- vendor's component descriptions.

Additional information about specifications, in relation to instruments, is to be found in Sydenham (1983b, 1986) and in references given there.

7.4 COMPUTER-AIDED DESIGN

7.4.1 Role of computers

Much of the design and manufacture of instruments is highly suited to the use of computer-based design tools. It is to be expected that their use will increase, for they offer great advantages compared with the traditional method.

The highly competitive nature of the instrument industry, impatience of clients and designers who do not wish to waste time going over established paths, and the need for better, more reliable and lower-cost systems will each force the design to go this way.

Another key reason is that computers can contain expertise, making it available instantly so that users can carry out a large amount of their own design. Because applications differ in each design case, the need to provide users with this form of knowledge-based interface is highly appropriate.

Although many relevant software tools have been available for several years the adoption of computer-aided design (CAD) methods has been slow for the smaller design operations such as instrument design.

This is now changing. The cost of an adequate personal computer with plotter and printer peripherals, the degree of standardization of software, the level of graphics, the availability of software and changes in attitude brought about by more awareness have created the design climate where to not use CAD is to waste too much resource for too little result.

The key limitation for CAD-based instrument development is currently a shortage of appropriate software tools for, as has generally always been the case, instrument development has so far derived its tools from other larger discipline pursuits.

7.4.2 Some relevant software tools

The following summary provides examples of products. Usually many alternatives exist, each having their own features. The user must choose carefully allowing for his or her particular circumstances.

If starting from scratch the rule is to choose the software and then obtain a suitable machine on which it can be used. Unfortunately the choice of both together is difficult, for available software may dictate several machines being used. An added complication is that computer technology is fast changing, becoming old and less cost effective in as little as two years. The compensation is that their price has dropped dramatically and their power increased beyond comprehension and this trend shows little sign of slowing.

At the design administration level there are the *word-processing packages* that allow the user to prepare documents of high quality with comparatively little secretarial skill. Documents can be searched, altered and copied with ease. Examples are WORDSTAR, McWrite, The Word and CHIWRITER.

Keeping financial records and preparing budgets has been streamlined with *spreadsheet software*. These allow the user to set up tables in which calculations and automatic editing are provided as entries are made. Examples are Lotus 1-2-3 and MICROPLAN.

Managing projects can be made much easier with *planner* software that assists set up schedules, establish critical paths and plot planning time charts. Examples are McProject and PROJECT PLANNER.

Figure 7.7 is a planner chart for the development of an acoustic sound recognition sensory system.

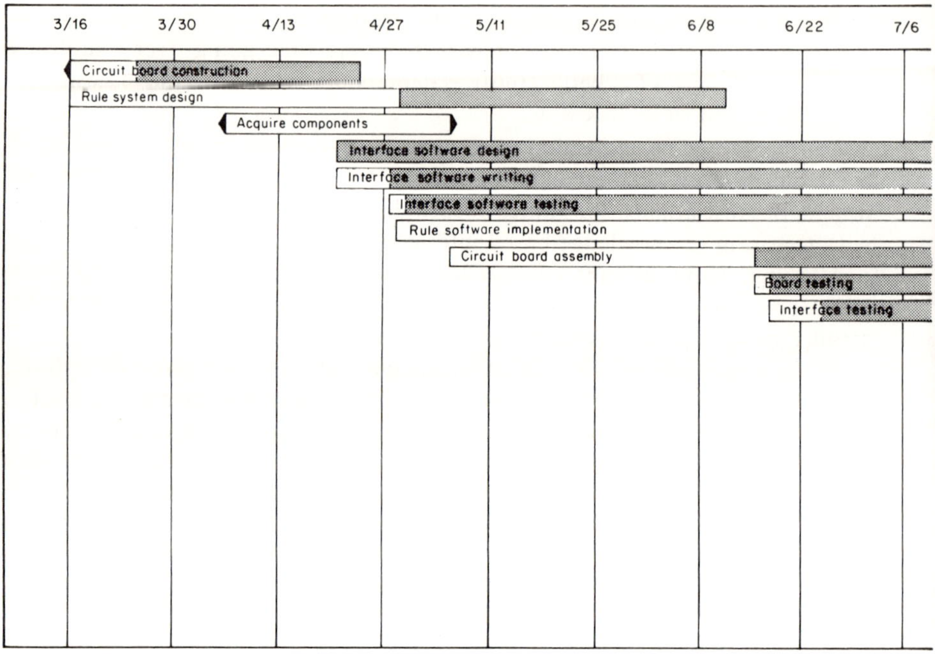

Figure 7.7 Project planning chart produced with McProject

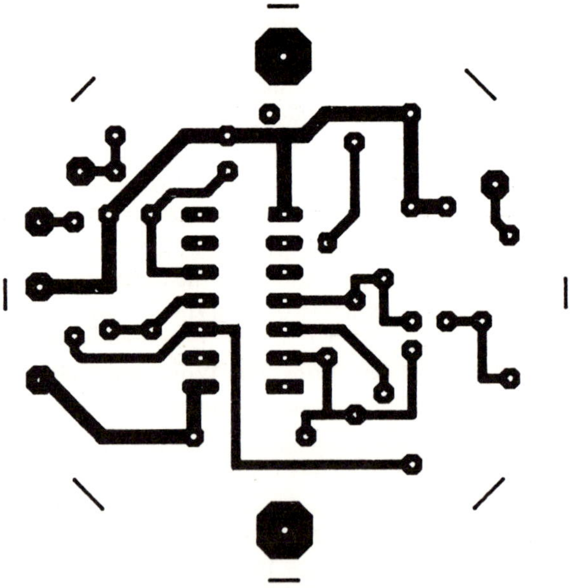

Figure 7.8 Printed circuit board produced using PROTEL tool

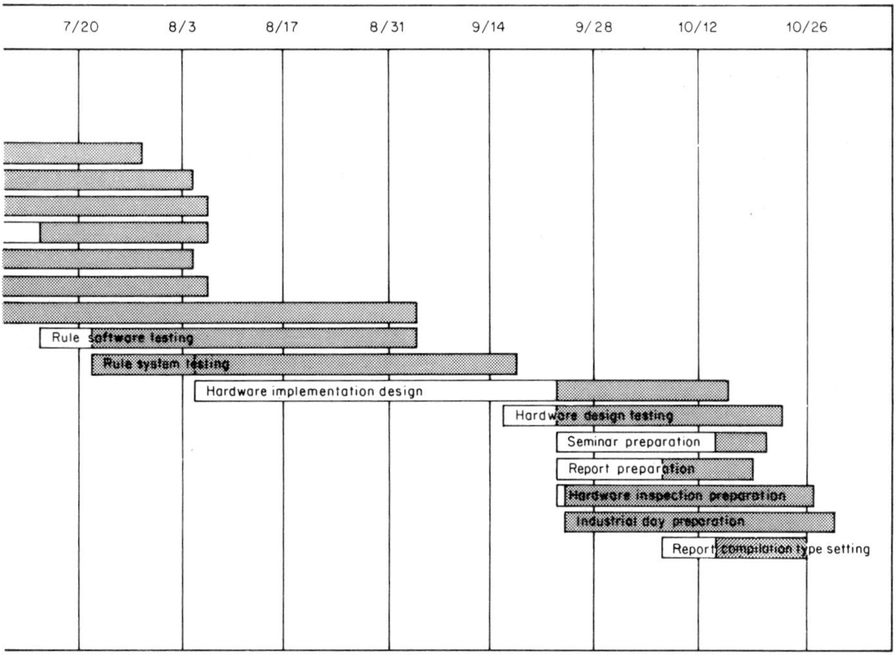

The above areas are well catered for because they align with the practice of general business practice. The following are more specialized and, therefore, less well supplied.

In the electronic systems field there are numerous relatively low level, but very capable, packages for *laying out* printed circuit boards. Figure 7.8 is the plotted layout produced for the electronic circuit of a strain-gauge based pressure sensor. It was produced using PROTEL. Another personal-computer level board-layout tool is SMARTWORK. More expensive systems exist which provide more power – an example is Racal's REDCAP.

Board-layout software is available to high levels of capability but can need mainframe computers. The trend is for users themselves, rather than draughtspersons, to develop the layouts. Once prepared the electronic file is then check plotted. Files can then be sent over data links to the board maker. The need for craft in drawing has largely been removed.

Development of circuitry design and operation are assisted by many packages, known as *schematic capture* and *simulation* software, by which the user enters a suitable circuit onto the screen. This is then analysed for performance and signal plots provided. A survey has been published (Jain and Harpas, 1987).

The problem with the above simulation packages is that the user has to have a well-developed knowledge of electronic engineering in order to be able to input a suitable circuit.

The trend is now to develop software that moves progressively more toward use by a person with *applications knowledge*, the software supplying the

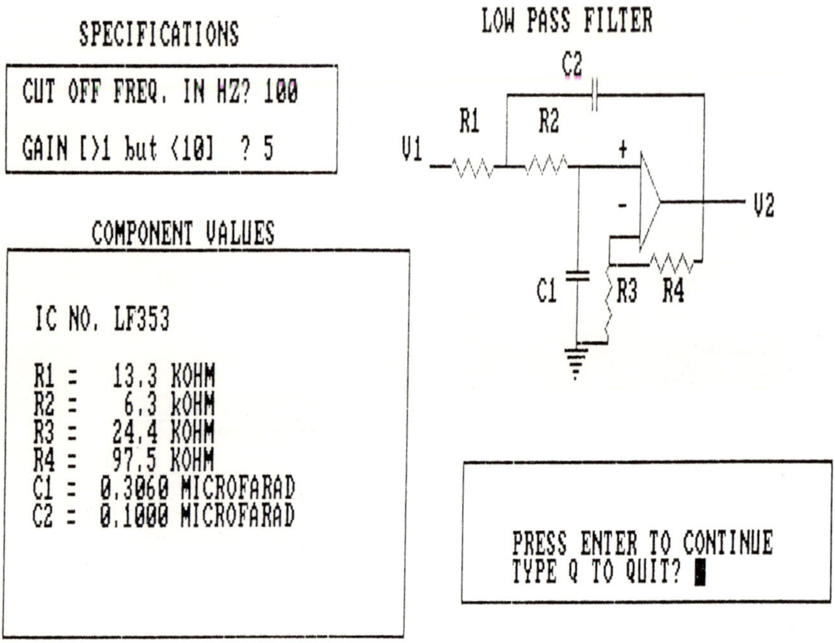

Figure 7.9 Sample display from AJITA software. The user answers applications questions to obtain a suitable circuit

electronic systems expertise. Answers to questions about operational need, rather than engineering functional descriptions lead, in minutes, to circuit board layouts (Figure 7.9) carrying appropriate circuits that deliver the necessary functionality.

The general availability of draughting software has made mechanical component design easier. However, this software often needs a large computer and well-trained operators. An example is given in Figure 7.10. It was produced with PALETTE software. This component has been set up with an *independent program interface* (IPI) that allows the necessary variables to be specified by an electronic data file from a previous design stage. Once defined in this way the plotted drawing will be of the specific case in hand – which can vary over a considerable range of form and shape.

Moving down to the *in-development* stage of emerging software that is relevant to instrument engineering, mention must be made of MEDIEM. This package (see Section 5.4.2) is used to model multi-energy domain systems such as sensors and other transducer stages. It is based upon the mathematics of energy systems and allows a sensor to be simulated for both static and dynamic response before it is manufactured. The alternative finite-element methods (FE), described in Section 5.4.3, are also highly pertinent.

Another instrumentation support tool is SPECRITER, used to produce a professional sensor system operational specification in a matter of minutes by an applications-oriented person (see Cook, 1988).

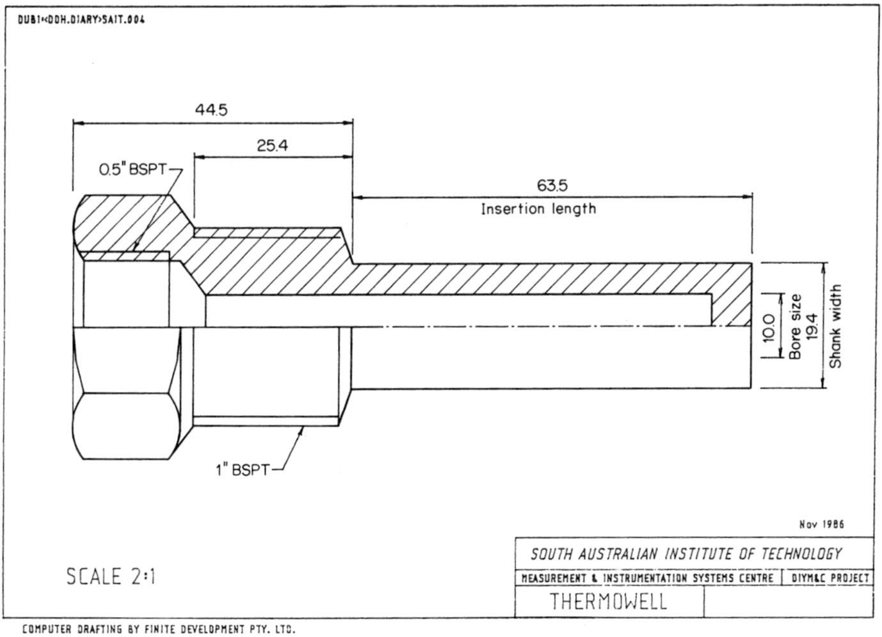

Figure 7.10 Thermowell manufacturing drawing produced with PALETTE software. The specific design drawn is decided by several variable values

A full CAD system for the design of instruments is being developed that embodies many of these tools into a coherent system (Sydenham *et al.* 1988).

It is to be expected that CAD tools will be increasingly used to design instruments, for this methodology provides applications persons with the ability to prepare their own high-quality systems in times that are shortened from weeks to hours. Whether sufficient resources will flow into the development of tools for instrument work is yet to be seen. It has already been found that the effort is well placed, for this methodology appears to be able to bring about the first revolution in how instruments are conceived, made and used.

7.5 PROTOTYPING

7.5.1 Need for test models – the prototype

Having developed what appears to be a sound design concept and specified the actual parts and their organic dispositions it is necessary to build the first specific *artifact* that represents this design. This is generally referred to as the *prototype*.

It is an exacting step in the life of the instrument for this is where many aspects come together as a hopefully coherent assembly. Failure of just one part will prevent successful operation. Meticulous care with all parts is, therefore, needed to produce a satisfactory prototype. The product cannot progress until the prototype is deemed satisfactory.

It is usual practice to break down the building task into separate modules that are each fabricated and tested in isolation before joining into the whole. Certain aspects may also need to be first proved with special rigs and tests. Sound design identifies the difficult problem areas for solution before more routine, known, technological aspects are undertaken.

Careful records should be made as each sub-assembly is manufactured and tested. These can become vital at a later stage, especially if later versions do not perform as they were expected.

Prototypes should not be broken up or destroyed until it is certain their role is no longer needed – often they are valuable as historical items. They will often be needed to see how they performed. They are the best record there can be of that stage of the work. The process of building and testing is one of learning; many prior assumptions will be revealed as incorrect. Tests that should have been made may not have been carried out, and so on.

The development of prototype instruments is best carried out by technicians and engineers who have this kind of experience. It is not as straightforward as production-line manufacture and assembly. The prototype team need a high level of autonomy and flexibility. They need a *general-purpose workshop* suited to this kind of work. Use of facilities and operators from existing production lines is not effective.

Firstly, the methods used to make parts in such low volume (typically 2–5 in the first building operation – one only is not recommended, because it leaves no comparison object as it is invariably altered) need to be highly labour and skill intensive. For example, frameworks must be made 'by hand' as injection moulding jigs and press tools are too expensive to entertain at this stage.

Secondly, there will be need to improvise and modify as the building operation progresses. Instruments seldom are given the level of up-front resources that allow in-depth design studies such as are associated with, say, a large-scale consumer-product enterprise where even a single connection can attract extensive design attention.

Special skills are needed to obtain good finishes in panel lettering and paint aspects. Testing procedures are often worked out as testing commences. Additional features are added as the prototype grows.

There may be need for subsequent *marks* of prototypes, especially if the volume of manufacture is large. In larger-scale production another stage, the *pre-production* model, is used before finalizing the design.

The prototype stage is generally one were change can be accommodated. As the design cycle progresses it becomes essential that the design becomes *frozen* – it is too expensive then to make changes.

7.5.2 Prototyping in the various regimes

The high cost and, often, long time delays in use of production manufacturing techniques to make parts does not suit the development of the prototype. Such methods are only justified when exact detail is established and their cost can be amortized over a large volume of instruments.

Prototypes are very much about proving ideas to be sound and workable. To this end it is often feasible to use certain versatile *prototyping technologies* that allow a system to be built in a reasonable time, and at a more affordable cost. These methods are rarely suited to final manufacture: modification is needed to take a successful prototype to the next manufacturing step.

In the electronic regime, general-purpose, printed circuit boards can be used to produce a circuit without needing to make a dedicated printed circuit board. Figure 7.11 shows some examples. Tracks are joined and broken as needed in the parallel strip boards. In another method (*wire wrap*) wires are wrapped around connection posts with a special wiring tool. Components and connecting wires are pushed into connecting holes in the *plug-in* boards. Once wired these are energized and tested with conventional electronic laboratory instruments.

Larger, dedicated-purpose, *prototyping systems* are available for such areas as analogue, communications and digital systems development.

To some extent much of the development can be carried out with computer-based design tools. These cannot, however, replace final systems testing.

The form of these electronic prototyping methods is very close to the final form of the product. Componentry is easily altered and adjusted.

Recently marketed are systems that go one step further on with the manufacturing process. After a printed circuit board has been laid out, on screen, using a CAD tool, the track geometry information is used to selectively expose photo-resisted printed board stock placed into the same flat-bed plotter that was used to *test plot* the art work. Here the pen is replaced with a radiation source.

After etching the board (elsewhere) it is placed back into this special plotter which then has a drill placed in the pen holder. Drilling for components is then carried out under computer control. With such systems a prototype board can be designed and made in a matter of hours.

Turning now to the development of the mechanical aspect, there are marketed proprietary systems for this purpose. They are expensive but are soon justified, as the labour cost to make fine mechanical assemblies is high. These systems are fast to use, providing high levels of mechanical complexity and precision.

That shown in Figure 7.12 appears to be the most suitable for this purpose. Others sold, as toys, can be pressed into service to some extent but do not provide the same degree of rigidity, precision and closeness to a final production form.

Apparatus sold for servomechanisms (such as Feedback Pty Ltd, UK) and instrument systems (such as J. J. Lloyd Instruments Pty Ltd, UK) teaching and development are also relevant, depending upon prototyping need.

Whilst these mechanical systems can replicate the essential performance needed, the form of the prototype is far from how it will be produced. For this reason extensive draughting and design must follow their use to convert their operation into a manufacturable instrument.

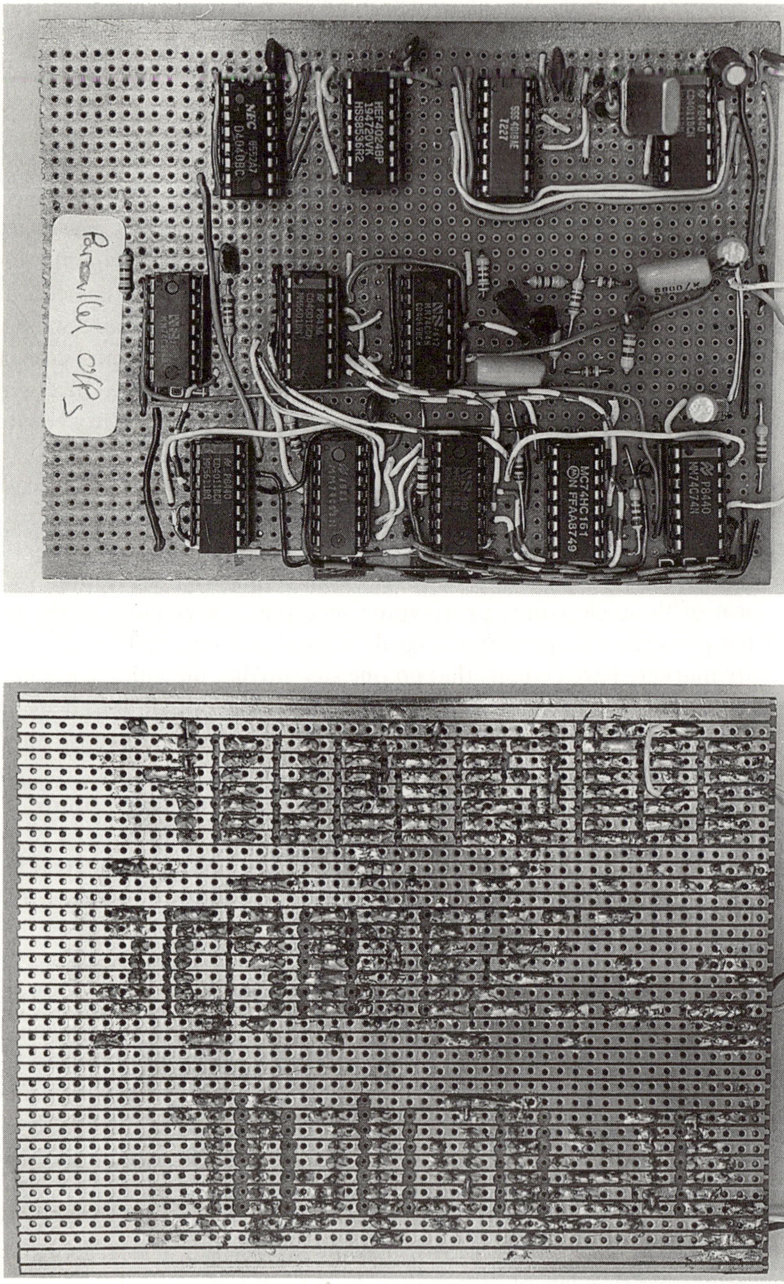

(a)

Figure 7.11 Examples of prototyping boards for electronic circuit use. (a) parallel copper strips printed board; (b) wire-wrapped board; (c) plug in board

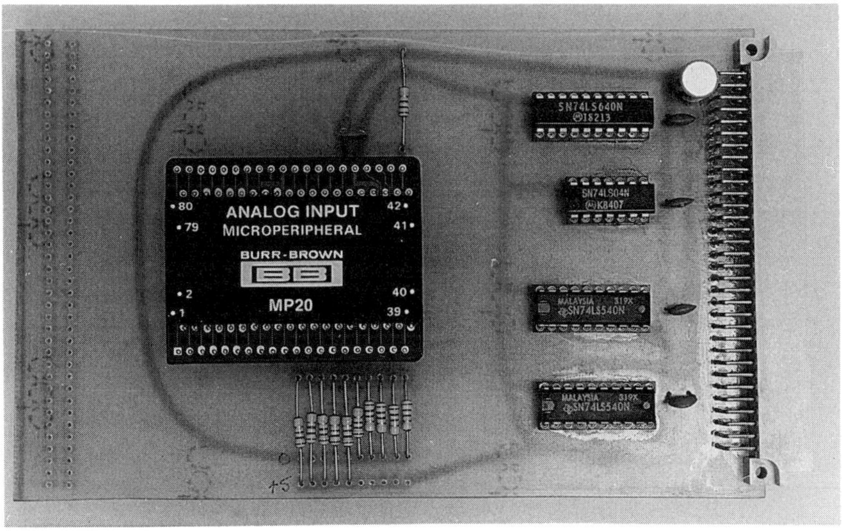

(b)

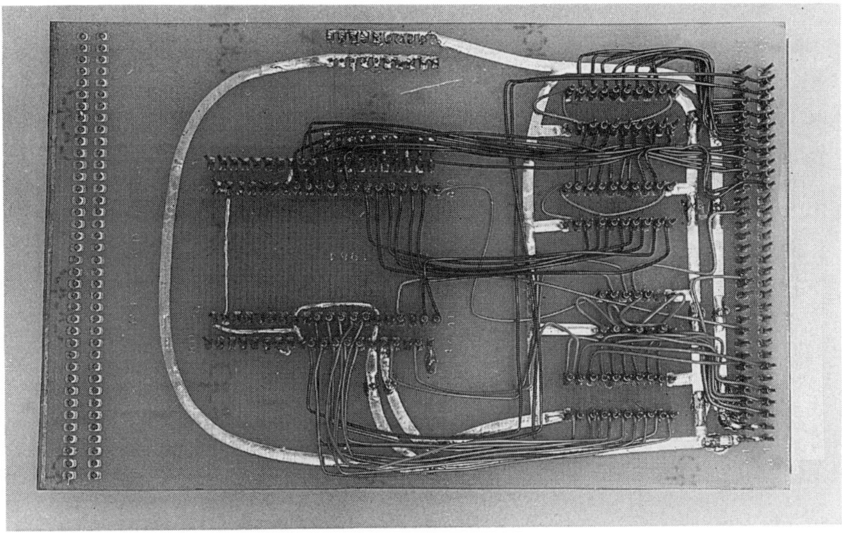

Figure 7.11 (*continued*)

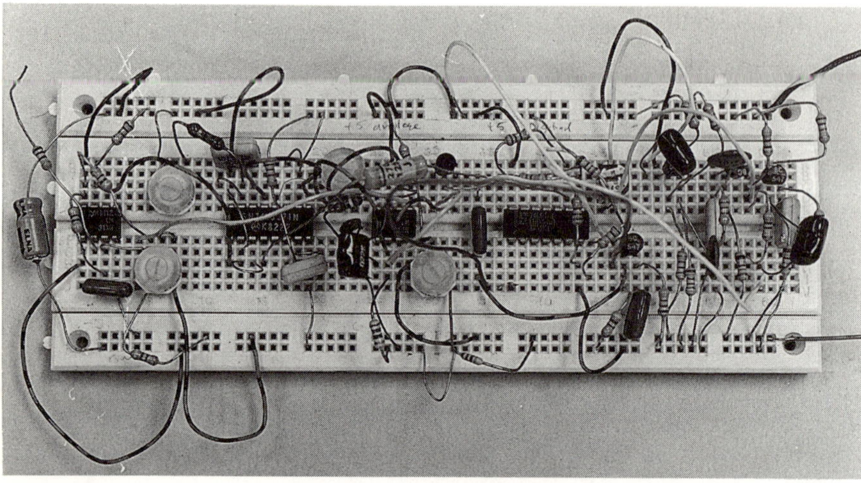

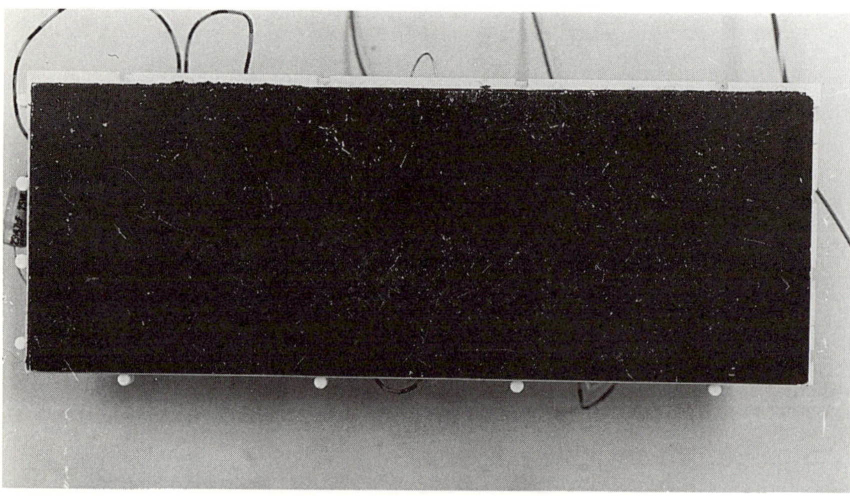

(c)

Figure 7.11 *(continued)*

Figure 7.12 Mechanical prototyping constructional system. (Courtesy FAC, Sweden.)

Radiative systems can be modelled using optical bench systems made to support adaptable component holders and standard size components. Figure 7.13 shows one form of this kind of system. This apparatus can be very expensive and is usually only available to large programs.

A certain degree of low-cost, in-house, construction can be used in optical instrument development but the positioning and controlled movement of components usually requires high-precision manufacture. If the instrument is simple then it is usually more economic to make a custom structure. Mechanical constructor kits can be used.

The cost of good optical components will be high where effective development is involved. Use of *educational quality* componentry is not recommended as they are not made to the optical quality needed. Section 6.4 explains the importance of quality optical componentry.

Pneumatic and hydraulic systems can also be prototyped relatively easily, for their final manufacturing form uses similar items. The use of easily made connections and standardized fittings gives high flexibility in development.

Air is obviously cleaner to use; oil spills in hydraulic systems need to be contained. The former finds more application in information aspects of instruments. Pneumatic systems are still in use but generally this technology is fading out of use for processing information.

Fluid systems, however, still find instrument use as necessary auxiliary devices, examples being to blow dirt from optical windows, as actuators in position balance devices and so on.

The reader is referred to Table 6.1 for a comparison of the prototyping in each of the information energy regimes.

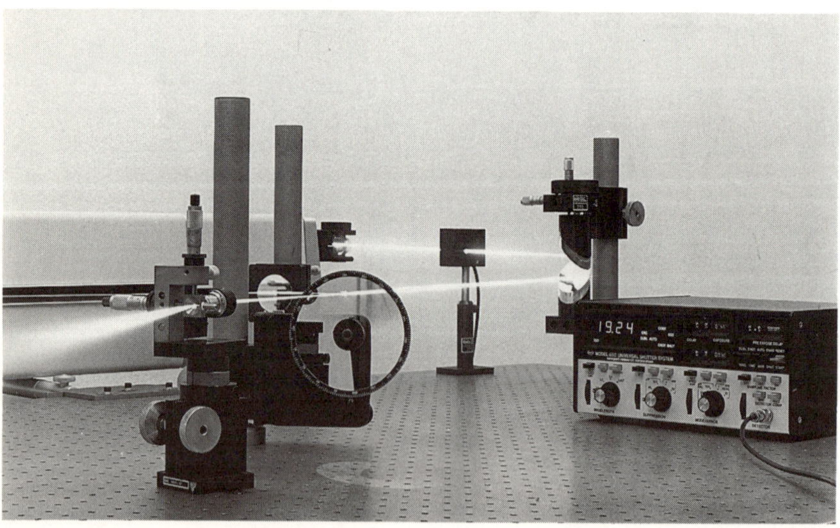

Figure 7.13 Example of optical prototyping system. (Reproduced by permission of Newport Corporation.)

The development of an optical sensor is explained in Example 7.5 as that needed use of fine mechanical, optical and electronic components.

Example 7.5 Prototyping an optical encoder system

One form of *incremental optical shaft encoder* consists of a transparent disc of plastic on which are formed, by photographic means, radial lines at uniform spacing and equal opaque–transparent widths. A typical system has 1000 lines on a disc size of 50 mm diameter.

By placing a fixed segment of grating close and parallel to the rotating disc, but not quite touching, *Moiré fringes* are produced when it is viewed through the two layers. To a sensor, seeing through the two gratings, these move past the field of view at the rate of one fringe movement for each grating cycle moved (1/1000th of a rotation for a 1000-line disc).

The advantage of the Moiré method is that the viewing area is enlarged to a practical 2 mm diameter allowing practical light-sensing diodes to be used. It also averages defects over as many as fifty lines or so.

By the use of at least two sensors, set to be one quarter of a cycle apart, it is possible to detect direction of rotation and produce four pulses per movement of a fringe, that is, a 1000-line grating produces 4000 pulses per revolution.

The plastic disc and reading reference segment were machined on a lathe. The bearing and its support system were built up using a constructional kit that allowed running tolerances and rigidity to be held within $\pm 25\ \mu m$. This was done without workshop support except for the manufacture of a hub to hold the disc. A small d.c. motor was connected to drive the disc for test purposes. The optical source and sensor were held in place by suitable mountings (made with the constructional kit) that were made adjustable for testing purposes. A synchronizing optical trigger system was added to define a datum angular location. Assembly of this part took two days, including the machining.

The electronic circuits were assembled using proprietary boards that allow components and wires to be plugged into small slots. Circuitry was powered with a laboratory-adjustable power supply. This activity took two days.

This done, the system was then ready to use in a learn-and-test capacity. Testing of dynamic performance was carried out using the drive motor. Static accuracy was measured using a correctly coupled angle transducer of known accuracy, obtained on loan.

The prototyped arrangement, once proven, was transformed into an *engineered* model for final use, an activity that took about two months. Engineering drawings were needed along with parts procurement. The body and cover had to be made from metal stock using saw, lathe, mill and drill operations. Parts were plated. The electronic circuit was converted into a printed circuit board, shaped to fit into the encoder outline. A report was compiled after testing was completed.

To have attempted to build the final version immediately would have required many modifications and taken many weeks of effort. Sensitivities of system elements could not have been tested so easily due to the fixed nature of mechanical parts and circuit values; in the prototype method these were easily varied.

Whilst this example is of a simple sensor the same methods are used to develop large systems – the limit being the availability of prototyping equipment.

7.5.3 Into production

The development of the prototype, as the first step into replication, must make allowances for its transition into a mass-produced item.

Componentary used must be obtainable in large numbers. It is preferable to have them potentially available from more than one source (*double sourced*) as this helps ensure supply. Procurement of large numbers of an item can take months. The pricing structure is better but more constrained in options.

The production assembly methods may dictate a different component package. There may be need for special jigs and tools for their preparation before assembly. Tooling for moulding plastics and pressing metal components will be needed.

Often the producer of the instrument would prefer to use componentary items and design aspects different from the prototype for good manufacturing reasons. This can cause delay.

Because of these, and many more, parameters it is wise to begin a relationship with a manufacturer during prototype development. Much time and other resource is lost by not integrating production early in the design cycle. The now critical importance of ever-shortened *lead-time-to-market* of instrument products means the whole process from idea to market must be highly efficient and productive.

It is often too late to wait for the completion of a satisfactory prototype before finance generation, manufacturing and marketing aspects begin.

The next step after the development of the laboratory-tested prototype stage are *field trials*. Several more units are constructed, again by costly hand methods.

These are placed in typical uses for which the instrument is being targeted. Records are kept by the users who are also interviewed after the trial period.

It is normal to expect some problems to be found that need correction. The extent and impact of these will range from disasters wherein the whole product concept is now seen to be untenable (numerous examples exist in the instrument industry, many of which do not get to be known outside manufacturers' premises) to those needing small changes that can easily be accommodated at this stage.

It is reasonable to assume that the better the design team, its support and its devotion to quality of design, the less the risk of traumatic failure. However, it is a fact that as product conception must rely on the making of many assumptions there is always a finite chance of a failed product.

Reasons that can bring about failure include:

- vendor unable to finally supply adequate sub-assemblies;
- emergence of a better competitive product at the same time;
- collapse of the intended market;
- incorrect marketing plan;
- changes to government regulations; and
- realization (all too late) that the instrument concept is just not right.

Collapse of the adopted specific design at the prototype stage may be very disappointing but it is nowhere near as costly to fail then as it will be later. Furthermore, it is easy to do a redesign or make modifications as only the prototype is involved. Often a change in direction can make effective use of the work already done.

It is most unwise, but often happens, to proceed to manufacture knowing that flaws exist in the marketing or design. There is little evidence that products improve themselves or that changes can be made more easily later. The cost of having to *recall* a 'thousands off' product to make even the smallest of changes can easily exceed all profit for that product. The loss of *market confidence* caused by a failed release is hard to overcome.

So why bother with instrument production? The simple answer is that, overall, manufacturing enterprises can be successful. Instrumentation is well known for being able to provide very lucrative rewards – if its supporters and designers get it right and have some luck!

Prototyping of instruments has been covered in more depth in Sydenham (1986). Relevant material is also given in Sydenham (1983b, 1985a). Workshop detail, as opposed to design management, is well covered in Moore *et al.* (1983).

Unfortunately the apparently small nature of this apparently diverse industry has not provided the incentive for in-depth books and courses to emerge on the development of instruments in particular. Readers needing greater depth must make best use of material provided in a more general context.

TUTORIAL EXAMPLES

Note to the reader: A design process takes many months to traverse. It is not appropriate for users of this text to attempt a full design during a two-semester teaching period.

As the majority of this chapter is concerned with decision-making and management concepts, not hardware design in detail, the use of studies of an available simple instrument is recommended. Common use of the same example will allow more meaningful discussion. The questions suggested can be used to form reports done individually, or as the basis of group discussion.

7.1 Using Figure 7.1 as a basis, and for a simple instrument of your choice, develop a suggested path of key decisions by which the idea was taken to the finally used specific product. Then add, to that decision path, an alternative route that could have been used to obtain an instrument that provided a similar operational performance.

7.2 An order has been won to design and make an instrument that measures the rotational angle of a person's elbow joint during rehabilitation of their health after an accident. It is intended that the first order for twenty systems will be used to develop a wider market. A new operation to carry out this task is to be developed.

Prepare a short document, for presentation to the group leader, that outlines the resources that will be needed to carry out this task placing the report in context with local availability.

7.3 For an instrument of your choice develop an operational requirements specification and then a functional and technical requirement statement.

(Use of an existing simple instrument example by the full class is recommended as the answers can then be used in group discussion to develop this aspect.)

7.4 To develop the use of tree diagrams construct one showing the generic possibilities that exist that might have been used to design the existing instrument of your choice. Write brief comments on the pros and cons of four of them.

7.5 Investigate the availability of CAD tools that are available to you to assist design of instruments. Suggested parameters for consideration are available software, where it is provided, on what machines, what it does and how it is used. Prepare a short report on its use to design the instrument of your choice.

7.6 Using an available instrument report or an actual instrument that incorporates mechanical, electronic and optical regimes in its operation, discuss how each regime aspect could have been prototyped using developmental system available locally. If time and availability permit form parts of a prototype replica.

REFERENCES

Bosman D, 1978, Systematic design of instrumentation systems. *J. Phys. E.: Sci. Instrum.*, **11**, 97–105.

Cook S, 1988, Automatic generation of measuring instrument specifications. *Measurement*, **6**, 4, 155–60.

Finkelstein L, 1977, Instrument science: Introductory article. *J. Phys. E.: Sci. Instrum.*, **10**, 566–72.

Hugill A, 1984, A gravimeter with automatic readout. PhD Thesis, Flinders University of South Australia.

Jain L K and Harpas P, 1987, PC software for electronic circuit design in printed circuit form. *CAE J.*, August, **5**(4), 148–52.

Marshall R C, 1978, Ideas to order. *Electronics & Power*, January, 54–7.

Moore J H, Davis C C and Coplan M A, 1983, *Building Scientific Apparatus – a Practical Guide to Design and Construction.* Addison-Wesley, Reading, Mass.

Sydenham P H (Ed), 1983a, *Handbook of Measurement Science*, Vol 2, Wiley, Chichester.

Sydenham P H, 1983b, Engineering in instrumentation. *Celebration Compendium*, Central Scientific Instruments Organisation, Chandigarh, India.

Sydenham P H, 1985a, *Quality in Scientific Instruments*, Australian Scientific Industry Association, Melbourne.

Sydenham P H, 1985b, Structured understanding of the measurement process. *Measurement*, **3**, 115–20, 161–8.

Sydenham P H, 1986, *Mechanical Design of Instruments*, ISA, Research Triangle Park, North Carolina, USA.

Sydenham P H, Skinner A and Biejer, R J, 1988, Do-it-yourself measurement and control CAE package. *Measurement and Control*, **21**, 69–75.

CHAPTER 8

Selection and testing of instrumentation

8.1 CRITICAL APPROACH TO INSTRUMENTATION

8.1.1 Introductory remarks

Moving through the life cycle of an instrument this chapter now addresses how to select existing or offered designs, how to assess their viability and how to install them to procure the knowledge that is sought about the selected process.

It is equally as important to study the installation as it is to design a measuring instrument; any weak link in the serial engineering process of placing an instrument into an application can destroy its effectiveness. To have an instrument in place that is not fulfilling the desired role can lead to inefficiency and waste of resources. All instrument engineers can recall many cases of incorrect data being captured only to find, much later after it could not be repeated, that it was all a waste of time. Example 8.1 is such an example that would have been avoided by a simple test of the measuring apparatus.

Example 8.1 How to be fooled by appearances – a greenhouse controller

A greenhouse temperature sensor plotted daily temperature fluctuations on a recorder for over a year. When a routine service was then carried out it was found that it had not been recording greenhouse temperature but the recorder's internal temperature compensation signal!

Investigation showed that the circuits had been modified when purchased, in an attempt to change the sensitivity and span of the system from 40–400 °C to 0–40 °C as the required latter range was not available. Unfortunately it had not been changed by an expert. The change made had only altered the zero point to 0 °C without increasing the sensitivity. Inside the controller, supplied with the temperature sensor, was placed a temperature-sensitive resistor that compensated for influence temperature changes in the controller location – which was not in the greenhouse.

The year-long records were of the temperature of the control room, not of the greenhouse. As they had the expected appearance of daily fluctuations and at the expected amplitude they were considered correct. A simple test of heating the temperature sensor would have shown up the fault. The moral is that an installed sensing system should be rigorously calibrated at regular intervals especially at installation and after repair.

Another example illustrates another well-known trap. Example 8.2 is about another kind of often met error - of unclear and systematic thinking.

Example 8.2 The source of a problem can be overlooked – an aircraft vibration problem that wasn't

Many months were spent redesigning an aircraft engine gearbox to eliminate mechanical resonances in drive shafts that had been indicated from vibration monitoring tests. Despite considerable effort and rebuilding of parts the vibration could not be eliminated.

By *chance* the output of the vibration meter was observed with its *supply turned off* only to find the supposed 'resonance' was still present - the observed signal, being interpreted as vibration, was actually stray-noise pick-up from the electrical mains supply.

Whereas an instrument might have been found to be working properly at the time of first installation there can be no guarantee that it will continue to do so after that. Regular recalibration and assessment of performance is necessary.

The repercussions of incorrect measurements can cost lives and money. For example, drift in the calibration of an in-process diameter gauge can gradually take manufactured metal parts out of allowable specification so that the machining system is then making scrap. Other instances could be a jumbo jet, or a submarine, guided to its peril by an incorrectly set guidance system. Submarine gyroscopes are recalibrated as often as each ten days - even in storage!

The nature of the instrument output data - its similar appearance even when quite wrong - easily give the appearance of being *on track*. Systematic testing is needed to ensure that correct data, and hence required information, is being produced. In large and expensive measurement projects, such as a seismic sea-bottom profiling exercise, at least one person has the assigned task to continuously observe the data and carry out tests to ensure it is sound.

Stein (1967) contains many examples of calibration and testing situations where error has arisen due to inadequate consideration of the whole system being used. Example 4.10 discusses the *loading effect* of using a bulky thermometer that is physically too large for the task.

8.1.2 Guilty until proven innocent

The attitude taken toward measuring instruments must be that they are not functioning correctly in their actual application until *practically* proven to be so.

Furthermore it cannot be assumed that once working correctly they will *continue* to do so for ever. There must exist well-substantiated reasons that support confidence that they will hold *calibration* over the chosen *calibration interval* to the next recalibration.

Designers, and especially users, must adopt the view that all instruments are not necessarily up to their task until proven to be so. In the words of Peter Stein, they must be considered to be

'guilty until proven innocent'.

Various authors have written on the problems of calibration and the related traceability concept: for example, see Daneman (1975), Stein (1967), and Chapter 31 in Sydenham (1983a). Another source of information is the regular *NCSL Newsletter* of the National Conference of Standards Laboratories, USA.

Having shown, by way of several practical examples, how it is important not to accept instruments and their output data on face value this chapter now develops the necessary understanding for selection and installation.

8.2 SPECIFICATION OF SELECTION AND PERFORMANCE

8.2.1 Assessing what is offered

Let us assume that a measurement problem has been identified and a good operational specification developed that describes the task. Original design or purchase can now commence. In cases of purchase the user must make a sound assessment of what is offered. This section addresses the problems of system description, in the printed form, for this is the usual mechanism by which acquisition *decisions* are made. Unfortunately, too often, critical acceptance is not carried out with instrumentation products.

Practical reasons dictate that system description must be succinct yet comprehensive – two aims that are in conflict. On the one hand the *reader* of the document expects to be fully informed about the instrument's capability without paying the cost of the supplier's providing this potentially large amount of documentation.

The *supplier*, on the other hand, is being kept down to a price by market forces and finds that survival can be obtained by giving shorthand statements of many system description parameters and by omitting many others.

A balance has emerged in this industry which now generally gives too little real information. This has arisen because purchasers do not question the supplied information enough, do not always properly test the instrument or are not prepared to pay the price of extra information.

It has become a *buyer-beware* situation. This is to be expected, for the effort needed to properly understand each application is usually more than can be expected of the supplier within the cost of each individual purchase.

To shorten printed descriptions and conversation times many 'shorthand' statements have been adopted that describe much larger phenomena, often with a single word – see Section 4.3. Interpretation of the content of reports and pamphlets needs a deeper understanding than might appear to be the case.

Another factor that must be considered is that documents are targeted at certain readers. A *publicity flyer* announcing a new product release is likely to be rather glowing in its description. It might well be describing an instrument still to be made and tested. Its presentation probably has some degree of *artist's licence*. A *media press report* is highly likely to have the technical facts incorrectly stated.

This is not to say that suppliers intend to be dishonest but more that the reader must be aware of the different purposes of documents in the descriptive aspects of instruments.

A *business-plan description* would need to be optimistic, for a degree of overstatement is often necessary to sway investors into innovative paths.

A preliminary *patent specification* will usually have more innovation embodied into it than actually eventuates in the final patented form.

Description of a part used in a *mass-produced product* must be very carefully developed, for the cost implications of failure to comply can be great to both supplier and to the purchaser.

It is therefore important for the purchaser to take a critical view to assessment of the description of an instrument.

A particular problem of instrumentation is that they are mostly small product-run items and therefore often have little *history* of use. Laboratory testing can only take performance evaluation so far. Field testing often shows up faults that were not expected even by the best-intentioned designer.

Some authorities have the policy of only purchasing so-called *proven* systems and often only accept the *lowest price*. These dictates can severely limit development of a national instrument industry and they do not always guarantee what was expected. The special nature of instruments often needs *local involvement* in design and manufacture to obtain best results.

8.2.2 Assessment checklist

The experienced person will instinctively know which parameters to investigate when assessing an instrument proposal but many persons who select and evaluate instrument purchases and uses, unfortunately, do not have that experience.

The following checklist is provided to assist the assessment process. It can be extended by the reader to suit a particular industry. Prolific assessment duties in a group suggests the preparation of proforma style documents, preferably in a computer-based form, or questioning software using knowledge-based methods, for speeding up the acceptance process. The problem is that even the experienced assessor will not remember all facets and after considerable duty of this kind will become tired of the procedure.

These suggested points are in addition to ensuring the instrument meets the declared performance as published in the data sheets.

- Does the instrument *excessively load* the observed situation?
- Is it *compatible*, if not what extra interfacing is needed?
- What is the *detection sensitivity*? Is it adequate or is amplification needed in its actual use?
- Is it able to *respond fast enough* to changes in observed variables?
- How well will it work in the face of the various *influence* variables existing in its working environment?
- What is the *real cost* when all additional charges are added for installation, spares, calibration, freight, etc.?
- What *calibration interval* is appropriate? How will it be calibrated?
- What is the essence of the *transduction principle* used? Is it sound and effective?
- Is it *well made*?
- Has its *declared performance been verified* anywhere, and how (data sheets, published papers and reports, existing users)?
- Is the *documentation adequate* for its lifelong needs?
- What *power supplies* are needed?
- What are the *maintenance requirements*?
- What is the *spare parts* situation?
- Is any special *support* equipment needed?
- How is it *calibrated*, where and how long does that take?
- How *well known* and reputable are the makers?
- Is the *technology* old, current or advanced? Each has its pros and cons.
- What is the *delivery time*? It could be many months if not an *off-the-shelf* product.

8.3 DECISION MAKING AND SELECTION

8.3.1 Selection of parameters

Whilst measuring instruments may appear to be relatively simple systems total assessment would often involve far too many parameters to be practicable. A process must be used to extract the most important features working downward from these few to more as the case becomes refined. The choice of parameters is a subjective task but it is made more objective through subsequent record keeping and by using *systematic operations* to incorporate them into a decision-making framework. Some easy-to-apply methods for getting started are now suggested.

Slip writing

The human brain is able to recall facts more capably and rapidly if one's thoughts are not constrained to processing the facts as they emerge from the subconscious. Further, as facts emerge in this way they can stimulate other related facts.

In *slip writing* a group of appropriately trained and experienced people (or a lone individual) are each given a note pad with small pages and a pen.

The problem is then presented to the group. An example statement might be 'What do you consider to be the important parameters for evaluating the several weather-station measurement systems offered to us?'

The group then rapidly jot down all facts that come to mind, one per page, on their pad. After about five to ten minutes their memories will begin to dry up. The session leader collects all pages and sorts them into classes.

This simple and quick method is surprisingly powerful. The *consensus* of the group is reached and many parameters will be identified. The sifted material then becomes the basis of a more detailed study.

In slip writing there is no opportunity for the members of the group to trigger innovative thoughts in other members' minds – but it does avoid domination by an 'ideas leader'.

Brainstorming and thinktank sessions

Another methodology, *brainstorming* or *thinktank* sessions, is to set up a team of experts who can represent the facets of experience needed for an holistic study of the problem. They each know of the problem in advance of meeting and can interact during the session. A session secretary records their ideas such that all can see the latest thoughts at a glance – large sheets of paper on the walls or several blackboards are suggested.

A session would last an hour or two. The result is a large amount of key information linking relevant facts to each other. The participants have been encouraged to think deeply *motivated by the ideas of others*. Weak ideas will be argued out and strong ones supported. The secretary then prepares a report from the recorded material – paper sheets are better than use of blackboards as they can be taken away from the meeting room. Example 8.3 is the account of a brainstorming session carried out to seek ways to detect sand from castings.

Example 8.3 Brainstorming session on how to detect sand in castings

A major manufacturer of aluminium castings requested assistance in finding a way to detect the core sand retained after casting had taken place.

The engineering consulting group drew together a team comprising a metallurgist/moulding expert, electrical engineer, physicist and chemical technologist.

The above statement of the problem, plus some samples of castings with parts of core still in them, were supplied to the team in advance of a meeting.

The meeting took around two hours, after which the paper sheets were used by the leader to prepare a report that itemized each alternative, along with comments that had arisen about each. Figure 8.1 shows a tree diagram prepared as the summary of the session. The numbering facilitated comments on each option.

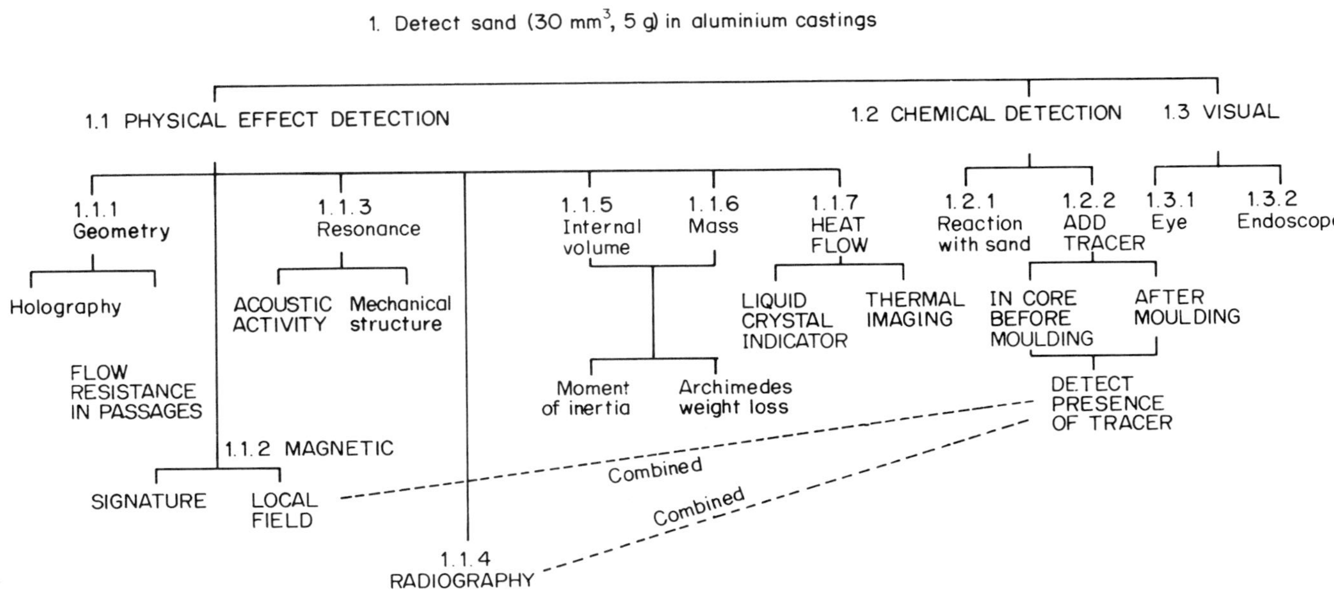

Figure 8.1 Chart summarizing ways to detect sand in castings. It was developed by a two-hour brainstorming session

The report recommended the first three methods considered to be worthy of further study and gave suggestions for their verification.

A useful book describing many methods for generating ideas is Smith and Ainsworth (1985).

Getting relevant information

The above methods all assume someone in the group already has the knowledge needed. When assessing instruments that is often not the case.

Review of past and current designs, study of manufacturers' sales literature, personal ideas, library material searches, comparison with analogous situations, and theoretical and scenario modelling are each possible sources of knowledge. The nature of each is expanded in Sydenham (1986).

8.3.2 Use of utility analyses

Utility analysis was briefly introduced in Section 7.2.2 as a method of reticulating a design task. It also has considerable relevance to assessment of instrument performance.

It is now explained in terms of an example; how to select an instrument, from several offered for sale, is covered in Example 8.4.

Example 8.4 Use of utility analysis to select an oscilloscope

A laboratory has to decide which cathode ray oscilloscope (CRO) to purchase. Sales leaflets of several are obtained and from their study it is clear that it is not easy to decide which is the obvious purchase. The manufacturers each have persuasive ways of making theirs appear to be the best with slightly differently expressed performance and each extolling different virtues.

The answer to this choice dilemma is to conduct a *utility analysis.*

The first step is to nominate 5–20 key performance and operational parameters (only four are used here for brevity). This can be done by a slip-writing session that is carried out once one has become generally familiar with the instrument class from the brochures and existing similar instruments.

For each parameter in turn (here volume V, weight W, sensitivity S and frequency limit F) a *utility curve* is drawn that expresses how the usefulness of the parameter varies with its variation – see Figure 8.2(a). For example, a CRO unit weighing $W = 20$ kg has (our decision here) zero utility, whereas a 1 kg unit would have 100% utility. The choice of limits and how the parameter varies in between limits is a matter of judgement for each particular application.

Whilst these curves are subjective statements, they are seated in sound experience and provide a record of the basis on which later decisions are made. When the task can justify the action they can be made more objective by conducting trials and tests.

The next step in a utility analysis is to decide the *relative weighting* that should be applied for each parameter – they will not all be as important. There exist here two

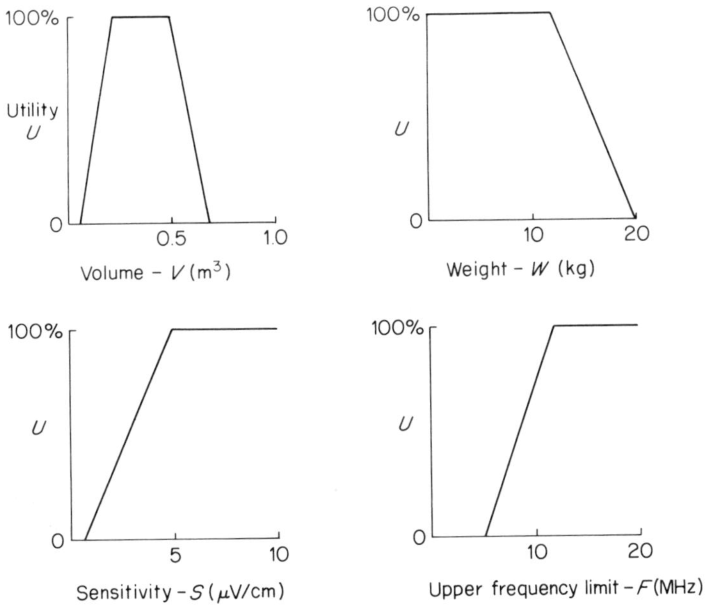

(a)

Parameter	V	W	S	F	Figure of Merit $\Sigma U \times C$ horiz.
Weight coeff. C	0.50	0.80	1.00	1.00	
CRO 1-U	0.23	0.38	0.83	1.00	2.25
CRO 2-U	1.00	1.00	0.60	0.92	2.82
CRO 3-U	1.00	1.00	0.90	0.85	3.05

(b)

Figure 8.2 Steps in utility analysis of which oscilloscope is the best buy: (a) four utility curves (more are normally used); (b) tabular calculation leading to best option (CRO 3)

choices; either to scale the vertical axis of each utility graph accordingly or else apply a *weighting coefficient* C to each as a value is taken from the curves (as done in the example – see Figure 8.2(b)).

Having created the set of utility curves and decided the weightings to be applied to each parameter the user is then ready to begin evaluation of the CRO units offered, three in this example.

For each CRO unit the utility value is read from the appropriate curve according to its particular data for that parameter. For example CRO2-U, in Figure 8.2(b), has a weight W offering a utility $U = 1.0$ and sensitivity S utility of $U = 0.60$. These values are recorded in the table (Figure 8.2(b)) as shown in the horizontal columns.

The table is then reduced toward the need to know which is the best choice by adding (horizontally) the utility values for each CRO unit (after multiplication of each by the weighting coefficients) to establish a utility *figure of merit* for each. The *best* option is the highest score.

The figure-of-merit score can be obtained by many other mathematical manipulations of the raw tabular data. That shown in Example 8.4 is usually a satisfactory method for cases where only a small number of utility parameters are used. The processing options change the emphasis of individual data such as the effect of *outlier* values. In this case, however, unlike much of measurement data the few samples and precision of the data usually mean that outlier values cannot be ignored.

It will often be sufficient to go no further than the simple level of sophistication given in the above example. Whilst there is room to debate the best utility analysis procedure an important point is that whichever method is used every step and decision is recorded for later study. The method also highlights strong (high parameter score) and weak points (low parameter score) of designs.

Cost-to-purchase is better not considered as a utility analysis parameter but as an ordinate of a plot of figure-of-merit score versus cost. This graph rapidly places the various designs in perspective. This will often show that best design performance is not always obtained by purchasing the more expensive unit. The example illustrates this point if you accept that the parameters chosen are the most important and only ones relevant!

Extensive literature exists on decision making of this form – for example Sidall (1972) – but it is generally directed at applications where considerable effort can be applied such as in warfare outcomes, large business developments and where the thrust of a new technology by a giant corporation should be directed.

8.4 QUALITY ASPECTS OF INSTRUMENTS

8.4.1 What is quality?

When assessing a design or a product a most important aspect is the quality offered.

Quality of a product or service is *its fitness for the intended purpose defined by its set of primary functions.* Quality must be assessed by much more than how well it is made or looks; a vital aspect is how well it performs its purpose. This section addresses the quality aspect of instrumentation.

In general quality functions divide into two distinctly different groups. The *use functions* refer to its properties related to carrying out its performance role. The *esteem functions* are concerned with such issues as its attractiveness as a wanted possession and the feeling of well-being it engenders when in use. Such human factors are often vital to effective operation and commercial success of instrumentation.

If *fitness* is to be used as the dominant measure, then there must be a standard with which to make comparison, and hence judgements, about quality.

Overall quality assessment will be the combination of *qualitative* (non-numeric) and *quantitative* (numeric) aspects. Whilst the latter give more apparent comfort because numerical quantities are involved, the *subjective issues are equally as vital to the success of an instrument.*

8.4.2 Quality in relation to instruments

A measuring instrument quality must primarily be judged in terms of how well it *extracts* the information needed from the observed situation and how reliably it will go on doing so under defined conditions.

Metrological performance is usually an obvious quality aspect to assess including tests of such parameters as its accuracy, repeatability, reproducibility, discrimination and uncertainty. There is, however, much more to quality than these factors. How well it holds its calibration, how good is the documentation and many more facets need to be addressed. The task of holistic instrument performance considerations is called *instrument evaluation* – this is covered below in Section 8.5.

Quality is the concern of a wide range of people who would be associated with the instrument. The main viewpoints will be those of

- the *designer*;
- the *original equipment manufacturer* (OEM);
- the *full system manufacturer*;
- the *purchaser*; and
- the *user*.

These are considered in turn.

Designer's viewpoint of quality

Designers strive to implement an instrument on a sound physical principle. The basic idea (a measuring instrument is the embodiment of a physical relationship that transduces one energy variable into another) should not need excessive compensation for extraneous influencing effects. It should be well understood either experimentally and theoretically. It should be as simple as possible. Use of the right principle can greatly reduce the complexity of the instrument, give best performance and at the same time be economic.

OEM's viewpoint of quality

The original equipment manufacturer (OEM) knows that the overall quality of the full system being built relies on the quality of the subsystems used. This organization should ideally test subsystem instruments as they are selected.

As an example, the instrumentation used in a new model motor car is developed by OEMs for supply to the car makers. Stringent quality requirements must be met if the supplier is to be chosen for such large orders.

Full-system manufacturer's viewpoint of quality

For the full system to meet requirements all subsystems must work when connected together. The full-system supplier is closely concerned with quality for the purchaser directs complaints to this organization, not to the OEM concerned. The cost of replacing even a trivial, say, electrical resistor under a *product recall* condition can often financially break the whole-system supplier.

Purchaser's viewpoint of quality

The user of an instrument is not always the purchaser. For example, cable detectors used by telecommunications technical staff would be purchased by a purchasing section on their collective behalf. Common sense would suggest purchasers would always rationally decide which is the right instrument to buy. Unfortunately the complexity and time needed to adequately specify instruments before purchase and test them after delivery, coupled with their often low value, means they do not always get proper attention by the purchaser.

A manufacturer with a 'good name' is often selected in the belief that this will ensure adequate quality but that is too often not true – see Section 8.5.1.

User's viewpoint of quality

The constant, and increasing, pace of change in information-technology products makes it hard for users to be as familiar with instrument facility and capability as they need to be.

Users often do not have an adequate understanding of the measurement need before purchase and only learn about its full capability (or lack of capability) after delivery. To combat this problem they often over-specify the called for performance and facilitates to allow them to be more certain it will be useful to them.

Users are those who appreciate the esteem functions and also the ergonomic aspects. This is where the final verdict about an instrument is reached. Even if it works well, bad styling, poor documentation, poor introduction into the workplace, lack of communication and perceived fear of its consequences can be formidable barriers to its success.

Subjective quality aspects of instruments have not attracted much published literature but more detail is available in Chapters 28 and 29 of Sydenham (1983a); in a series of seminar papers Sydenham (1983b), Hewlett-Packard (1981) and in papers published in the *NCSL Newsletter* such as Iervolino (1983), Schultz (1981) and Moss (1978).

Attention is now directed to the *objective* aspects of quality. Areas that lend themselves to such considerations are those concerned with a system's

- *reliability*;
- *maintainability*; and
- *availability*.

Economic considerations are also necessary because changes in the above factors generally mean trade-off in cost to manufacture and cost of ownership.

An understanding of the principles and methodology of reliability estimation and cost of ownership issues is now presented.

8.4.3 Reliability theory

Reliability is the ability of a system or component part of a system to perform a needed function, under defined conditions, for a defined time or on a defined occasion.

Reliability is formally described in terms of the *failure rate* λ of a system, subsystem or component. It has been observed experimentally that most systems and components, large or small, display the 'bathtub' curve of reliability which is illustrated in Figure 8.3. Electronic systems follow this curve more closely than mechanical systems.

The curve may be divided (roughly) into three parts which are, respectively

- 'burn-in' period (in which odd manufacturing faults and substandard components cause a number of early failures);
- 'useful life' (during which the failure rate is low and roughly constant);
- 'wear out period' (where failure rates again increase, doing so more with passing time).

During the *useful life* it is assumed that *only random failures* occur, and hence the rate is constant for a given system, subsystem or component. As failure rates are very low it is usual to refer more commonly to the *mean time between failure* (MTBF), symbol m, usually expressed in %/1000 h. The MTBF and the failure rate are simply related by

$$\lambda = 1/m \tag{8.1}$$

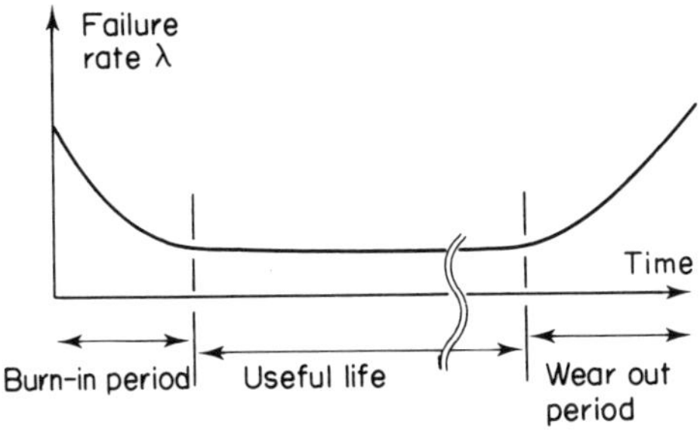

Figure 8.3 The bathtub failure rate curve

Over the *useful life* only, the reliability $R(t)$ can be defined as the probability of survival of the item at time t and will, therefore, take a value between 0 and 1. Because the failures are assumed (quite reasonably) to occur randomly it can be shown quite simply that the reliability will decay exponentially with time according to

$$R(t) = e^{-\lambda t} = e^{-t/m} \tag{8.2}$$

Alternatively if we have N_0 items at the commencement of their useful life, then after a time t there will be $N(t)$ items still serviceable, given by

$$N(t) = N_0 e^{-\lambda t} = N_0 e^{-t/m} \tag{8.3}$$

Figure 8.4 illustrates the relations of Equations (8.2) and (8.3).

It is important to note that after the duration of the MTBF, the point $t = m$ in Figure 8.4, there is only a 37% probability that the item will still be serviceable; or alternatively only 37% of the production total will still be serviceable. Indeed at 0.1 MTBF, when $t = m/10$, the reliability $R(t)$ has already dropped to 90%.

Composite reliability

From the foregoing theoretical ideas it is possible, and extremely useful, to calculate the reliability of a composite system from the individual reliabilities of each of the component parts. The interconnection of the component parts must first be considered; in most simple systems (especially so the instrument system) they are connected *serially* in the sense that if any one link in the chain of blocks should fail, the whole system fails.

In this case if we have n components each with failure rates $\lambda_1, \lambda_2, \ldots, \lambda_n$, then the system failure rate λ_s will be given by

$$\lambda_s = \lambda_1 + \lambda_2 + \cdots + \lambda_n = \sum_{i=1}^{n} \lambda_i \tag{8.4}$$

This is illustrated in Figure 8.5(a).

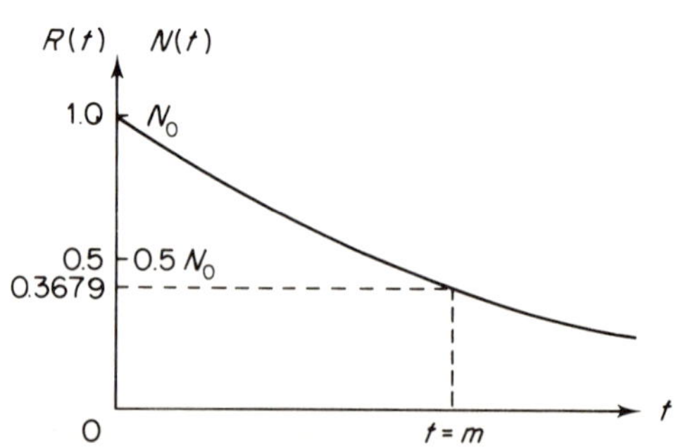

Figure 8.4 Exponential decay of reliability $R(t)$ and serviceable population $N(t)$ with time

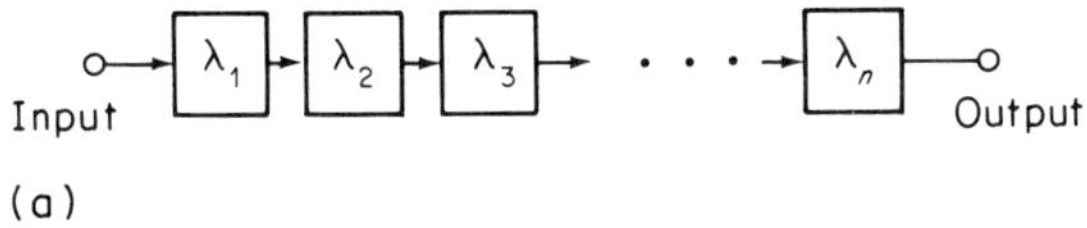

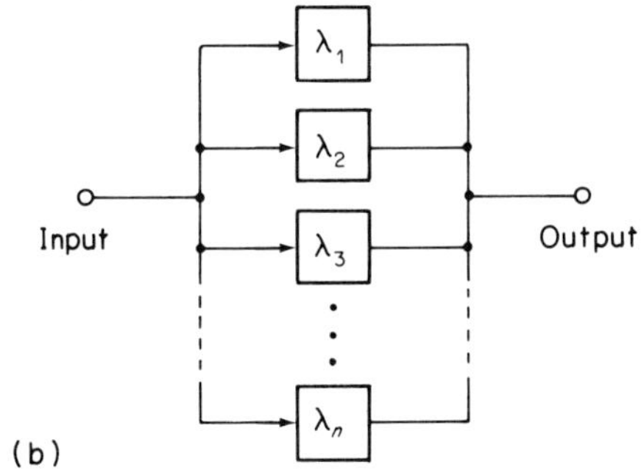

Figure 8.5 Composite reliability of systems comprising: (a) *n non-redundant* component parts; (b) complete *n-fold* redundancy

It follows that the reliability $R_s(t)$ of the composite system (that is, the probability of *no* failure) will be the *product* of the individual component reliabilities

$$R_s(t) = e^{-\lambda_1 t} e^{-\lambda_2 t} \cdots e^{-\lambda_n t} = \prod_{i=1}^{n} R_i(t) \tag{8.5}$$

Equation (8.5) is the *product law of non-redundant composite reliability.*

The effect of the product law is quite severe, particularly with large, complex, systems. Consider a system with a million essential (non-redundant) components, each being individually very reliable and having an MTBF of $m_i = 10\,000\,000$ h. Hence all have $\lambda_i = 10^{-7}$ per hour, so the system failure rate will be

$$\lambda_s = 10^6 \times 10^{-7} = 0.1 \text{ per hour}$$

giving a system MTBF of only $m_s = 10$ h.

Similarly it is clear that if we require high reliability for such a complex system, then the individual reliability must be *extremely* high.

8.4.4 Achieving high reliability

Faced with the problems outlined above there are four major ways in which the reliability of complex systems (which invariably mean the electronic processing parts of information machines) may be enhanced. These are

- simplification;
- subsystem redundancy;

- derating of components;
- use of microelectronics.

According to the circumstances all of these approaches may be employed and each is now considered.

(1) *Simplification.* Although it may be thought that this is obvious many systems, large and small, are often unnecessarily complex for their required task. It is natural human tendency to over-design, over-specify and incorporate additional features 'just in case . . .'. In a lot of cases the additional features serve only to increase the cost and lower the reliability, if they are not actually employed.

At the subsystems level simplification usually involves reducing the component count. In electronic systems, however, individual component reliabilities can vary widely from very high-reliability solid-state components to lower reliability, electromechanical parts such as switches, relays and connectors. Multi-way cable connectors and printed circuit board edge connectors are common sources of low reliability and should only be used where essential.

Mechanical parts of instruments are particularly unreliable (compared to electronic components) as they are often made by small batch processes and have not been designed for the highest-reliability performance due to cost constraints. A dominant difficulty in properly estimating system reliability where mechanical parts, transformers and sundry electronic electromechanical parts are used is getting accurate failure rate data to carry out the design optimization.

(2) *Sub-system redundancy.* Reliability may be increased by providing duplicate (or triplicate, or more) subsystems essentially *operating in parallel* such that all subsystems must fail before the overall system goes unserviceable. This is illustrated in Figure 8.5(b) in which there are depicted n redundant subsystems.

In this case the system MTBF will be given by the sum of the component MTBFs, that is

$$m_s = m_1 + m_2 + \cdots + m_n \tag{8.6}$$

or

$$\lambda_s = \frac{1}{1/\lambda_1 + 1/\lambda_2 + \cdots + 1/\lambda_n} \tag{8.7}$$

The subsystems are essentially identical so that the $m_s = nm_i$ and hence it can be shown that the system reliability for n identical subsystems of reliability $R_i(t)$ will be given by

$$R_s(t) = 1 - (1 - R_i(t))^n \tag{8.8}$$

Study of Equation (8.8) shows that it is possible to make great gains in reliability by the use of redundant systems – but at the cost of supplying extra replicated componentry, space to put it and more systems hardware to connect the several systems to become a common operation.

For the foregoing to be applicable it is essential that the redundant systems be *truly independent* as required by the underlying probability theory. For example, if they are electronic systems each must possess a separate power supply or, if mechanical, a separate motor.

Examples of duplicated and triplicated instruments are to be found in aircraft flight systems, in nuclear power station monitoring systems and in space-craft instrumentation.

(3) *Derating of components.* Every component or subsystem has specifications which include a number of *ratings*; for example a 100 Ω resistor may be rated at 5 watts, a 5 volt power supply for a logic circuit board may be rated at 2 amperes, and a servomotor or actuating solenoid may be rated at 5 watts. Rating may also refer to properties of a component other than its *power* capability, examples being the *working voltage* of a capacitor, the *over pressure* of a pressure gauge and other quantities related to such influence parameter variables as the humidity, vibration or temperature. In each case the figure quoted is intended to be an *operational* guide only.

Figure 8.6 illustrates how the reliability of a component might decrease as it is made to 'work harder' and used to an increasingly higher (or lower) level of *working stress* than its quoted value.

It follows from Figure 8.6 that reliability may be increased, and sometimes maximized, by using components well below their rated value and by deliberate specification of a component rated well in excess of the task required. Components are formally *derated* by a factor of 2 or 10 when higher reliability is required, and factors of 100 or more may be used in satellite and space-craft applications where reliability is paramount – imagine designing instruments for use in a Mars, or deeper space, probe where they must operate for tens of years without failure.

Similarly, to maintain reliability, derating may be necessary when components are used in exceptional and demanding circumstances – lead–acid batteries, for example, are severely derated for use in low ambient temperatures, resistors are usually run at half their rated power rating because the declared rating actually runs the component at an elevated temperature giving a surprisingly short life.

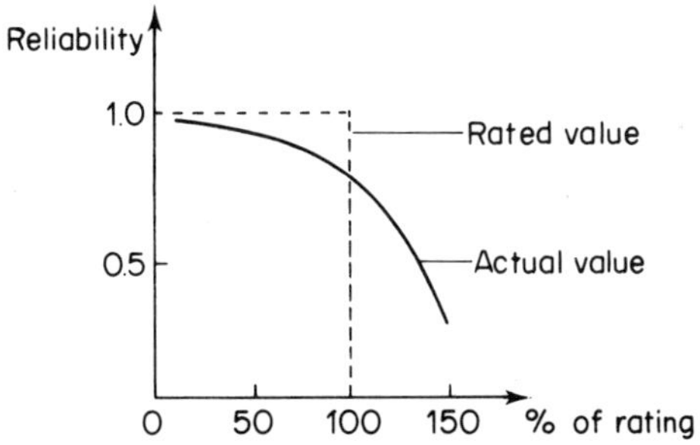

Figure 8.6 Reliability of a component with increasing stress

There are also *absolute maximum ratings* for certain components which must not be exceeded. These are not related to reliability but refer to the point where they will be destroyed. Logic circuits, for instance, made with transistor-transistor logic (TTL) technology are designed to be powered with 5.0 volts, but usually have an absolute maximum rating of 5.15 volts.

(4) *Use of microelectronic circuitry.* This technique is of particular value to information-machine systems and is one of the major reasons that instrumentation is dominated by electronic and microelectronic technology.

Microelectronic systems are loosely defined as those with a device or component density exceeding 3 devices/cm^2. Compared with modern integrated circuits this is a very low component density, but it has been shown that as density is increased, and particularly as the uniformity of manufacture is increased, the reliability is also increased. This is illustrated in the tree diagram of Figure 8.7 which sets out the available microelectronic technologies. It is beyond the scope of this text to delve more deeply into this aspect but such knowledge is well established.

Putting the above theory into practice, for assessment and adjustment of the reliability of a system, is reasonably easy to carry out. (Software is appearing on the market to assist the task.)

Example 8.5 shows how to estimate the reliability of an electronic circuit board (that does not have redundancy). The calculation given there is the easiest to carry out as it only requires component catalogue data plus failure rates. This method, called the *parts count procedure*, does not use *actual* stress levels so can be only a rough, but usually more conservative, estimate. If the board has

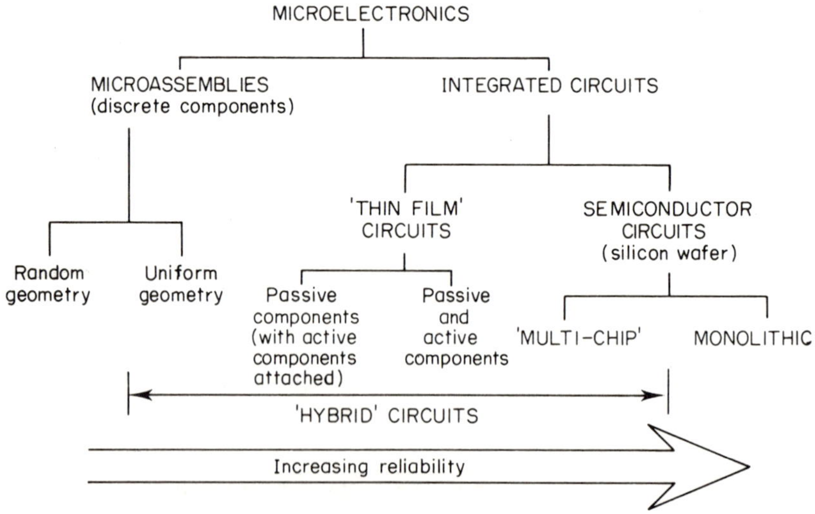

Figure 8.7 Scope and relationship between the various microelectronic technologies

already been made, or simulated with software, then much more realistic stress levels can be used to obtain a closer estimate.

Example 8.5 Calculation of estimated reliability of an electronic circuit board

An electronic circuit has been designed and laid out in printed circuit board form. It is necessary to estimate the chance of failure and its reliability before the design is approved for routine manufacture.

One way to decide how reliable it will be could be to make several boards and subject them to tests. Whilst this is the best solution it takes months to achieve. Calculation from basic component data, however, will give an estimate that takes only a few minutes to carry out.

The first step is to prepare a table listing similar kinds of parts, their strength ratings and number, as shown in Table 8.1.

For each class the typical failure rate is now added. Obtaining specific data is not always possible but available data banks and manufacturer's product data can allow a reasonable estimate. In the case of large-volume products more accurate data should be available. Data for electronic components is readily available through several data banks but that for other components, such as motors, gearboxes, sensors and actuators, is not so easily found. It is often necessary to make an estimate of failure rates.

The table is then calculated, as shown, to arrive at an overall failure rate from which the MTBF can be calculated. In this example the failure rate is 34 367 failures per million hours, from which the MTBF is calculated to be 29 098 h. As the data and methodology is not as accurate as the figures suggest results should be rounded back to only three significant places, that is a failure rate of 34 400 per million hours and MTBF of 29 100 h.

Table 8.1 Calculation of failure rate of a system formed of parts. (*Note*: All failures are per million hours)

ITEM	QUANTITY	BASE FAILURE RATE	QUALITY FACTOR	FAILURE RATE
PWB	1	0.320	10.0	3.200
Connector PCB	2	0.120	1.0	0.240
Connector	2	0.250	1.0	0.500
Resistor (use 1)	1	0.012	1.0	0.012
Resistor (use 2)	9	0.470	3.0	12.690
Capacitor (use 1)	1	0.055	3.0	0.165
Capacitor (use 2)	21	0.030	3.0	1.890
Transistor	1	0.063	10.0	0.630
Lin device	2	0.061	20.0	2.440
Dig device	18	0.035	20.0	12.600
			Total failure rate:	34.367

It must be stressed that reliability estimates are only as good as the test data supplied and give indications only. The presented example makes allowance for an expected stress factor but in operation the stress level will vary continuously and in unexpected ways. This practical aspect makes the task of reliability prediction and control only an *estimation* discipline, working with groups of a product. The single item therefore cannot be entirely guaranteed - only a probability of failure can be given.

The literature on system reliability is vast. Suggested as a start (works related to instruments) are Jensen (1982), Klaassen and Van Peppen (1986), Millar (1980), O'Connor (1981), Simpson (1976), Smith (1972), Chapter 28 in Sydenham (1983a), a chapter in Sydenham (1983b) and MIL (1974), the much used source document for serious involvement.

8.4.5 Economic issues

In the above example the cost of the assembly could be reduced by using lower wattage rated resistors, or lower peak rated voltage capacitors. This would, of course, reduce the reliability by a known amount. To make such judgements an appreciation of the cost regime of an instrument is needed.

The *total cost of ownership* is formed of both *acquisition cost* plus many other *support*, *installation* and *operating costs.* A generalization of the cost cycle for an instrument is given in Figure 8.8.

Each step involves an associated cost as the following check list of cost of ownership (adapted from H. Blanks' chapter in Sydenham (1983b) of an instrument highlights.

- initial acquisition of instrument;
- acquisition of supporting components for installation;
- acquisition of spares;
- acquisition of replacements;
- acquisition of contingency spare instrument;
- acquisition of service tools and calibration plant;
- promotional charges;
- regulatory costs for approvals, etc.;
- charges for holding spares, service tools, test rigs, etc.;
- scheduled maintenance and recalibration;
- unscheduled repairs;
- income lost when instrument being repaired, etc.;
- losses due to spoilt product, or closed service due to instrument failure;

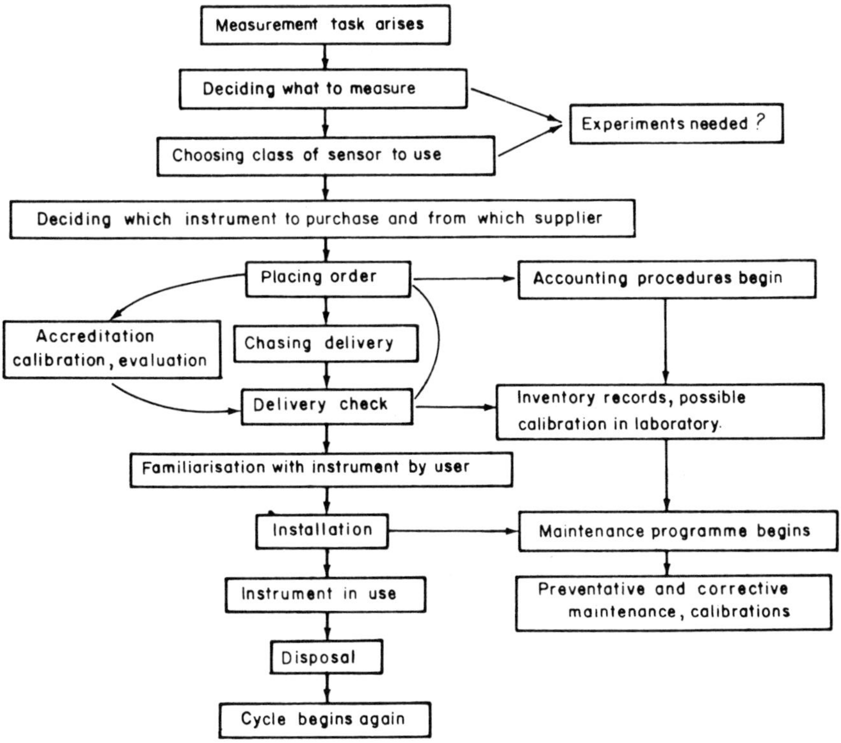

Figure 8.8 Cost cycle of a measuring instrument from need to removal. (HBMS, figure 3.3)

- damage cost caused by failed instrument;
- legal liability of failed instrument outcome;
- training costs and lost product during breaking period;
- technical, engineering, administrative overhead.

Ownership cost is a variable quantity, for many parameters can be traded off. For example, the cheapest system to purchase is often the most expensive to operate. On the other hand, if the very best of everything is purchased the initial cost will be the largest but the running cost will probably be the smallest. Economists have developed ways to analyse this situation; it is often found that the minimum total cost results from a compromise between the two extremes (see Figure 8.9).

Unfortunately there will not be found a unique solution to any given case because there exist many optimization targets that do not coincide. Some of the optimization situations that will be met are represented in the hypothetical set of alternatives given in Figure 8.10 – taken from Blanks in Sydenham (1983b).

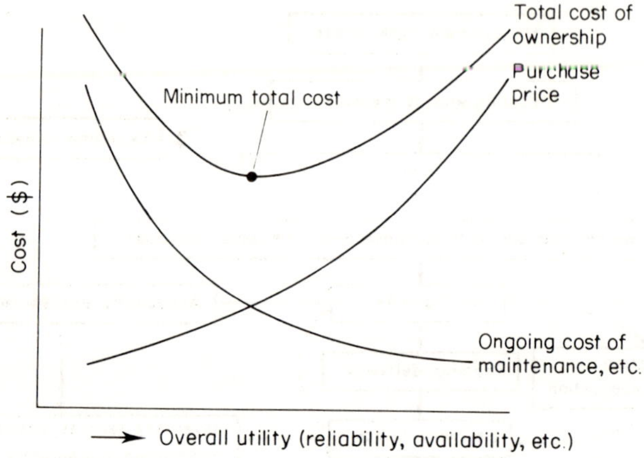

Figure 8.9 Cost of ownership is a composite of many variable costs – see text

Figure 8.10 Comparison of often met, design cost alternatives: (a) maximum effectiveness, irrespective of cost; (b) maximum effectiveness, subject to cost constraint; (c) minimum cost, irrespective of effectiveness; (d) minimum cost, subject to constraint on minimum effectiveness; (e) maximum effectiveness/cost ratio, irrespective of any constraints; (f) maximum effectiveness/cost ratio, subject to a constraint on minimum effectiveness; (g) maximum effectiveness/cost ratio, subject to constraints on both cost and minimum effectiveness. (From Sydenham, 1983b. Reproduced by permission of Henry Blanks.)

8.5 INSTRUMENT EVALUATION

8.5.1 Need and purposes of instrument evaluation

A definition of instrument evaluation is:

> The investigation and testing of the performance of instruments and related equipment in accordance with an agreed programme based on the manufacturer's specification, relevant national and international standards and specific user requirements

(from Chapter 31, Section 2 of Sydenham (1983a).

This definition explains the methodology by which a manufacturer or prospective purchaser of a specific instrument obtains *operational conformance* information about the equipment when it is operated under agreed environmental conditions and also about peripheral factors such as the quality of the construction and documentation.

Evaluation of instruments can be a wise investment, to both manufacturers and purchasers, because the quality of instruments has been found in many surveys and tests to be far short of expectations.

Section 31.2.7 of Sydenham (1983a) lists the results of 700 evaluations made by one of the largest such testing agencies, now called SIREP. The breakdown of statistics include 25% unsatisfactory as received for evaluation, 27% outside specifications under reference conditions and 24% that failed whilst being evaluated – and these were for equipments that could be selected for testing before submission.

A definitive USA instrument testing experience (Moss, 1978) came to similar conclusions.

Wherever the cost of incorrectly sensed information becomes significant then a case is made for the typical US $50 000–100 000 evaluation cost for an instrument to be worthwhile. This cost level can arise as a loss in minutes in military combat operations, in a month's operation of a power station not up to expected efficiency or several months of accumulation in the fiscal metering of a crude oil pipeline flow. Unfortunately the accumulated cost of poor measurements in scientific development is too diffuse to allow monetary forces to routinely apply in research costing considerations.

If an instrument is not *fit for its purpose* then it produces *false data* and *financial losses.*

Evaluation is carried out at many levels by many different agencies. During manufacture tests will be carried out, but it is unreasonable to expect self-evaluation to always be truly unbiased. An independent specialist agency is to be preferred. These can be consultant groups such as campus companies and other research laboratories with the appropriate facilities.

Several groups exist across the world that specialize in instrument evaluation (see Section 31.2.3 of Sydenham, 1983a).

8.5.2 Procedures for instrument evaluation

Procedures are typical to those of the SIREP service, as shown in Figure 8.11. The full cycle will take several weeks, to months, to complete depending on the

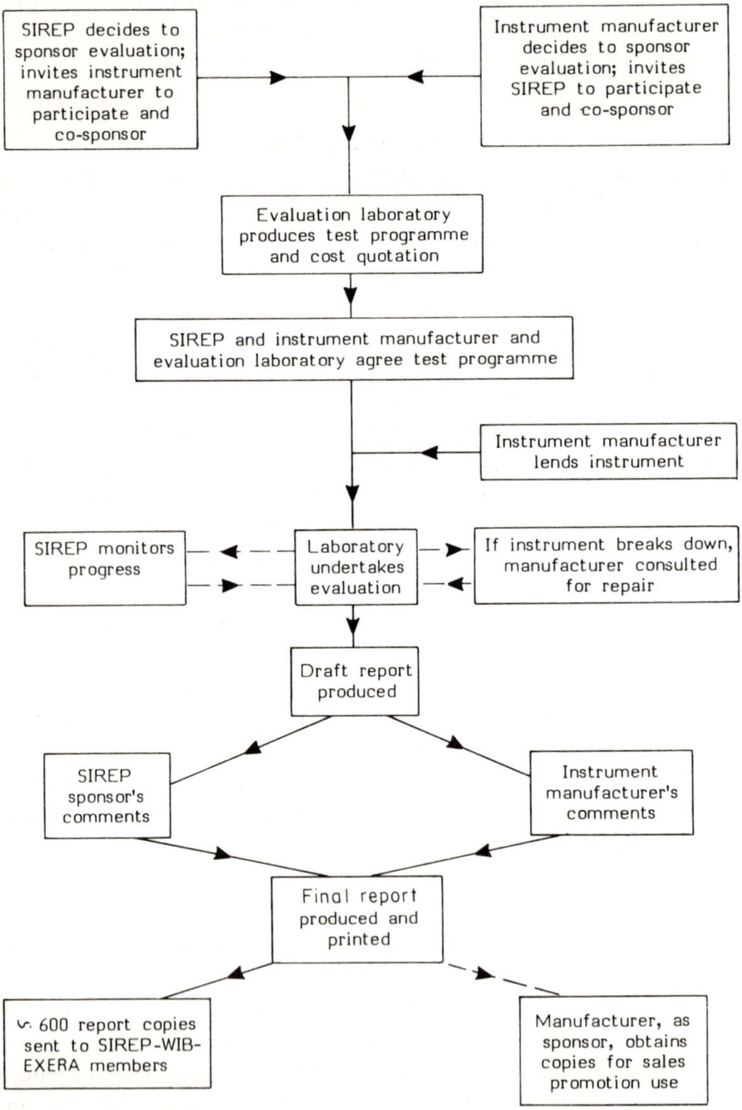

Figure 8.11 The SIREP instrument evaluation procedure. (HBMS, figure 31.1). (Reproduced by permission of SIREP International Instrument Users' Association.)

extent of the agreed tests. More than one unit will often be needed. A confidential report is issued to the sponsors. Manufacturers have the opportunity to react and make changes as part of the process.

8.5.3 Suggested test parameters

An abridged list of suggested evaluation parameters is as follows:

Tests under reference conditions

Basic accuracy
Warm-up time
Short-term drift
Response time
Frequency response
Dead zone
Long-term drift

Tests under simulated environmental conditions

Climatic tests
Power-supply variations
Power-supply interruptions
Overload
Series-mode interference
Common-mode interference
Flash test and insulation resistance
Output terminal isolation
Magnetic field interference
Mounting position
Load variation
Cross sensitivity
Sample temperature compensation
Vibration
Drop tests
Driving rain
Dust test
Salt spray

Assessment of construction and literature quality

Standards of construction and finish
Packaging
Maintenance
Safety
Documentation

The software CAE tool, SPECRITER (Cook, 1988), effectively encapsulates a large checklist incorporated into a procedure whereby the user adds the required information specific to any given instrument need, the system then producing a comprehensive printed specification document. It is based on the MIL-STD requirements for such documents and is a useful aid to setting up evaluations.

8.6 INSTALLATION AND COMMISSIONING

8.6.1 Installation a critical stage

Being within a serial chain from establishment of need to collection of flowing information, the installation phase of the life cycle of an instrument must also be taken seriously, otherwise bad data may be collected and the system may fail prematurely. Example 8.1, a real case, is worth re-reading at this stage.

The installation stage is where the instrument is fully tested in its final circumstance – prior to this all modelling and testing was simulated to expectations.

Instrument requirements often tend to be arranged as afterthoughts in large system planning and thus may be forced into harmony with the system during installation. They are often a modification of a lesser cost production item.

On receipt they should always be checked through prior to installation. Transport conditions of vibration, shock, extremes of temperature, pressure and relative humidity can play havoc on a pretested item. As has been explained in Section 8.5.1, they are too often found to be faulty or out of adjustment.

Advance planning is needed to prepare for installation. Special skills and tooling may be needed – such as welding in a thermowell insert flange for a temperature sensor in a chemical vat. The plant may need to be shut down to make the installation – in some cases a *time window* to do this may only be of a few hours' duration in a several-year period. The machine needing the instrument may need to be made available during its mobile life, as would be the case of installing new destination signs in a fleet of buses. It is not uncommon for these *shut-down losses* to exceed the purchase cost of the instrument.

Systems into which instruments are installed may need modification for successful installation, examples being addition of flow straighteners for flow meters, shielding of electric current carrying busbars when visual display monitors are placed near them and addition of stilling tubes in fluid-level sensing.

Installation should be carried out by suitably trained and competent persons. It might be carried out by the supplier, the user or by a contracted agency. Which it is needs clarification as it can be an expensive operation.

Instruments are often specialized items made to order and have taken considerable time to order and receive. Their less commonly met complexity and sophistication means they are not understood adequately by routine technical staff. Unexpected requirements can render a system out of calibration if special needs are not adhered to. For example, the windings of an electric current measuring transformer must not be left open-circuited when it is energized, excessive forces should not be placed upon bolts of load-cell mounts and impacts on housings during installation can give rise to zero offset in fine mechanical aspects of, say, pressure gauges.

When the system is complex and must be put into service rapidly, as in the commissioning of a ship of an oil refinery, operators must be trained well before installation. Suppliers of *turnkey* projects usually offer a training package.

In large-value projects an *installation-specification* document will be developed to guide this step.

Systematic testing schedules

Once the system is installed, the following simple tests should always be carried out to ascertain if it is ready for detailed assessment and calibration. As well as checking that the system is properly conditioned this testing will see if unwanted noise has found its way in.

By ensuring that the various input signal and possible noise sources are selectively and sequentially controlled to known values it is possible to ascertain if the system is generally operating correctly.

Sensing stages will operate with either analogue or digital signal forms - many begin as analogue (the sensed real world is largely that) possibly being converting into digital form before transmission or recording is carried out. The analogue type is typified by the general system shown in Figure 8.12. As a guide digital systems, as this primary stage of a sensing arrangement, carry out pseudo-analogue functions and can often be considered in an analogue sense at the block-diagram level.

At the first stage, point A in Figure 8.12, the sensing signal has an energy form of the observed variable. This has to be transduced into the more convenient electrical representation. Testing requires availability of a suitable physical system to provide original signals for driving the sensing system through its operational range. Where these are not available during testing, then simulation with special test rigs and alternative signals must be used (see later).

Where the sensor stage is of the *modulating* block type, a *biasing* power supply (for example, as with a magneto-resistor or light-dependent resistor) or it might be *self-generating* (for example, a photovoltaic light illumination level sensor, or thermocouple).

The output of the primary stage, point B, may need *interrogation* to extract the measurement quantity. A commonly met circuit is the Wheatstone bridge, which needs its own stable power supply to provide the *excitation* for operation of the differential action of the bridge. Bridges can use either d.c. or a.c.

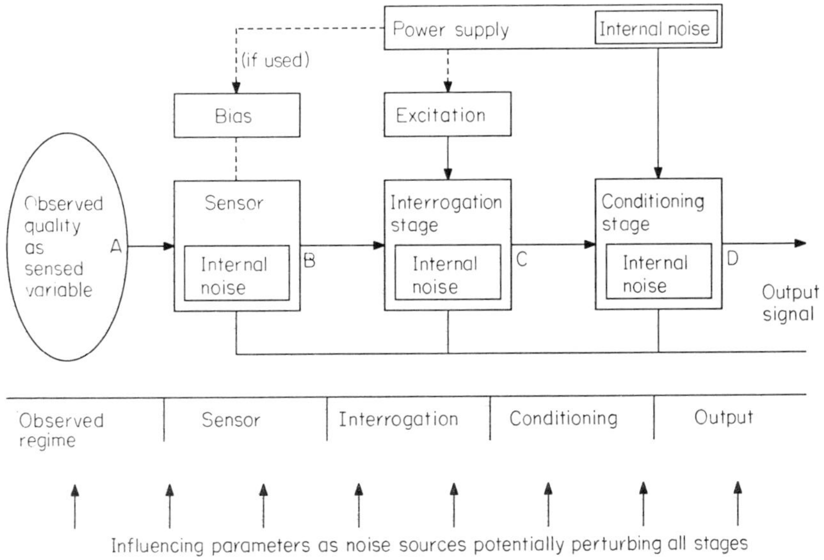

Figure 8.12 The measuring instrument as a 'black-box' with one required sensed input

excitation, the latter being variously known as *phase-sensitive detection, synchronous detection, homodyne* and *lock-in detection.*

The signal at point C, in Figure 8.12, closely represents the sensed variable but usually is both too small in magnitude and incorrect in form. There is usually need to *condition* the signal.

Power for the bias and excitation is usually obtained by further processing of the less stable 'main' power supply. Some sensors will have their own in-built battery supplies.

Noise may enter the system through any of the many energy paths formed between the sensing hardware and its environment. It can be induced in the connections and the componentry by electromagnetic induction and also by any number of influence parameter transducer effects (such as piezoelectric effects in vibrating leads). It can find a way through the power-supply paths into the various stages.

The first step is to establish a block diagram for the system along the lines of Figure 8.12. It should be drawn and labelled showing all potential *signal ports* (that is, both wanted and unwanted).

Next, by measurement, ensure that all ports of the system have expected signal magnitudes and frequencies. An oscilloscope should be used to show up higher-frequency effects that are integrated out in d.c. metering. A bridge with an open-circuit arm caused by a poor connection of the sensor into the circuit, or failed sensor element, will show an abnormally high output voltage. Excessive a.c. *hum* pick-up may saturate amplifiers and power-supply controls or cover over the sensing signal.

An excursion of the sensing input at point A should cause a correlated output change, at point D, at the expected speed and magnitude.

Next, set the sensing input at zero sensed quantity – for example, by shorting the terminals in the electrical case; by clamping movement in displacement sensors; by venting the pressure-sensing port to ambient in relative pressure gauges or to the second sensing port in differential versions; by removing all forces on load cells. Under this condition the output signal will be the system *offset* plus the effect of other unwanted signals. The signal at point D, in Figure 8.12, should be small and trimmable to be close to zero. If not, then unwanted signal is finding its way through the system.

If the offset is not small, then removal of the power supply, or shielding from changing influence inputs, will help establish the source of the unwanted signal. Once an adequately small zero has been established the system should be retained in that state for an adequately long period to test its *stability*.

When excitation is used this should also be removed without changing any other port values. If significant signal still exists at point D then the system is faulty; the source should be tracked down and removed.

When the physical variable, at point A, cannot be varied in practice during development and testing (supplying a varying 1000 tonne force calibration signal or a rapid and large pressure pulse are examples) the input transducer stage can be replaced with a *dummy equivalent electrical network*. In the case of

Wheatstone bridge sensors used in strain gauges, light meters, load cells, and more, a simple calibrated and highly stable variable resistor can replace the sensor in the bridge arrangement. This cannot establish if the sensor transduction stage is operating satisfactorily but can confirm, and help set up, subsequent stages. In some cases it will not be possible, or the best approach, to test the sensor by a direct method. Indirect testing, coupled with a mathematical model to transform a measured variable into a wanted one, can be used.

A particular trap worth mentioning is that arising when observing virtually unchanging signals such as occur when recording the stability of a process. If all is performing well, then no change is expected. However, a faulty sensing system or a lead-off of the recorder can give a perfectly unchanging record. To ensure the system is monitoring properly the recorded variable should have a feature that shows it is working, such as use of enough system gain to record a varying signal or addition of a regularly injected calibration signal.

Another trap is where an unexpected influencing variable is present. An appreciation of the physical principles on which the sensor operates is needed both at its primary level and in terms of the secondary effects. Where these are gross effects they are soon detected. However, they tend to be often only a small proportion of the signal and must be found by accurate systematic measurement at the various stages of the information pathway. For example, the metal-resistance strain-gauge system is well known to be affected by its ambient temperature, but often the sensor assembly can be magnetoresistive when mounted on certain materials. With thermocouple sensors it is highly important to track all connectons around the 'hot-cold junction' circuit, looking for unexpected thermo junctions produced at terminals and switches.

Small influencing effects can arise virtually anywhere in the system. Radio-frequency pick-up in leads, thermoelectric effects, piezoelectric generation and many more can provide frustrating challenges at times!

Some sensors use digital techniques virtually from the outset. Examples are position indicators that count pulses generated as magnets move past proximity sensors and the optical encoder discs producing pulses as a shaft rotates. In these cases a digital pulse generator of a programmed digital signal sequence should be used to verify that the electronic parts of the system are operational.

Much of the above suggested testing procedures still apply but the digital nature of the signal is less easy to interpret, especially during rapid changes. Observation of digital system states is conventionally achieved using logic analysers which operate with fixed clock frequencies. This form of tool can be applied to data-processing stages but is less useful for sensor signals where the output frequency is variable from zero hertz upward. The use of an appropriate digital-to-analogue converter (DAC) can restore the analogue time and magnitude information needed during the initial system set-up stages.

8.6.2 The commissioning phase

Commissioning is the process whereby the supplier finalizes installation and initial operation, handing over the full system to the purchaser. For a high-cost project there may be a case for formalized commissioning of the instrumentation.

A *commissioning specification* explaining the procedures to be used, who shall carry them out and what performance is to be met, will be drawn up for the larger contracts. These often include a *penalty clause* for failure to be operational by a set time.

Some instruments are so costly as to be a major system in themselves - for example a giant optical telescope, or a remote sensing platform in space.

The commissioning function may need to be conducted in the manufacturer's premises or in a mutually agreed special test facility. It would not be feasible to commission, for example, a balloon-borne radio telescope *in situ* because it is clearly impracticable to take automatic test equipment into space. Extensive ground-based simulation is used instead.

The commissioning tests will be attended by representatives of the parties involved - the maker, the user, the purchaser, the banker, etc. Once agreed parameters are verified and recorded a contractual document is signed, passing the responsibility for operation and payment over to the client. The document must clearly state then the liability of the supplier.

For low-cost instruments there is usually no formal commissioning stage, the purchaser being left to undertake adequate testing work before the various warranty or guarantee periods expire.

8.6.3 Influence parameters

An instrument has to work in practice in a certain environment. The physical variables of this environment can affect the operation of the instrument. The influence parameters concerned at the place of installation need to be verified as being as expected, for if not the instrument may give erroneous results when extremes occur. The performance of an instrument is usually influence effect limited. These variables have a habit of being at their extremes when the instrument has to be working to its best.

The main influence variables are temperature, pressure, relative humidity, vibration and electromagnetic interference (EMI); others may be relevant.

The variables involved may not be normal ambient environment expectations. A temperature recorder for a flown goods container may only have to record small excursions of temperature but its working environment may be excessive if its electronic circuitry is not located along with the goods. A borehole geophysical instrument has to work in extreme pressures and temperatures - active and passive isolation is often necessary.

It is prudent to look for unusual influence effects that may be present. For example, a load cell used to weigh the plated amount of an aluminium electrolysis cell may need correction for the magnetostrictive effect contracting the load cell (caused by the very large electric currents and resultant magnetic fields).

As influence effects are not constant over time it is necessary to estimate their magnitude. Confirmation of the estimates might be appropriate using long-term recorders.

8.6.4 Calibration aspects

Associated with the calibration are two costs – that of making it and that of not making it. The first is relatively easy to ascertain and depends on the instrument's complexity, ease of adjustment, degree of automation, procedure needed, staff capability, loss of production whilst out of service, transport costs to the test laboratory, spares holdings and paperwork. This all builds to be a significant cost with the consequence that few instrument users have it fully carried out.

The costs of such deficiencies are application dependent. The major costs will be losses incurred due to lost system operational time, production of scrap or non-standard product or, in scientific research, unnecessarily long research times in establishing new theories. Errors caused by lack of calibration are often detected at the worst times, resulting in excessive losses.

The perspective for viewing the resource needs of calibration is to regard it as a form of insurance, as illustrated in Example 8.6.

Example 8.6 Calibration is a form of insurance

A critical process parameter in the production of a special alloy is the temperature at which it is poured into casting moulds – this decides the crystallization process and hence final properties (which finally emerge after two weeks of a many-stage process).

For three decades Bob has done this job. He keeps record cards for each product variety, which tell him at which temperature he must pour each one. He measures temperature, which has to be within 1 °C at nominally 1100 °C, with a proprietary *disappearing filament pyrometer*.

Pyrometers make use of the heat radiation from the observed subject. The above-mentioned type needs calibration because the *emmissivity* of the surface actually being sensed varies with each case. The pyrometer has a calibration certificate somewhere in its well-worn box sitting in the battered metal cabinet next to the furnace. It was last calibrated when it was purchased some twenty years ago.

All is fine – until, one day, the pyrometer is dropped! After years of satisfactory operation the situation suddenly becomes a disaster.

First, there is no back-up instrument on hand, so production ceases. In his country there is no adequate calibration facility for pyrometers (he needs that to a high precision) so repair will be a lengthy process taking months.

In an attempt to overcome the problem a new pyrometer is purchased. It takes two months to arrive. It is put into service only to find, after much lost product (and customers' orders), that each casting lot is a failure. What has gone wrong?

What has happened is that the performance of Bob's original instrument had drifted markedly from the true *traceable* calibration it started with so many years beforehand.

His carefully kept record-card temperatures were made to an unknown, time-varying, temperature scale - one that could not be reproduced without performing time-consuming product tests.

How should they have operated their metrology administration?

There should have been a traceably calibrated *laboratory standard* pyrometer held by the firm, or one known to be readily available for *transfer* calibration through a local *certified* calibration laboratory.

Bob's everyday working instrument, a more robust, less accurate, instrument should have been the only one used on a regular basis. Periodically it should have been compared with the standard instrument and the accuracy departures recorded on a *calibration chart.*

It would then have become clear whether the routine instrument was holding its calibration or needed an overhaul. Furthermore, at any time the routine instrument could then be replaced with another having the *known* scale. Hence *calibration is like insurance* - if you do not pay your premiums then do not expect support when needed; if you have paid the premium then losses are minimized.

Having set up a traceable system the next consideration is how often should an instrument be calibrated? This time period is called the *calibration interval.*

No objective procedures allow this to be formulated, as it depends on the instrument, its application and importance. Some guiding rules, however, to help establish this follow.

The instrument maker will often suggest an interval for the instrument. This is used to set up a declared *calibration programme* when the instrument is placed into service for the first time. This programme will declare *recall* dates. It is wise, initially, to use a shorter interval, lengthening it as confidence builds in the product.

Two methods can be used - on a *time-in-service* or a *calendar-time* basis. The impact of *smart* instruments, with various levels of in-built calibration ability (but this can never replace the need to compare it with the declared standard) will change attitudes and procedures.

Whereas an instrument might go on being proved to be within calibration limits this must not lead to the interval being stretched to very long periods. In normal practice the interval should rarely become longer than 1-2 years. Corrosion, physical damage, and other stochastic factors out of the designer's control, are facts of life that can take a system out of calibration by just enough to cause trouble, but go unnoticed.

8.6.5 Measurement assurance planning

The quality of calibration practice can drift downward unless specific action is taken to ensure the accuracy of the processes involved is maintained. Within factories, test and scientific laboratories, national standards and international physical measurement laboratories, special programmes, called *measurement assurance programmes* (MAPs), are used for this purpose.

They make use of *peer review* to compare the performance of the various laboratories. In a *round-robin* programme a set of objects representing a type and range of measurand is sent around a laboratory circuit. Each laboratory makes the defined measurements, reporting their work to a central organizer. The results are then compared. These have been done for the units of voltage, small and large length, mass, and others.

8.6.6 Management of existing systems

The measurement system is installed and commissioned. Is that the end of necessary attention? Unfortunately not! It would be most desirable if sensing systems could be used on a *fit-and-forget* basis. The reality is that several good reasons exist that make this requirement unreasonable.

The first and most usual restriction is largely a question of cost. Given enough development and testing resources extremely reliable and long-life sensors can be built - but the cost is often prohibitive. In those fields where the cost of failure is high (such as in aerospace, medical and military endeavours) or the demand huge (such as in the automotive industry), measurement systems development will be well supported. In everyday, more routine, semi-custom needs, designs will often be minimal to keep their cost down to what the market is prepared to pay.

Research into sensing systems that include *intelligence* is gradually reducing the cost of systems whilst giving better performance. Tomorrow's sensor systems will be most sophisticated, yet cheap. But for now it is necessary to remember that measurement systems will need considerable care and possibly much re-work after commissioning.

A check list of factors that may be the reason for revisiting the system is as follows:

- it is found that the sensor is not up to the need;
- updating may be needed to follow regulatory changes;
- calibration cannot be relied upon to remain constant;
- physical damage and corrosion arise;
- the response of the process it is monitoring may be upgraded requiring sensor improvement;
- changing technology often renders the unit unserviceable;
- changing process control or experimental observational needs can call for an improved design;
- the user/purchaser often learns what is really needed after the first has been installed;
- working environment of the system can be made less protective;
- the output of the system has to be changed to suit a new processing arrangement.

For more detail refer to Chapter 30 of Sydenham (1983a).

TUTORIAL EXAMPLES

Note: These exercises can be developed to suit a particular case for which data is readily available.

8.1 Discuss, as a group, the reasons why the thermometer of Example 4.10, Section 4.5.2, might not provide a sound measurement of the temperature of a cup of coffee. Aim to develop scientifically based reasons that take heuristic reasoning onto mathematical modelling so that the magnitude of effects might be assessed.

8.2 List systematic testing steps for verifying that an installed measuring system is working correctly. How would you decide how long can pass before the instrument should be recalibrated?

8.3 For an instrument of your choice prepare an evaluation plan suited to the availability of one of the following sources:
(a) sales brochure;
(b) handbook;
(c) instrument.

8.4 Using data given in various sales catalogues conduct a utility analysis to decide which product to purchase. Suggested products for consideration might be:
(a) electronic signal generator for laboratory teaching of low-frequency linear circuits;
(b) thermometer for monitoring temperature in a controlled environment room and providing an output for recording purposes;
(c) weighing scale for kitchen use;
(d) power-plug outlet timer unit.

8.5 Using an available electronic circuit board calculate its expected MTBF. (The failure rates given in Table 8.1 can be used for the exercise but it is better to find a source related to the actual product used in the exercise.) Suggest how it could be improved, detailing the changes and resultant gains.

8.6 Estimate the various costs associated with the purchase, installation and operation of an instrument of your choice; one of the instruments suggested in Exercise 8.3 could be used.

REFERENCES

Cook S, 1988, SPECRITER *Measurement*, **6**(4), 155–60.

Daneman, H L, 1975, *Some Aspects of Traceability*, HLD Associates, Sante Fe, New Mexico, USA.

Hewlett-Packard, 1981, *Improving Productivity and Quality Through Electronic Testing*, Hewlett-Packard, 02-5952-7610 I.E., Palo Alto.

Iervolino J L, 1983, The Productivity–Quality Connection. *NCSL Newsletter*, **23**(1), 41–47.

Jensen, 1982, *Burn-In: An Engineering Approach to the Design and Analysis of Burn-In Procedures*, Wiley, Chichester.

Klaassen K B and Van Peppen J C L, 1986, 'Reliability in Instrumentation', *J. Phys. E.: Sci. Instrum.*, **19**, 577–87.

MIL, 1974, *Military Standardisation Handbook, Reliability Prediction of Electronic Equipment*, MIL-HDBK-217B (20-9-1974), Dept of Defense, USA.

Millar A, 1980, Maintenance constraints on system design, *Measurement and Control*, **13**, 5–8.

Moss C, 1978, Quality of precision measurement equipment. *NCSL Newsletter*, **18**(1), 23–8.

O'Connor P D T, 1981, *Practical Reliability Engineering*, Heyden, London.

Schultz D, 1981, Quality, the key to productivity. *NCSL Newsletter*, **22**(1) 18–25.

Sidall J, 1972, *Analytical Decision Making for Engineering Design*, Prentice-Hall, Englewood Cliffs.

Simpson A, 1976, *Testing Methods and Reliability in Electronics*, Macmillan, Houndsmill.

Smith D J, 1972, *Reliability Engineering*, Pitman, London.

Smith N and Ainsworth M, 1985, *Ideas Unlimited*, Nelson, Melbourne.

Stein P K, 1967, *Traceability – the golden calf*, Pub. no. 7. In *Measurement Engineering*, P K Stein Engineering, Arizona.

Sydenham P H, 1983a, *Handbook of Measurement Science* Vol 2, Wiley, Chichester.

Sydenham P H, 1983b, *Quality and Scientific Instruments*, Australian Scientific Industry Association, Melbourne.

Sydenham P H, 1986, *Mechanical Design of Instruments*, ISA, Research Triangle Park, North Carolina, USA.

INDEX